I0064850

Handbook of Metals and Alloys

Handbook of Metals and Alloys

Edited by
Ricky Peyret

WILLFORD PRESS

www.willfordpress.com

Published by Willford Press,
118-35 Queens Blvd., Suite 400,
Forest Hills, NY 11375, USA

Copyright © 2019 Willford Press

This book contains information obtained from authentic and highly regarded sources. Copyright for all individual chapters remain with the respective authors as indicated. All chapters are published with permission under the Creative Commons Attribution License or equivalent. A wide variety of references are listed. Permission and sources are indicated; for detailed attributions, please refer to the permissions page and list of contributors. Reasonable efforts have been made to publish reliable data and information, but the authors, editors and publisher cannot assume any responsibility for the validity of all materials or the consequences of their use.

Trademark Notice: Registered trademark of products or corporate names are used only for explanation and identification without intent to infringe.

ISBN: 978-1-68285-575-1

Cataloging-in-Publication Data

Handbook of metals and alloys / edited by Ricky Peyret.
    p. cm.
Includes bibliographical references and index.
ISBN 978-1-68285-575-1
1. Metals. 2. Alloys. 3. Metallurgy. 4. Materials science. I. Peyret, Ricky.
TA459 .H36 2019
620.16--dc21

For information on all Willford Press publications
visit our website at www.willfordpress.com

WILLFORD PRESS

Contents

Preface

Over the recent decade, advancements and applications have progressed exponentially. This has led to the increased interest in this field and projects are being conducted to enhance knowledge. The main objective of this book is to present some of the critical challenges and provide insights into possible solutions. This book will answer the varied questions that arise in the field and also provide an increased scope for furthering studies.

A metallic element, compound or alloy is characterized by high structural strength, opacity to light, optical luster, malleability, ductility, and good thermal and electrical conductivity. Most pure metals are combined with one or more metallic or non-metallic components in varied proportions to achieve desired characteristics of structural hardness, resistance to corrosion, suitable color and luster. This combination of metals and other components results in the formation of alloys. The most commonly used alloys of iron are steel, stainless steel, cast iron, etc. are used in a variety of industries like construction, automobile industry, etc. Since the study of metals is fundamental to all areas of engineering. This book aims to expound some key concepts and theories related to the science of metals and alloys in detail and provide extensive information on their practical applications. With state-of-the-art inputs by acclaimed experts working in this field, this book is meant for materials scientists, engineers, experts and students.

I hope that this book, with its visionary approach, will be a valuable addition and will promote interest among readers. Each of the authors has provided their extraordinary competence in their specific fields by providing different perspectives as they come from diverse nations and regions. I thank them for their contributions.

Editor

A Case Study for the Welding of Dissimilar EN AW 6082 and EN AW 5083 Aluminum Alloys by Friction Stir Welding

Sefika Kasman [1,*], **Fatih Kahraman** [2], **Anıl Emiralioğlu** [2,3] and **Haydar Kahraman** [3,4]

[1]   Izmir Vocational School, Dokuz Eylul University, Izmir 35160, Turkey
[2]   Department of Mechanical Engineering, Faculty of Engineering, Dokuz Eylul University, Izmir 35140, Turkey; fatih.kahraman@deu.edu.tr (F.K.); anilemiralioglu@gmail.com (A.E.)
[3]   Graduate School of Natural and Applied Sciences, Dokuz Eylul University, Izmir 35140, Turkey; haydar.kahraman@deu.edu.tr
[4]   Department of MetallurgicalandMaterialsEngineering, Faculty of Engineering, Dokuz Eylul University, Izmir 35140, Turkey
*   Correspondence: sefika.kasman@deu.edu.tr

Academic Editors: Halil Ibrahim Kurt, Adem Kurt and Necip Fazil Yilmaz

Abstract: The aim of this study is to investigate the effect of keeping constant the tool rotational speed to the welding speed ratio (υ ratio) on the mechanical properties of the dissimilar friction stir welding of EN AW6082-T6 and EN AW5083-H111. Two different pins shaped as triangular and pentagonal were associated with the constant υ ratio. From the tensile test results, it was found that the υ ratio does not create an evident change in the weld joint strength. The small cavity- and tunnel-type defects were observed at the nugget zone and located on the advancing side of the pin. These defects caused a decrease in the strength and elongation of the weld joint. The most important inference obtained from the experimental results is that if the υ ratio is kept constant, the weld joint strength for each weld does not correspond to a constant value.

Keywords: dissimilar friction stir welding; pin shape; mechanical properties

1. Introduction

It is well known that aluminum alloys have inherent and versatility properties such as resistance to corrosion, good formability, a good strength to weight ratio, low density and electrical and thermal conductivity. These properties make it a high-demand material compared to the steel alloys in industrial application, particularly in the automotive, shipbuilding, packaging, construction and architecture fields, etc. With the growth of demand for aluminum alloys in a wide variety of applications, the welding and manufacturing process of aluminum alloys requires special knowledge and experiences. The welding of aluminum alloys by fusion welding techniques produces some defects such as pores, loss of some elements, hot cracking, stress corrosion cracking, and mismatch between the filler alloy and the workpiece material in dissimilar welding, causing the loss of strength of the weld joint [1]. To overcome these problems, the solid-state welding techniques are the best alternatives for the welding of aluminum alloys.

Aluminum-magnesium alloys are non-heat-treatable alloys which provide good mechanical properties, corrosion resistance, and good workability and weldability. These excellent properties make it an attractive material in a wide range of construction and structural applications in the automotive and shipbuilding industries. Aluminum alloy EN AW5083 is one of the aluminum-magnesium alloys and it has high mechanical strength and fusion weldability [2]. The aluminum-magnesium-silicon alloys are heat-treatable alloys and have a medium strength with excellent corrosion resistance. These alloys are being used in automotive parts, especially in body sheets to decrease their weight. EN AW6082 alloy is one of the aluminum-magnesium-silicon alloys and it has very good weldability, but its strength is lowered in the weld zone [3]. Many studies have focused on the friction stir welding (FSW) process of aluminum alloys and it is well known that some of aluminum alloys cannot be welded by the fusion welding process. Therefore, in order to achieve more information about the FSW process, critical factors such as the tool design, rotational speed and welding speed have been subjected to investigation. There are many studies related to the FSW of EN AW5083 and EN AW6082, but the dissimilar FS welding of these alloys is limited. Also, in most of the studies, the tool pin shape is a straight, cylindrical threaded profile. There are some studies in the literature on dissimilar FS welding of EN AW6082 and EN AW5083. Peel et al. [4] performed studies to determine the effect of the tool rotational and welding speed on the microstructure of welding zones and weld properties of dissimilar FSW EN AW5083–EN AW6082 joints. Donatus et al. [5] focused on identifying corrosion-susceptible regions of dissimilar friction stir welds of EN AW5083-O and EN AW6082-T6. Donatus et al. [6] performed dissimilar FSW studies on EN AW5083-O and EN AW6082-T6 to investigate the material flow in the friction stir welds. Steuwer et al. [7] investigated the effect of the welding parameters on the residual stress profiles on the same welds. Apart from these studies, Leitão et al. [8] conducted two-stage experimental studies in order to determine the effect of FSW parameters on each material and the weldability of dissimilar alloys at high temperature. Sun et al. [9] studied the dissimilar friction stir welding of ultrafine-grained 1050 and 6061-T6 aluminum alloys to understand joint characterization. Aval et al. [10] investigated the thermo-mechanical behavior and microstructural events of dissimilar FSW of AA6061-T6 and AA5086-O.

The present study is focused on the effect of tool geometry and the ratio of the tool rotational speed to the welding speed (v ratio) on the mechanical and macrostructural properties of dissimilar FSW of EN AW6082-T6 and EN AW5083-H111 alloys.

2. Friction Stir Welding

Friction stir welding is the alternative welding technique for some aluminum alloys that cannot be welded by the fusion welding techniques. The major difference is that FSW occurs below the melting point and it is located in the welding group occurring in the solid-state phase. The principles of the FSW process are that a tool consisting of a shoulder and a pin plunges into the butted surface of the plates and then the rotational and translational movement perform the welding process. The necessary heat for welding is provided by the tool shoulder and pin, the speed of the tool rotation and the movement. The friction between the tool pin and the workpiece generates the heat and, therefore, the pin shape has a particular effect and an increase in the tool rotational speed increases the heat. Also, the heat decreases with the increase in the tool movement speed or welding speed. In this context, a mathematical approximation is developed as in Equation (1) [11,12].

$$Q = (\alpha q)/W = 4/3\,\pi^2\,(\alpha\mu P.\text{TRS}.R^3)/WS = \beta.\text{TRS}/WS \qquad (1)$$

where Q is the heat input per unit length, α is the heat input efficiency, WS is the welding speed, and TRS is the tool rotational speed. For a welding condition, α, μ, P, R and β are the constant values. It is understood from Equation (1) that the tool rotational speed and the welding speed are the main factors for determining the per unit heat input. By the effect of the generated heat, the material around the pin flows from the front to the back of the tool and a deformation occurs from the tool movement. Then, the weld joint is completed by repeating this cycle.

3. Materials and Methods

Aluminum alloys EN AW5083-H111 and EN AW6082-T6 plates with a thickness of 5 mm, width of 150 mm and a length of 200 mm were used as the weld plates. The chemical composition and the mechanical properties of the plate materials were given at Tables 1 and 2, respectively. Before welding experiments, the butted surfaces of plates were cleaned from dust and any residue.

Table 1. The chemical composition of aluminum alloys used in the present study (wt. %).

Alloy/Elements	Cu	Si	Mg	Mn	Zn	Ti	Cr	Fe	Al
EN AW6082-T6	0.23	0.98	1.02	0.6	0.21	0.01	0.03	0.6	Bal.
EN AW5083-H111	0.056	0.093	4.19	0.53	0.09	0.007	0.083	0.243	Bal.

Table 2. Mechanical properties of aluminum alloys.

Alloy	Ultimate Tensile Strength (MPa)	Elongation (%)
EN AW6082-T6	293.21	8.12
EN AW5083-H111	325	17.6

The dissimilar FSW were performed with a universal milling machine (Russia). For the welding experiments, a backing plate was used to place and clamp the plates. A schematic FSW experiment and components was shown in Figure 1. The welding direction was set as the parallel to the rolling direction of plates and the tool was tilted at the 2° to the normal direction of the plates. A sample figure for FSW experiments was given in Figure 1. The welding tool was made of DIN EN 1.7131 steel and geometrically shaped as a triangular and pentagonal shown in Figure 2. The tool properties were listed in Table 3. With the aim of analyzing the effect of the constant ratio of the tool rotational speed to the welding speed "υ" (TRS (rpm)/WS (mm·min^{-1})), an experimental layout was conducted given at Table 4. As shown in Table 4, the υ ratio was kept fixed at 10.

Figure 1. A sample for FSW experiment.

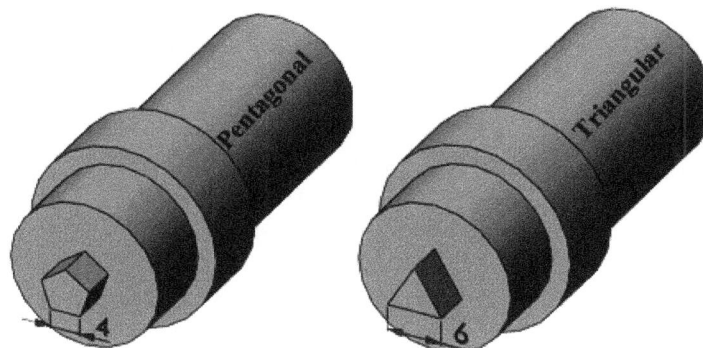

Figure 2. The tool pin profiles used in the FSW experiment.

The macrostructural analyses were performed on the cross sectioned perpendicular to the weld direction. Each sample was subjected to the standard grinding and etching process. The etching is performed by the modified Keller solution. Following the surface preparations, the optical macroscopic analyses were done.

A joint fabricated by the FSW process was characterized with the ultimate tensile strength (UTS, MPa) and percentage of elongation (ε, %). The tensile tests were carried out on 500 kN hydraulic testing machine (Shimadzu, Kyoto, Japan) with a cross-head speed of 2 mm·min^{-1}. The samples were prepared according to the ASTM E8-M. At least two samples were tested and the average of those tests was used in the final decision and analyses.

Table 3. FSW process parameters and their levels.

Parameters (Unit)	Symbol	Levels			
Tool Rotational Speed (rpm)	TRS	400	500	630	800
Welding Speed (mm·min^{-1})	WS	40	50	63	80
Tool Rotational Speed-to-Welding Speed	υ		10		
Pin shape	PS	T		P	
Fixed parameters					
Tool shoulder diameter (mm)	D		20		
Tool tilt angle (°)	α		2		
Dwell time (s)	t		20		

Table 4. FSW experimental layout with responses (UTS, ε, %).

Exp. No.	Process Parameters				Welding Performance				
	PS	TRS	WS	υ TRS/WS	UTS (MPa)	ε (%)	Efficiency 6082	5083	Defect
1	T	400	40		164.36	0.49	56.06	50.57	Defective
2	T	500	50		188.68	4.6	64.35	58.06	Defective
3	T	630	63		198	4.7	67.53	60.92	Defective
4	T	800	80	10	198.48	4.26	67.69	61.07	Sound weld
5	P	400	40		180.59	4.14	61.59	55.57	Defective
6	P	500	50		192.27	4.312	65.57	59.16	Defective
7	P	630	63		181.96	0.39	62.06	55.99	Defective
8	P	800	80		187.85	4.24	64.07	57.80	Defective

4. Results and Discussion

The present study focused on the effect of fixing the ratio of the tool rotational speed to the welding speed "υ" on the microstructure and mechanical properties of dissimilar FS welding of EN AW6082 to EN AW5083. In total, eight FSW experiments were performed and the results were evaluated as below.

4.1. Macrostructure and Microstructure Investigations

A weld joint is characterized by three distinct zones and a sample is shown in the Figure 3. The zones are affected by the tool rotational speed, welding speed and tool pin shape. The investigation of the macrostructure for each weld joint in the context of the present study was performed on the cross-section of the weld joints shown in Figure 3. As shown in Figure 3, the shape of nugget zone (NZ) and thermomechanically affected zone (TMAZ) is highly affected from the FSW parameters and pin shape. The pin shape shows the predominant effect on the NZ shape compared to the υ ratio. The onion ring–type structure was detected all in the NZ. The visibility of onion rings in the NZ produced by a pentagonal-shaped pin is more distinctive and the wideness of NZ is higher compared to the triangular-shaped pin. The reason could be the dimension of the circumference of the pin. The flow of deformed and elongated grains in the TMAZ is affected from the tool pin shape and generated heat. The width of the TMAZ is bigger at the condition of Exp. 3,4,7,8. It is attributed to the higher heat input. As seen in Equation (1), an increase in TRS increases the heat input due to higher frictional heat [12] and this causes an increase in width of the TMAZ. Exp. 3,4,7,8 were performed with a higher TRS.

Figure 4 shows a sample microstructure taken from the weld joint of Exp. 6 and 8. The regions and borders of the welded metals can be clearly detected due to different etching behavior. The flow of the material in the nugget zone shows differences according to the value of TRS and WS. The shape of the onion rings is similar for the entire weld joint and almost has an elliptical form. When the onion ring is sectioned as vertical, it is seen that a part of the onion ring closer to the joint root is composed from EN AW6082. The reason is due to the rotation of the tool and the placement of EN AW6082. An onion ring is composed of a dark and a bright ring. These rings grow outward from the center and are placed in the nugget zone as they are ordered consecutively. The composition of the onion rings depends on the tool rotational speed. The welding speed determines the rotation times of the tool per minute. The material forged by the tool and is moved from one side to the other side which is repeated many times by the effect of tool rotational speed. This effect causes the forming of onion rings.

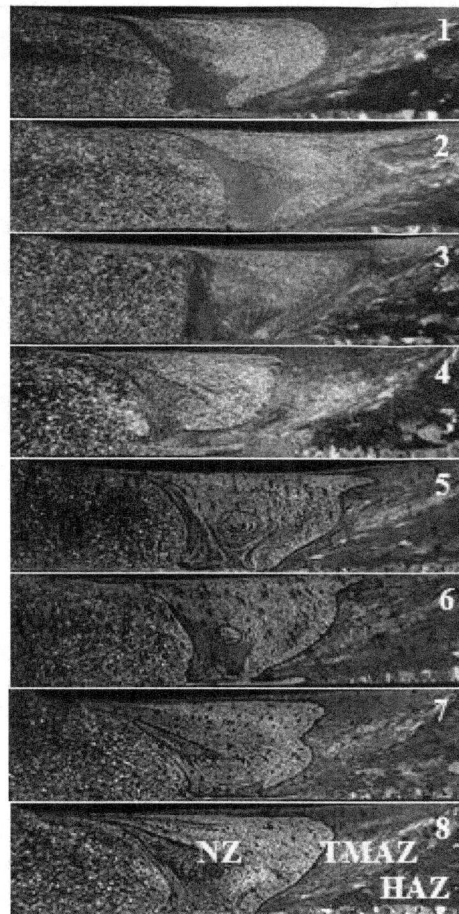

Figure 3. Cross-section of each weld joint.

Two defective welds shown in Figure 5 were taken from the joint fabricated by the triangular-shaped pin at the TRSs of 400 and 500 rpm and WSs of 40 and 50 mm·min^{-1}. Tunnel- and cavity-type defects were clearly detected from the weld joint and similar defects were also observed at the other joints. The defects were located at the side of the pin tip and EN AW6082. There is more than one small defect and these caused a decrease in the mechanical properties of the weld joints. The reason of formed defects is the amount of heat input; in some cases, a higher or insufficient heat input produced defects, and especially insufficient heat input resulted in inadequate material flow which led to tunnel- or cavity-like defects. In the present study, the small clustered defects at the root of weld joint were clearly observed and appeared at both the advancing and retracting side. The possible reason for these defects is insufficient heat input.

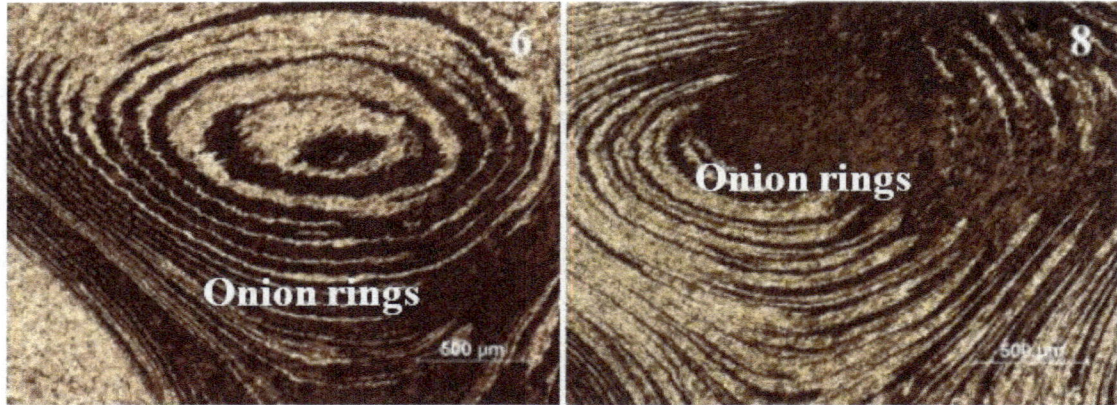

Figure 4. Nugget zone structures consisting of onion rings.

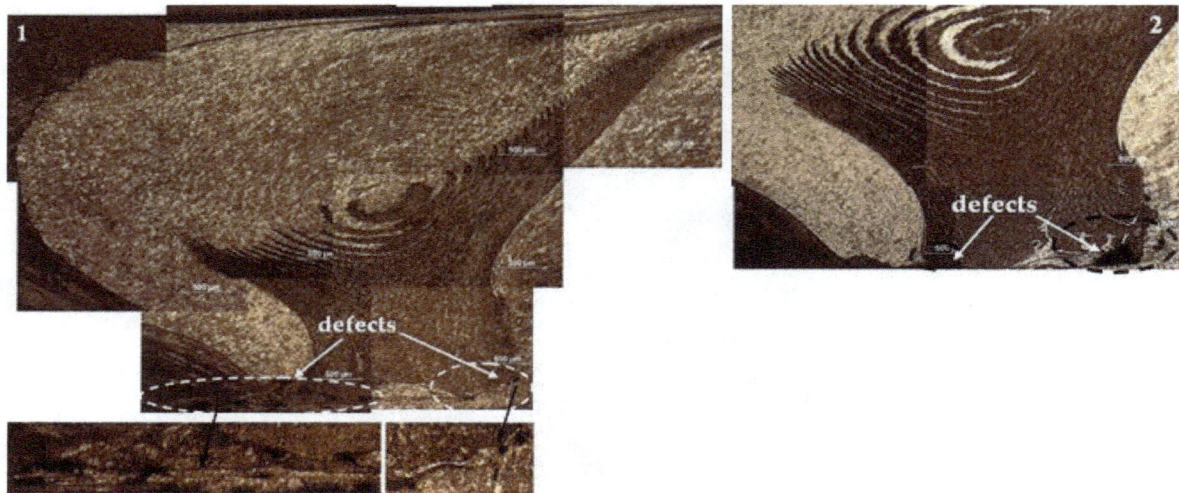

Figure 5. Samples for defective weld joints (Exp. 1 and Exp. 2).

4.2. Mechanical Properties

In order to determine the mechanical properties of each weld joint, the tensile tests were conducted and the results were associated with the process parameters. The ultimate tensile strength (UTS) and percentage of elongation (ε, %) were considered as the process responses. The tensile test results were given in Table 4 and Figures 6 and 7. Also, the fracture location of each weld joint is shown in Figure 8. As shown in Figure 8, except the welding of Exp. 1 and Exp. 7, the fracture of the joints was located at the heat affected zone (HAZ) and the side of EN AW6082. The fracture for the welding of Exp. 1 and Exp. 7 was located at the nugget zone. As mentioned before, except Exp. 4, all the weld joints contained many small defects such as cavities and some of the joints as in Exp. 2 had a tunnel-type defect. From Figure 6, the effects of the defects on the UTS can be clearly seen and all the UTS values are lower than those of the base materials. The UTS value for the triangular-shaped pin increased with the increase of the tool rotational speed independent of the welding speed. However, this tendency was not observed for the pentagonal-shaped pin. The UTS for Exp. 1 was the lowest (164.36 MPa) one of all in the weld joint and it corresponded to 56.06% efficiency for EN AW6082. The ε value in Figure 7 was verified by the UTS result. The value of ε was 0.49 and, also, the fracture location of this joint was at the NZ and we observed that there was no significant elongation compared with the other joints. A similar result was observed at the welding of 630 rpm and 63 mm·min^{-1} (Exp. 7) produced by the pentagonal-shaped pin. The fracture location was also at the NZ. According to the UTS results, the welding with the highest strength was performed at the condition of 800 rpm and 80 mm·min^{-1} by the triangular-shaped pin, and an important issue was observed relating to the UTS values for both of Exp. 3 and Exp. 4: the results are almost the same. The difference between the two UTS values is small enough to be ignored. It can be taken as an important inference for the strength and the elongation of the welded samples; the results are smaller than the base metals' UTS and ε. The welding efficiency was determined by comparing the UTS values for all the weld joints with the base metals. The results were listed in Table 4. It is obvious that the lowest efficiency was obtained from the defective weld joints. The efficiency for joint strength ranged from 55% to 68%. As an association between the UTSs of weld joints and UTSs of base materials, a result was reached: defects cause the decrease of the UTS and ε of FSW samples compared to those of base materials.

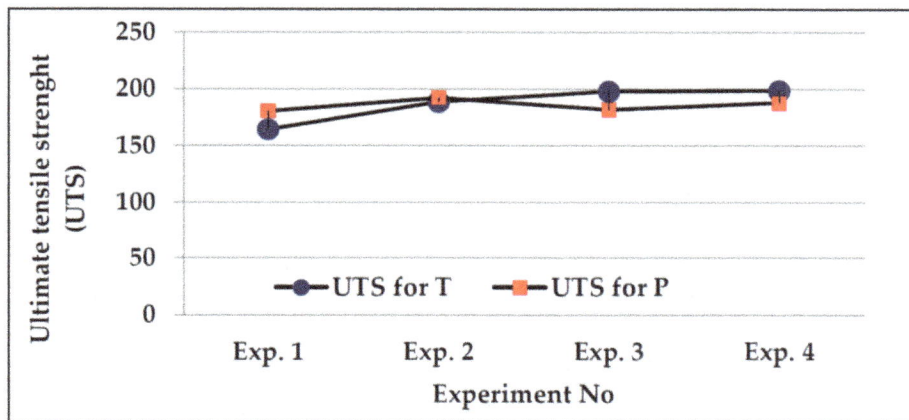

Figure 6. The comparative UTS results for both tool pins.

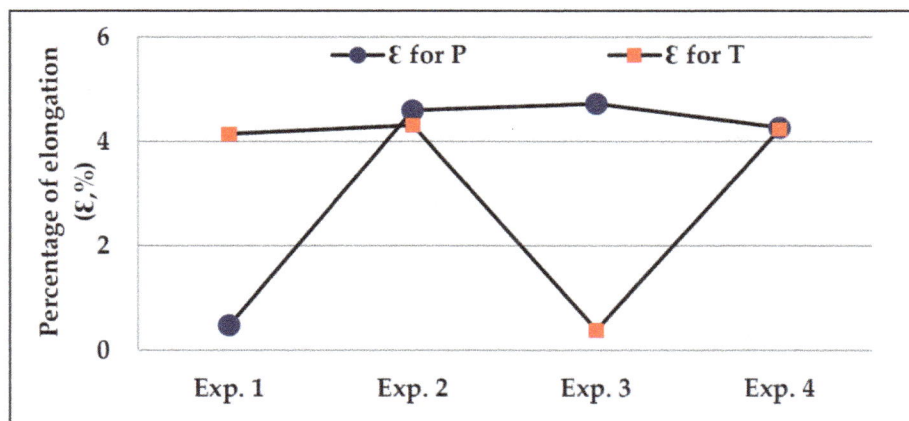

Figure 7. The comparative ε results for both tool pins.

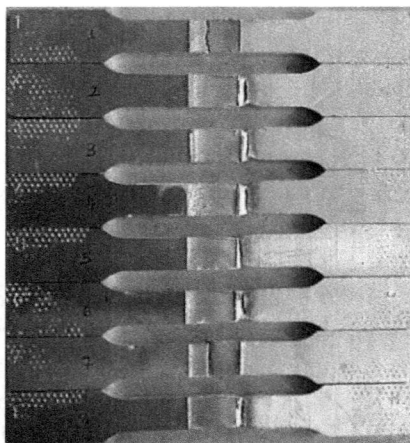

Figure 8. The fracture location of weld samples after tensile tests.

5. Conclusions

The present study focused on the effect of the ratio of the tool rotation speed to the welding speed (v ratio) on the tensile properties and macrostructural alteration for dissimilar friction stir welding of EN AW6082 to EN AW5083 aluminum alloys. Two different pin shapes were associated with a constant v ratio. The following conclusions were drawn.

- The pin shape has a significant effect on the tensile properties and microstructure. The strengths of the weld joint fabricated by the pentagonal-shaped pin was smaller than those of triangular-shaped pin for Exp. 7 and Exp. 8.
- The highest tensile strength was obtained from the weld joint fabricated with a triangular-shaped pin and the UTS and ε values were 198.48 MPa and 4.26%, respectively.
- The efficiency for joint strength is ranged from 55% to 68%. The variation trend changes depending on both defects in the weld joint and the strength of the base material.
- Except for the fourth weld joint, the others contained small cavity and tunnel defects.
- The fracture of each joint except in Exp. 1 and Exp. 7 was located at the side of EN AW6082.
- The nugget zone profile was affected by the pin shape. The nugget zone of each weld contained onion rings. The shape of the onion rings was dependent on the value of the tool rotational speed and welding speed.
- The effect of a constant v ratio on the profile and structure of the nugget zone is dependent on both the welding speed and tool rotational speed.
- At a lower tool rotational speed and welding speed for each tool pin shape, lower UTS values were obtained. The UTS increased as the tool rotational speed and the welding speed increased, while keeping the v ratio constant for the triangular-shaped pin.

Acknowledgments: The present study is supported by a scientific project research (No. 2011.KB.FEN.045).

Author Contributions: Sefika Kasman and Fatih Kahraman designed the experiments; Anıl Emiralioglu and Haydar Kahraman performed the experiments with Sefika Kasman and Fatih Kahraman; all authors analyzed the data and contributed to written of paper.

Conflicts of Interest: The authors declare no conflict of interest.

References

1.   DebRoy, T.; Bhadeshia, H.K.D.H. Friction stir welding of dissimilar alloys—A perspective. *Sci. Technol. Weld. Join.* **2010**, *15*, 266–270. [CrossRef]

2. Aalco Metal Limited, Aluminium Alloys. Available online: www.aalco.co.uk/datasheets/Aalco-Metals-Ltd_Aluminium-Alloy-5083--0-H111-Sheet-and-Plate_149.pdf.ashx (accessed on 9 November 2016).

3. Aalco Metal Limited, Aluminium Alloys. Available online: www.aalco.co.uk/datasheets/Aalco-Metals-Ltd_Aluminium-Alloy-6082-T6T651-Plate_148.pdf.ashx (accessed on 9 November 2016).

4. Peel, M.J.; Steuwer, A.; Withers, P.J. Dissimilar Friction Stir Welds in EN AW5083-EN AW6082, Part II: Process Parameter Effects on Microstructure. *Metall. Mater. Trans. A* **2006**, *37*, 2195–2206. [CrossRef]

5. Donatus, U.; Thompson, G.E.; Zhou, X.; Wang, J.; Cassell, A.; Beamish, K. Corrosion susceptibility of dissimilar friction stir welds of EN AW5083 and EN AW6082 alloys. *Mater. Charact.* **2015**, *107*, 85–97. [CrossRef]

6. Donatus, U.; Thompson, G.E.; Zhou, X.; Wang, J.; Beamish, K. Flow patterns in friction stir welds of EN AW5083 and EN AW6082 alloys. *Mater. Des.* **2015**, *83*, 203–213.

7. Steuwer, A.; Peel, M.J.; Withers, P.J. Dissimilar friction stir welds in EN AW5083-EN AW6082: The effect of process parameters on residual stress. *Mater. Sci. Eng. A* **2006**, *441*, 187–196. [CrossRef]

8. Leitão, C.; Louro, R.; Rodrigues, D.M. Analysis of high temperature plastic behaviour and its relation with weldability in friction stir welding for aluminium alloys AA5083-H111 and AA6082-T6. *Mater. Des.* **2012**, *37*, 402–409. [CrossRef]

9. Sun, Y.; Tsuji, N.; Fujii, H. Microstructure and mechanical properties of dissimilar friction stir welding between ultrafine grained 1050 and 6061-t6 aluminum alloys. *Metals* **2016**, *6*, 249. [CrossRef]

10. Aval, H.J.; Serajzadeh, S.; Kokabi, A.H. Thermo-mechanical and microstructural issues in dissimilar friction stir welding of AA5086-AA6061. *J. Mater. Sci.* **2011**, *46*, 3258–3268. [CrossRef]

11. Hao, H.L.; Ni, D.R.; Huang, H.; Wang, D.; Xiao, B.L.; Nie, Z.R.; Ma, Z.Y. Effect of welding parameters on microstructure and mechanical properties of friction stir welded Al-Mg-Er alloy. *Mater. Sci. Eng. A* **2013**, *559*, 889–896. [CrossRef]

12. Mishra, R.S.; Ma, Z.Y. Friction stir welding and processing. *Mater. Sci. Eng. R* **2005**, *50*, 1–78. [CrossRef]

Theoretical and Experimental Nucleation and Growth of Precipitates in a Medium Carbon–Vanadium Steel

Sebastián F. Medina [1,*], Inigo Ruiz-Bustinza [1], José Robla [1] and Jessica Calvo [2]

[1] National Centre for Metallurgical Research (CENIM-CSIC), Av. Gregorio del Amo 8, 28040 Madrid, Spain; irbustinza@cenim.csic.es (I.R.-B.); jrobla@cenim.csic.es (J.R.)

[2] Technical University of Catalonia (ETSEIB—UPC), Av. Diagonal 647, 08028 Barcelona, Spain; jessica.calvo@upc.edu

* Correspondence: smedina@cenim.csic.es

Academic Editor: Hugo F. Lopez

Abstract: Using the general theory of nucleation, the nucleation period, critical radius, and growth of particles were determined for a medium carbon V-steel. Several parameters were calculated, which have allowed the plotting of nucleation critical time vs. temperature and precipitate critical radius vs. temperature. Meanwhile, an experimental study was performed and it was found that the growth of precipitates during precipitation obeys a quadratic growth equation and not a cubic coalescence equation. The experimentally determined growth rate coincides with the theoretically predicted growth rate.

Keywords: microalloyed steel; nucleation time; nucleus radius; precipitate growing

1. Introduction

During the hot deformation of microalloyed steels an interaction takes place between the static recrystallization and precipitation of nanometric particles, which grow until reaching a certain size. The complexity of this phenomenon has been studied by many researchers and is influenced by all the external (temperature, strain, strain rate) and internal variables (chemical composition, austenite grain size) that participate in the microstructural evolution of hot strained austenite [1–8].

First of all, the precipitates nucleate and then increase in size by means of growth and coarsening or coalescence [9]. During coarsening, the largest particles grow at the expense of the smallest, and according to some authors this takes place due to the effect of accelerated diffusion of the solute along the dislocations, i.e., when the precipitation is strain-induced [10].

The present work basically addresses the precipitation in a vanadium microalloyed steel, distinguishing the growth and coarsening phases and determining which of these phenomena is really the most important for the increase in precipitate size. Calculations have been performed taking into account the general theory of nucleation [11,12].

Comparison of the calculated and experimentally obtained results helps to understand the precipitation phenomenon in its theoretical and experimental aspects, respectively.

2. Materials and Methods

The steel was manufactured by the Electroslag Remelting Process and its composition is shown in Table 1, which includes an indication of the $\gamma \rightarrow \alpha$ transformation start temperature during cooling (A_{r3}), determined by dilatometry at a cooling rate of 0.2 K/s, which is the minimum temperature at which the different parameters related with austenite phase precipitation will be calculated.

Table 1. Chemical composition (mass %) and transformation critical temperature (A_{r3}, 0.2 K/s).

C	Si	Mn	V	N	Al	A_{r3}, °C
0.33	0.22	1.24	0.076	0.0146	0.011	716

The specimens for torsion tests had a gauge length of 50 mm and a diameter of 6 mm. The austenitization temperature was 1200 °C for 10 min, and these conditions were sufficient to dissolve vanadium carbonitrides. The temperature was then rapidly lowered to the testing temperature of 900 °C, where it was held for a few seconds to prevent precipitation taking place before the strain was applied. After deformation, the samples were held for times of 50 s, 250 s, and 800 s, respectively, and finally quenched by water stream. The parameters of torsion, torque, and number of revolutions, and the equivalent parameters, stress and strain, were related according to the Von Mises criterion. Accordingly, the applied strain was 0.35, and the strain rate was 3.63 s^{-1}.

A transmission electron microscopy (TEM) (CM20 TEM/STEM 200KV, Philips, Eindhoven, The Netherlands) study was performed and the carbon extraction replica technique was used. The distribution of precipitate sizes was always determined on a population of close to 200 precipitates. This technique is widely used, but there are other competitive techniques such as quantitative X-ray diffraction (QXRD) [13].

3. Results and Discussion

3.1. Theoretical Model and Calculation of Parameters

The Gibbs energy for the formation of a spherical nucleus of carbonitride from the element in solution (V) is classically expressed as the sum of chemical free energy, interfacial free energy, and dislocation core energy, resulting in the following expression [14–16]:

$$\Delta G(J) = \frac{16\pi\gamma^3}{3\Delta G_v^2} + 0.8\mu b^2 \frac{\gamma}{\Delta G_v} \tag{1}$$

where γ is the surface energy of the precipitate (0.5 Jm^{-2}), ΔG_v is the driving force for nucleation of precipitates, b is the Burgers vector of austenite, and μ is the shear modulus. For austenite, $b = 2.59 \times 10^{-10}$ m and $\gamma = 4.5 \times 10^4$ MPa [9].

The equilibrium between the austenite matrix and the carbonitride VC$_y$N$_{1-y}$ is described by the mass action law [15]:

$$\Delta G_v \left(J \cdot m^{-3}\right) = -\frac{R_g T}{V_m}\left[\ln\left(\frac{X_V^{ss}}{X_V^e}\right) + y\ln\left(\frac{X_C^{ss}}{X_C^e}\right) + (1-y)\ln\left(\frac{X_N^{ss}}{X_N^e}\right)\right] \tag{2}$$

where X_i^{ss} are the molar fractions in the solid solution of V, C, and N, respectively, X_i^e are the equilibrium fractions at the deformation temperature, V_m is the molar volume of precipitate species, R_g is the universal gas constant (8.3145 J·mol^{-1}·K^{-1}), and T is the deformation absolute temperature.

The carbonitride is considered as an ideal mix of VC and VN, and the values of parameters (y; X_V^e; X_C^e; X_N^e) are determined using FactSage (Developed jointly between Thermfact/CRCT (Montreal, QC, Canada) and GTT-Technologies (Aachen, Germany)) [17,18]. The value of "y" in Equation (4) is the precipitated VC/VCN ratio, and "$1 - y$" is the VN/VCN ratio. On the other hand, $N_0 = 0.5 \Delta\rho^{1.5}$ is the number of nodes in the dislocation network, $\Delta\rho = (\Delta\sigma/0.2\mu b)^2$ is the variation in the density of dislocations associated with the recrystallization front movement in the deformed zone at the start of precipitation [9], and $\Delta\sigma$ is the difference between the flow stress and yield stress at the deformation temperature.

The atomic impingement rate is given as [10]

$$\beta'\left(s^{-1}\right) = \frac{4\pi R_c^2 D_V C_V}{a^4} \tag{3}$$

where D_V is the bulk diffusivity of solute atoms (V) in the austenite, a is the lattice parameter of the precipitate, and C_V is the initial concentration of vanadium in mol fraction. Here, the bulk diffusion coefficient (D_V) is replaced by an effective diffusion coefficient, D_{eff}, expressed as a weighted mean of the bulk diffusion (D_V) and pipe diffusion (D_p) coefficients, and used in the description of the precipitate evolution [10,19]:

$$D_{eff} = D_p \pi R_{core}^2 \rho + D_V \left(1 - \pi R_{core}^2 \rho\right) \tag{4}$$

where R_{core} is the radius of the dislocation core, taken to be equal to the Burgers vector, b.

The Zeldovich factor Z takes into account that the nucleus is destabilised by thermal excitation compared to the inactivated state and is given as [20]

$$Z = \frac{V_{at}^p}{2\pi R_c^2} \sqrt{\frac{\gamma}{kT}} \tag{5}$$

The flow stress increment ($\Delta\sigma$) has been calculated using the model reported by Medina and Hernández [21], which facilitates the calculation of flow stress. The dislocation density has been calculated at the nose temperature of the P_s curve corresponding to a strain of 0.35. When the austenite is not deformed the dislocation density is approximately 10^{12} m^{-2} [22]. The dislocation density corresponding to the curve nose will be given by $10^{12} + \Delta\rho$.

The incubation time (τ) is given as follows [23,24]:

$$\tau = \frac{1}{2\beta' Z^2} \tag{6}$$

The critical radius for nucleation is determined from the driving force and is given as [10]

$$R_c = -\frac{2\gamma}{\Delta G_v} \tag{7}$$

It is obvious that the nuclei that can grow have to be bigger than the critical nucleus, and in accordance with Dutta et al. [9] and Perez et al. [25] this value will be multiplied by 1.05. It must be noted that the factor 1.05 has little consequence on the overall precipitation kinetics.

The integration of Zener's equation for the growth rate will give the radius of the precipitates as a function of time [15]:

$$R^2 = R_0^2 + 2D_{eff}\frac{X_V^{SS} - X_V^i}{\frac{V_{Fe}}{V_p} - X_V^i}\Delta t \tag{8}$$

where V_{Fe} is the molar volume of austenite, V_p is the molar volume of precipitate, R is the precipitate radius after a certain time Δt, and R_0 is the average critical radius of the precipitates that have nucleated during the incubation period, coinciding with $1.05 R_0$.

The FactSage software tool for the calculation of phase equilibria and thermodynamic properties makes it possible to predict the formation of simple precipitates (nitrides and carbides) and more complex precipitates (carbonitrides), and the results can be expressed as a weight fraction versus temperature [26]. The FSstel database containing data for solutions and compounds [26] was used. The calculations were carried out every 10 °C in the temperature range between 740 °C and 1250 °C with the search for transition temperatures [27]. Figure 1 shows the fraction of AlN, VC, VN, and total VCN versus the temperature. In the model used (FactSage), the VCN fraction is the sum of VC and VN.

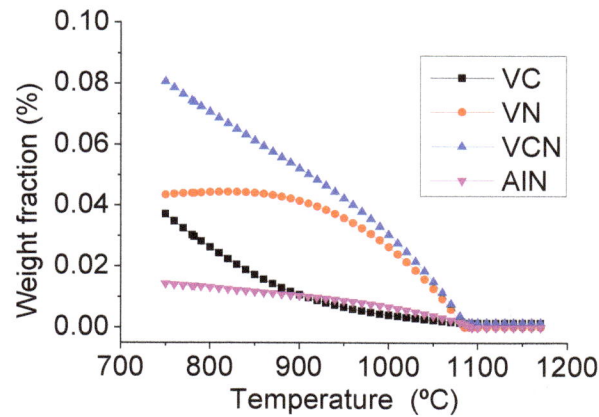

Figure 1. Equilibrium compounds predicted by FactSage.

The values calculated according to the above equations are shown in Figures 2–4. The incubation time (τ) is shown as a function of the temperature (Figure 2). As the effective energy was taken into account in Equation (3) instead of only the energy for bulk diffusion, the values of τ were lower. The critical radius (R_c) for nucleation was calculated from Equation (7) as a function of the temperature (Figure 3).

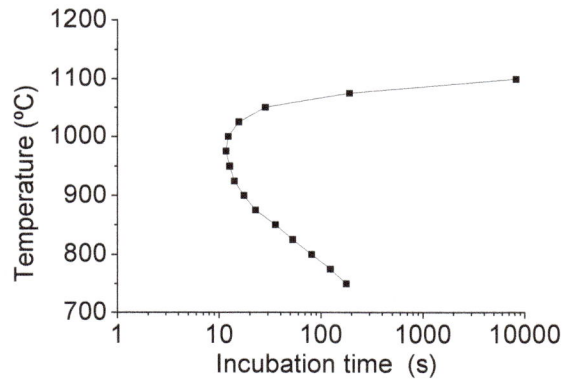

Figure 2. Incubation time vs. temperature (curve P_s) for VCN precipitates.

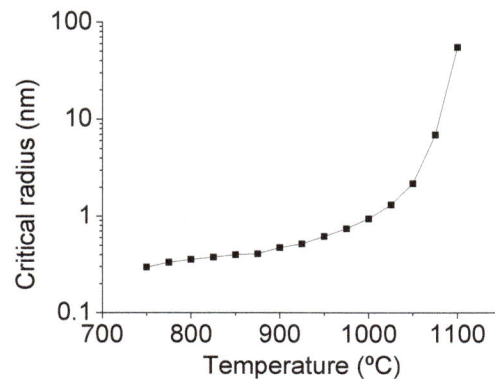

Figure 3. Critical radius size vs. temperature for VCN precipitates.

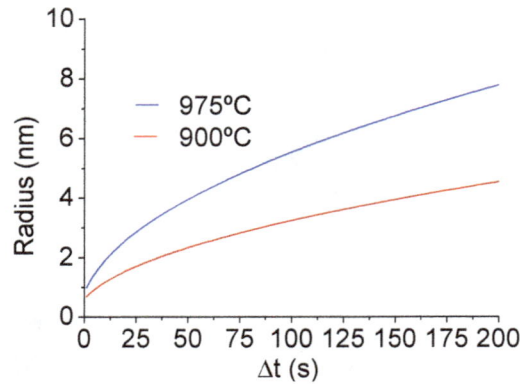

Figure 4. Precipitate size (radius) growth vs. time ($\Delta t > \tau$) for VCN precipitates. Equation (8).

Equation (8) has been applied to the precipitate growth, selecting temperatures of 975 °C (curve nose in Figure 2) and 900 °C. The Δt time starts to be counted after the nucleation period, which was 12 s at 975 °C and 17 s at 900 °C. The value of R_0 was 0.8 nm and 0.5 nm, both multiplied by 1.05, at 975 °C and 900 °C, respectively. The growth of nuclei is shown in Figure 4. The calculations have been performed using parameters extracted from literature [9,10,15,28,29] and shown in Table 2.

Table 2. Parameters used for calculations.

Parameter	Symbol	Value	Reference
Burgers vector	b (m)	2.59×10^{-10}	[9]
Shear modulus	μ (MPa)	4.5×10^4	[9]
Interfacial energy	γ (J·m^{-2})	0.5	[10,15]
Lattice parameter (VCN)	a (nm)	0.4118	[26]
Bulk diffusion of V	D_V (m^2·s^{-1})	$0.28 \times 10^{-4} e^{(-264,000/RT)}$	[27]
Pipe diffusion	D_p (m^2·s^{-1})	$0.25 \times 10^{-4} e^{(-210,000/RT)}$	[26]
Molar volume of VCN	V_P (m^3·mol^{-1})	10.65×10^{-6}	[15]
Molar volume of austenite	V_{Fe} (m^3·mol^{-1})	7.11×10^{-6}	[15]
Dislocation core radius	R_{core} (m)	5.00×10^{-10}	[10]

The term ΔG_v Equation (2) is important in the calculations carried out, whose values at 975 °C and 900 °C were -1.347×10^9 J·m^{-3} and -2.115×10^9 J·m^{-3}, respectively.

Dutta et al. [9] checked that the number of particles per unit volume decreased dramatically and this is not predicted by a simple growth model. Such a reduction in the density of particles per unit volume requires particle coarsening. The coarsening of precipitates by coalescence occurs once precipitation is complete, i.e., when the plateau has ended and recrystallization progresses again. Coalescence can be explained by the modified Lifshitz–Slyozov–Wagner theory (MLSW) [29,30], which predicts that while the basic $t^{1/3}$ kinetics of the LSW theory is maintained, the coarsening rate increases as the volume fraction increases, even at very small precipitated volume fraction values.

Coarsening is given by the expression:

$$R^3 = R_0^3 + \frac{8 D_V \gamma V_P^2 X_V^i}{9 R_g T} \Delta t \tag{9}$$

where R_g is the gas constant and the other parameters have already been explained.

Coarsening is the process by which the smallest precipitates dissolve to the profit of the larger ones, leading to a distribution peak shift to larger sizes. This phenomenon is particularly important when the system reaches the equilibrium precipitate fraction. It is to be noted that coarsening can occur at every stage of the precipitation process [15].

The most striking aspect of Equation (9) is that the mean radius cubed varies with time, as opposed to the squared radius in growth calculations. Coarsening is thus a much slower process than precipitate growth, as is reasonable given that the growth of one particle only occurs by cannibalistic particle growth [22].

Following Equation (9), nucleus growth was calculated for the temperatures of 975 °C and 900 °C. Once again, D_V has been replaced by D_{eff}, which is one order of magnitude higher. The result is shown in Figure 5, where it can be seen that coarsening does not take place at these temperatures and is therefore nil, unlike that shown in Figure 4. This is because the value of D_{eff}, or D_V, is very low at the mentioned temperatures and rises as the temperature increases. In other words, at higher temperatures coarsening by coalescence would also probably take place during precipitation and after this has ended.

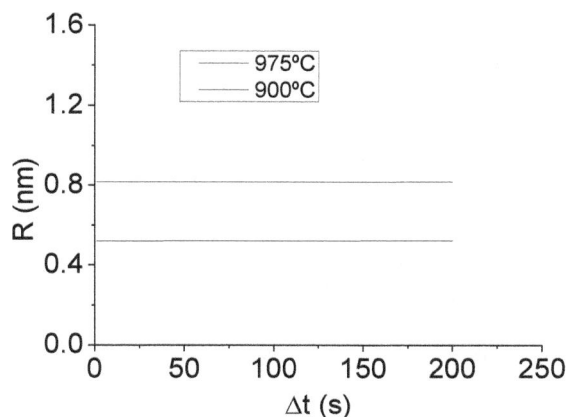

Figure 5. Precipitate size (radius) coarsening vs. time ($\Delta t > \tau$). Equation (9).

3.2. Evolution of Experimental Precipitate Size

The applied strain was 0.35, which was insufficient to promote dynamic recrystallization [31]. At high temperatures (above T_{nr}), the dislocation structure development governs the rates of recovery and recrystallization, which affect the austenite grain size, high temperature strength, and start of precipitation [32].

The resolution of precipitates using the carbon extraction replica technique by TEM is shown in Figure 6a–c, obtained on a specimen strained and quenched in the conditions indicated at the foot of the figure. The spectrum in Figure 6b shows the presence of V on the precipitate indicated by the red arrow in Figure 6a, and the lattice parameter determined from Figure 8c reveals a fcc cubic lattice with a value of $a = 0.413$ nm, which is identified, in accordance with the reference value found in the literature [22], as a vanadium carbonitride (VCN). The two unassigned peaks are Fe and Ni (6.4 keV and 7.4 keV, respectively). They may be due to grid impurities or residues from preparation in the evaporator. It is well known that the copper peak always appears in the carbon extraction replica technique and is due to the copper support grid.

On the other hand, the particle size distribution and determination of the average particle have been obtained by measuring an average number of 200 particles on the tested specimens. Figure 7 shows a log-normal distribution that corresponds to a holding time of 50 s at 900 °C, and a bimodal distribution corresponding to holding times of 250 s and 800 s, respectively, with a weighted mean size (diameter $\bar{D}$) of 11.2 nm, 19.2 nm, and 20.4 nm, respectively. Other authors have found similar sizes of V(C,N) precipitates [33].

(a)   (b)

(c)

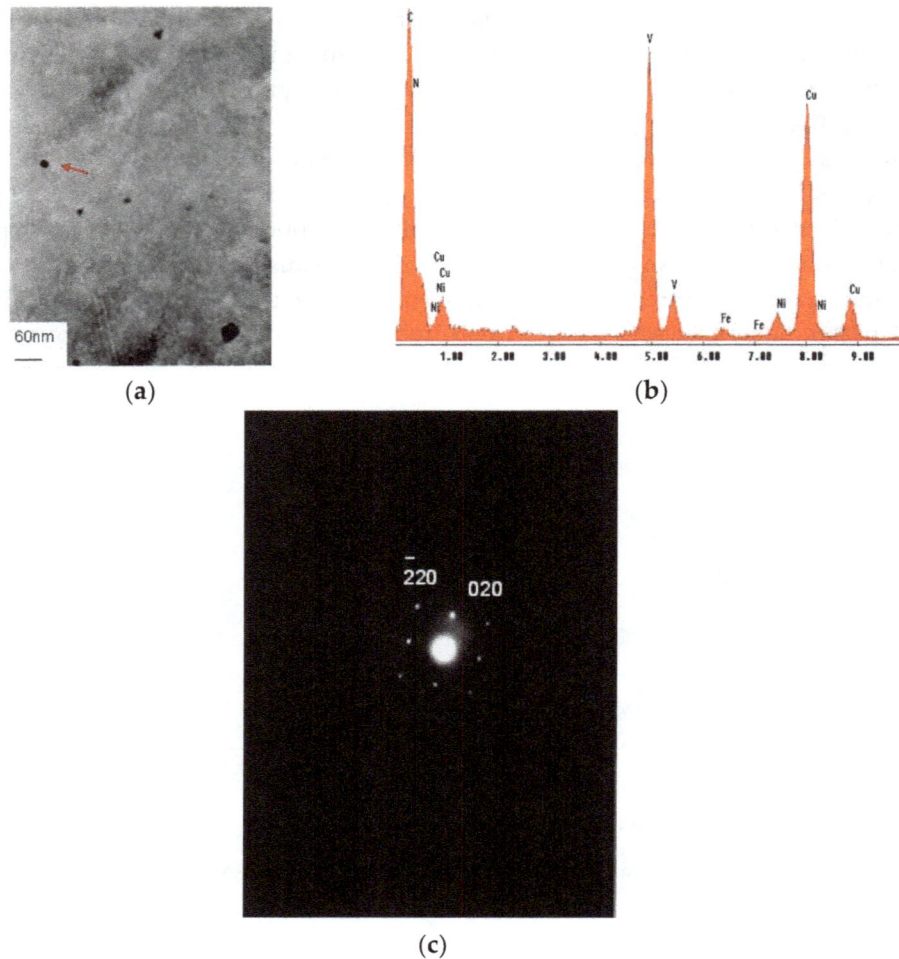

Figure 6. TEM images of steel used. (a) Image showing precipitates for specimen tested at reheating temperature of 1200 °C/10 min; Def. temp.= 900 °C; strain = 0.35; strain rate = 3.63 s^{-1}; holding time = 250 s; (b) EDAX spectrum of precipitate; (c) Electron diffraction image.

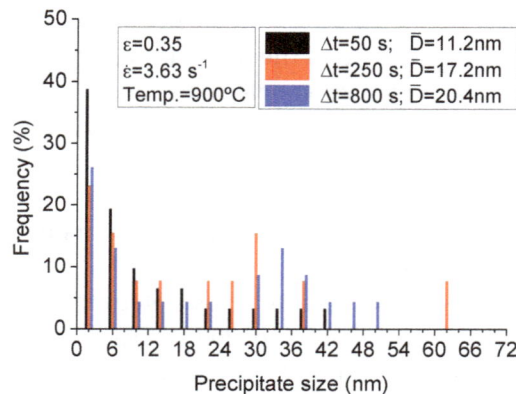

Figure 7. Size distribution of precipitates for specimen tested at reheating temperature of 1200 °C for 10 min, deformation temperature of 900 °C, and holding times of 50 s, 250 s, and 800 s.

If we accept that the precipitates grow once they are nucleated, then moment zero of growth coincides with the nucleation time (τ) and Δt will be the sample holding time after τ. Figure 8 shows a representation of the average precipitate growth at a temperature of 900 °C, where the size is given by the mean radius ($R = \bar{D}/2$). The holding times after the nucleation time were 50 s, 250 s, and 800 s, minus the value of 17 s corresponding to the calculated incubation time at 900 °C. At first sight there are seen to be two growth laws. Between the first two points, the nucleus grows notably; between the last two, it barely grows. These two laws obey Equations (8) and (9), respectively, and show that the precipitates grow during precipitation but that coarsening by coalescence does not take place. As coalescence does not occur, it has already been seen that Equation (9) is a constant at 900 °C (Figure 5). If a horizontal line is drawn from the final point, with coordinates at 783 s; 10.2 nm, to its intersection with the quadratic curve, the intersection point is determined to be at coordinates 275 s; 10.2 nm, which means that precipitation has ended 275 s after the nucleation time (17 s).

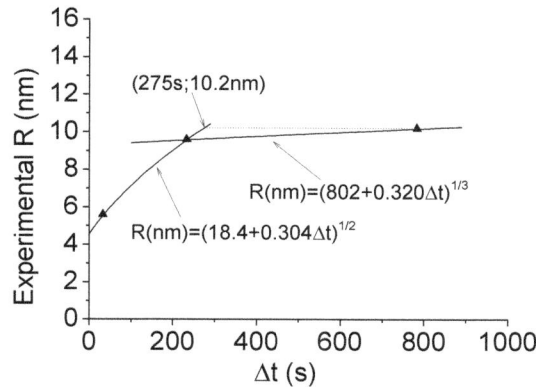

Figure 8. Experimental growth and coarsening of precipitates.

From Figure 8, in particular from the quadratic equation, it is possible to deduce the nucleus radius (R_0). This indicates a size of 4.3 nm when $\Delta t = 0$, corresponding to the nucleus size at 900 °C, which is greater than the nucleus radius calculated at the theoretical curve nose, which was approximately 0.5 nm. It is necessary to highlight the conceptual difference between the theoretical and experimental incubation time curves, respectively. The calculated and experimental nucleation times do not have the same meaning. The nucleation time calculated at any temperature means the time necessary for the formation of a number N of nuclei, of size R_0, whose growth continues when the precipitate radius reaches a size of $1.05R_0$. The experimental nucleation time refers to the time necessary for the pinning forces to exceed the driving forces for recrystallization, temporarily inhibiting the progress of recrystallization [34]. On the other hand, the carbon extraction replica technique does not easily detect precipitates with sizes of less than 1 nm.

The results confirm that the time of 275 s deduced for the end of precipitation, to which the 17 s of the nucleation time must be added, coincides approximately with the experimentally determined precipitation end time [35].

In other words, to explain the increase in the average precipitate size, it is sufficient to use the growth equation, and coalescence can be ignored. On the other hand, from the holding time of 50 s to the holding time of 250 s, the measured precipitate radius has increased from 5.6 nm to 9.6 nm, an increase of 4 nm. If the times of 50 s and 200 s are inserted in Figure 4, subtracting from both the calculated incubation time (17 s) at 900 °C, it is seen that the nucleus radius increases from 2 nm to 4.6 nm. Therefore, the calculated and experimental growth rates are similar.

It is thus deduced from both the experimental and theoretical points of view that precipitate growth between the start (P_s) and end of precipitation (P_f) can be predicted by the growth equation, as coarsening is not taking place, so it is not necessary to include the coalescence equation. The main difference between the calculated and experimental values is the nucleus size.

4. Conclusions

In conclusion, the calculated mean nucleus radius was approximately 0.5 nm and the experimental average radius was 4.3 nm at the temperature of 900 °C. This is due to the conceptual difference between the theoretical and experimental incubation time curves, respectively. The precipitate size growth rate between the experimentally determined P_s and P_f practically coincides with the calculated growth rate. Both are expressed by a quadratic growth equation in such a way that the coalescence of precipitates during precipitation is not taking place and can be ruled out. Since the effective diffusion coefficient (D_{eff}) parameter increases notably with the temperature, coarsening can take place at very high temperatures.

Acknowledgments: We acknowledge the financial support of the Spanish CICYT (Project MAT 2011-29039-C02-02).

Author Contributions: Sebastián F. Medina conceived and designed the work and also performed and interpreted the results; Inigo Ruiz-Bustinza and José Robla made calculations and drew graphs; Jessica Calvo did thermodynamic calculations and drew the corresponding graphs. All contributed to the writing of the paper, especially S.F. Medina.

Conflicts of Interest: The authors declare no conflict of interest.

References

1.   Gómez, M.; Medina, S.F.; Quispe, A.; Valles, P. Static recrystallization and induced precipitation in a low Nb microalloyed steel. *ISIJ Int.* **2002**, *42*, 423–431. [CrossRef]
2.   Medina, S.F.; Quispe, A.; Gómez, M. Strain induced precipitation effect on austenite static recrystallization in microalloyed steels. *Mater. Sci. Technol.* **2003**, *19*, 99–108. [CrossRef]
3.   Andrade, H.L.; Akben, M.G.; Jonas, J.J. Effect of molybdenum, niobium, and vanadium on static recovery and recrystallization and on solute strengthening in microalloyed steels. *Metall. Trans. A* **1983**, *14*, 1967–1977. [CrossRef]
4.   Kwon, O. A technology for the prediction and control of microstructural changes and mechanical properties in steel. *ISIJ Int.* **1992**, *32*, 350–358. [CrossRef]
5.   Luton, M.J.; Dorvel, R.; Petkovic, R.A. Interaction between deformation, recrystallization and precipitation in niobium steels. *Metall. Trans. A* **1980**, *11*, 411–420. [CrossRef]
6.   Gómez, M.; Rancel, L.; Medina, S.F. Effects of aluminium and nitrogen on static recrystallization in V-microalloyed steels. *Mater. Sci. Eng. A* **2009**, *506*, 165–173. [CrossRef]
7.   Gómez, M.; Medina, S.F. Role of microalloying elements on the microstructure of hot rolled steels. *Int. J. Mater. Res.* **2011**, *102*, 1197–1207. [CrossRef]
8.   Kwon, O.; DeArdo, A. Interactions between recrystallization and precipitation in hot-deformed microalloyed steels. *Acta Metall. Mater.* **1990**, *39*, 529–538. [CrossRef]
9.   Dutta, B.; Valdes, E.; Sellars, C.M. Mechanisms and kinetics of strain induced precipitation of Nb(C,N) in austenite. *Acta Metall. Mater.* **1992**, *40*, 653–662. [CrossRef]
10. Dutta, B.; Palmiere, E.J.; Sellars, C.M. Modelling the kinetics of strain induced precipitation in Nb microalloyed steels. *Acta Mater.* **2001**, *49*, 785–794. [CrossRef]
11. Russel, K.C. Nucleation in solids: The induction and steady state effects. *Adv. Colloid. Interface Sci.* **1980**, *13*, 205–318. [CrossRef]
12. Kampmann, R.; Wagner, R. *A Comprehensive Treatment, Materials Science and Technology*; Cahn, R.W., Ed.; VCH: Weinheim, Germany, 1991; pp. 213–303.
13. Wiskel, J.B.; Lu, J.; Omotoso, O.; Ivey, D.G.; Henein, H. Characterization of precipitates in a microalloyed steel using quantitative X-ray diffraction. *Metals* **2016**, *6*, 90. [CrossRef]
14. Fujita, N.; Bhadeshia, H.K.D.H. Modelling precipitation of niobium carbide in austenite: Multicomponent diffusion, capillarity and coarsening. *Mater. Sci. Technol.* **2001**, *17*, 403–408. [CrossRef]
15. Maugis, P.; Gouné, M. Kinetics of vanadium carbonitride precipitation in steel: A computer model. *Acta Mater.* **2005**, *53*, 3359–3367. [CrossRef]
16. Sun, W.P.; Militzer, M.; Bai, D.Q.; Jonas, J.J. Measurement and modelling of the effects of precipitation on recrystallization under multipass deformation conditions. *Acta Metall. Mater.* **1993**, *41*, 3595–3604. [CrossRef]

17. Salas-Reyes, A.E.; Mejía, I.; Bedolla-Jacuinde, A.; Boulaajaj, A.; Calvo, J.; Cabrera, J.M. Hot ductility behavior of high-Mn austenitic Fe-22Mn-1.5Al-1.5Si-0.45C TWIP steels microalloyed with Ti and V. *Mater. Sci. Eng. A* **2014**, *611*, 77–89. [CrossRef]

18. Mejía, I.; Salas-Reyes, A.E.; Bedolla-Jacuinde, A.; Calvo, J.; Cabrera, J.M. Effect of Nb and Mo on the hot ductility behavior of a high-manganese austenitic Fe-21Mn-1.3Al-1.5Si-0.5C TWIP steel. *Mater. Sci. Eng. A* **2014**, *616*, 229–239. [CrossRef]

19. Zurob, H.S.; Hutchinson, C.R.; Brechet, Y.; Purdy, G. Modeling recrystallization of microalloyed austenite: Effect of coupling recovery, precipitation and recrystallization. *Acta Mater.* **2002**, *50*, 3075–3092. [CrossRef]

20. Perrard, F.; Deschamps, A.; Maugis, P. Modelling the precipitation of NbC on dislocations in α-Fe. *Acta Mater.* **2007**, *55*, 1255–1266. [CrossRef]

21. Medina, S.F.; Hernández, C.A.; Ruiz, J. Modelling austenite flow curves in low alloy and microalloyed steels. *Acta Mater.* **1996**, *44*, 155–163.

22. Gladman, T. *The Physical Metallurgy of Microalloyed Steels*; The Institute of Materials: London, UK, 1997; pp. 28–56.

23. Perez, M.; Courtois, E.; Acevedo, D.; Epicier, T.; Maugis, P. Precipitation of niobium carbonitrides in ferrite: Chemical composition measurements and thermodynamic modelling. *Phil. Mag. Lett.* **2007**, *87*, 645–656. [CrossRef]

24. Radis, R.; Schlacher, C.; Kozeschnik, E.; Mayr, P.; Enzinger, N.; Schröttner, H.; Sommitsch, C. Loss of ductility caused by AlN precipitation in Hadfield steel. *Metall. Mater. Trans. A* **2012**, *43*, 1132–1139. [CrossRef]

25. Perez, M.; Deschamps, A. Microscopic modelling of simultaneous two phase precipitation: Application to carbide precipitation in low carbon steels. *Mater. Sci. Eng. A* **2003**, *360*, 214–219. [CrossRef]

26. FSstel Database. Available online: http//www.factsage.com (accessed on 4 December 2010).

27. Bale, C.W.; Bélisle, E.; Chartrand, P.; Degterov, S.A.; Eriksson, G.; Hack, K.; Jung, I.H.; Kang, Y.B.; Melancon, J.; Pelton, A.D.; et al. FactSagethermo-chemical software and databases—Recent developments. *Calphad* **2009**, *33*, 295–311. [CrossRef]

28. Mukherjee, M.; Prahl, U.; Bleck, W. Modelling the strain-induced precipitation kinetics of vanadium carbonitride during hot working of precipitation-hardened ferritic-pearlitic steels. *Acta Mater.* **2014**, *71*, 234–254. [CrossRef]

29. Oikawa, H. Lattice diffusion in iron-a review. *Tetsu Hagane* **1982**, *68*, 1489–1497.

30. Ardell, J. The effect of volume fraction on particle coarsening: Theoretical considerations. *Acta Metall.* **1972**, *20*, 61–71. [CrossRef]

31. Badjena, S.K. Dynamic recrystallization behavior of vanadium micro-alloyed forging medium carbon steel. *ISIJ Int.* **2014**, *54*, 650–656. [CrossRef]

32. Kostryzhev, A.G.; Mannan, P.; Marenych, O.O. High temperature dislocation structure and NbC precipitation in three Ni-Fe-Nb-C model alloys. *J. Mater. Sci.* **2015**, *50*, 7115–7125. [CrossRef]

33. Nafisi, S.; Amirkhiz, B.S.; Fazeli, F.; Arafin, M.; Glodowski, R.; Collins, L. Effect of vanadium addition on the strength of API X100 linepipe steel. *ISIJ Int.* **2016**, *56*, 154–160. [CrossRef]

34. Gómez, M.; Medina, S.F.; Valles, P. Determination of driving and pinning forces for static recrystallization during hot rolling of a Nb-microalloyed steel. *ISIJ Int.* **2005**, *45*, 1711–1720. [CrossRef]

35. Medina, S.F.; Quispe, A.; Gomez, M. New model for strain induced precipitation kinetics in microalloyed steels. *Metall. Mater. Trans. A* **2014**, *45*, 1524–1539. [CrossRef]

3

Vibration-Assisted Sputter Coating of Cenospheres: A New Approach for Realizing Cu-Based Metal Matrix Syntactic Foams

Andrei Shishkin [1,*], **Maria Drozdova** [2], **Viktor Kozlov** [3], **Irina Hussainova** [2] and **Dirk Lehmhus** [4]

[1] Rudolfs Cimdins Riga Biomaterials Innovations and Development Centre of Riga Technical University (RTU), Institute of General Chemical Engineering, Faculty of Materials Science and Applied Chemistry, Riga Technical University, Pulka 3, LV-1007 Riga, Latvia

[2] Department of Mechanical and Industrial Engineering, Tallinn University of Technology, 19086 Tallinn, Estonia; maria.drozdova@ttu.ee (M.D.); irina.hussainova@ttu.ee (I.H.)

[3] Sidrabe Inc., LV-1073 Riga, Latvia; vkozlovs@sidrabe.eu

[4] MAPEX Center for Materials and Processes, University of Bremen, 28359 Bremen, Germany; dirk.lehmhus@uni-bremen.de

* Correspondence: andrejs.siskins@rtu.lv

Academic Editor: Hugo F. Lopez

Abstract: The coating of hollow alumino-silicate microspheres or cenospheres with thin layers of Cu by means of vibration-assisted magnetron sputtering yields a starting material with considerable potential for the production of new types of metal matrix syntactic foams as well as optimized variants of conventional materials of this kind. This study introduces the coating process and the production of macroscopic samples from the coated spheres via spark plasma sintering (SPS). The influence of processing parameters on the coating itself, and the syntactic foams are discussed in terms of the obtained density levels as a function of sintering temperature (which was varied between 850 and 1080 °C), time (0.5 to 4 min), and surface appearance before and after SPS treatment. Sintering temperatures of 900 °C and above were found to cause breaking-up of the homogeneous sputter coating into a net-like structure. This effect is attributed to wetting behavior of Cu on the alumino-silicate cenosphere shells. Cylindrical samples were subjected to conductivity measurements and mechanical tests, and the first performance characteristics are reported here. Compressive strengths for Cu-based materials in the density range of 0.90–1.50 g/cm^3 were found to lie between 8.6 and 61.9 MPa, depending on sintering conditions and density. An approximate relationship between strength and density is suggested based on the well-known Gibson–Ashby law. Density-related strength of the new material is contrasted to similar findings for several types of established metal foams gathered from the literature. Besides discussing these first experimental results, this paper outlines the potential of coated microspheres as optimized filler particles in metal matrix syntactic foams, and suggests associated directions of future research.

Keywords: syntactic foam; spark plasma sintering; magnetron sputtering; fly ash; cenospheres; composite; metal matrix composite; metal matrix syntactic foam

1. Introduction

Metal-matrix syntactic foams (MMSF) using micro-scale filler particles for density reduction have recently been demonstrated to show attractive property profiles in combination with economic viability. The published literature encompasses melt- as well as powder-based processes and matrix materials ranging from magnesium to aluminum, titanium, zinc, iron, and steel, often in combination with either glass hollow spheres or cenospheres as additives. Cenospheres are a by-product of coal combustion and are available at low cost as so-called fly ash from coal-burning power plants. Effectively, they are a chemically inert [1] and temperature-resistant type of hollow ceramic micro beads which have successfully demonstrated their suitability for use as filler particle, even in conjunction with high temperature processes [2]. The present study looks at cenosphere adaptation by means of sputter coating in view of producing MMSF with optimized mechanical properties and reduced weight.

A comprehensive overview of MMSF methods and materials is offered by Gupta and Rohatgi [3]. Most processes investigated so far limit the effective range of porosities to an approximate maximum at around 50%. At the same time, several publications point at the fact that properties specifically under tensile load could benefit from improved interface strength between matrix and hollow particle, which is typically not achieved today [4,5]. The present study reverses the usual processing sequence by creating the interface between metal phase and hollow particles first and consolidating the resulting particles to form macroscopic bodies afterwards. The process employed for this purpose is vibration-assisted sputter coating of the hollow particles with metals. The expected benefits include increased interface strength, additional control over interface properties, and access to lower density materials through direct sintering of metal-coated spheres as done in the present study.

Sputter-coating of powders and even of hollow microspheres is not entirely new: Both Koppel et al. and Xiaozheng and Zhigang have reported electromagnetic absorption/shielding properties of Cu- [6] and Ni-coated [7] cenospheres in bulk condition [6] and embedded in a paraffin matrix [7]. Additional interest in this coating approach is based on the envisaged potential for performance optimization in conventional powder metallurgical syntactic foams through tailored interface definition. The main advantage of sputtering in this respect is its variability in terms of coating materials compared to conventional chemical powder coating methods, as well as the improved uniformity of the coating in terms of thickness and surface quality [8,9].

2. Materials and Methods

Cenospheres were acquired from Biotecha Latvia Ltd (Riga, Latvia). Their composition was chemically determined to be 53.8 ± 0.5 wt % SiO_2, 40.7 ± 0.7 wt % Al_2O_3, 1.4 ± 0.2 wt % CaO, and 1.0 ± 0.2 wt % Fe_2O_3, plus smaller amounts (below 1 wt %) of MgO, Na_2O, and K_2O. Bulk density of the particles prior to coating was 0.39 ± 0.006 g/cm^3 based on repeated Scott volumeter measurements according to ISO 3923-2-81.

Copper coating was conducted using an experimental magnetron sputtering setup at Sidrabe Ltd. (Riga, Latvia), in which the powder reservoir situated below the target was constantly agitated by a vibration source to ensure homogeneous coverage. The Cu sputter targets had a purity of 99.9%.

Consolidation of material samples for compression tests was done by spark plasma sintering (SPS) using KCE®-FCT HP D 10-GB equipment by FCT Systeme GmbH (Rauenstein, Germany). A graphite die of 21 mm diameter was filled with either 1 g (disc-shaped samples, nominal size of approximately 20 mm diameter and 2.5 mm thickness/height) or 5.5 g (cylindrical samples, nominal size approximately 20 mm diameter and 19 mm height) of Cu-coated cenospheres plus a graphite foil wrapping of 0.6 ± 0.2 mm thickness to prevent the sample from sticking to the die walls. Subsequently, the die was evacuated and the powder held at a constant pressure of 9.5 MPa throughout the process. Heating rates employed were 100 K/min for ramp-up and ~200 K/min during cooling. Sintering temperatures and times were varied in the ranges 850–1080 °C and 0.5–4.0 min, respectively. Sample core temperatures were acquired pyrometrically via a bore hole in the graphite punch. Prior to mechanical testing and density evaluation, the samples were machined in order to remove remainders of the graphite foil. This led to some variation in sample dimensions and weight which were naturally accounted for in the determination of both density and compression strength.

Imaging relied on a Zeiss EVO MA-15 scanning electron microscope (SEM) (Carl Zeiss AG, Oberkochen, Germany) and Image Pro 7 image analysis tool by Media Cybernetics (version 7.0, Media Cybernetics, Inc., Rockville, MD, USA, 2011).

Compression tests were performed on an Instron 8801 universal testing machine (Instron, Norwood, MA, USA) at room temperature and a constant quasi-static test speed of 0.01 mm/s on three samples per material variant.

Apparent density and porosity of sintered MMSF were evaluated according to Archimedes' principle (results are averaged over five samples per set of sintering conditions). Immersion in 2-prophanol (Sigma Aldrich, St. Louis, MO, USA, grade 99.5%) was employed to prevent Cu corrosion. The relative density of 2-prophanol at 22 °C (0.7834 g/cm^3) was taken into account in the calculation. Samples were encapsulated during immersion to avoid infiltration of the open porosity present between coated microspheres.

3. Results

Vibration-assisted sputter coating led to homogeneous and continuous metal layers in a thickness range of approximately 0.4–2.5 µm, as depicted in Figure 1.

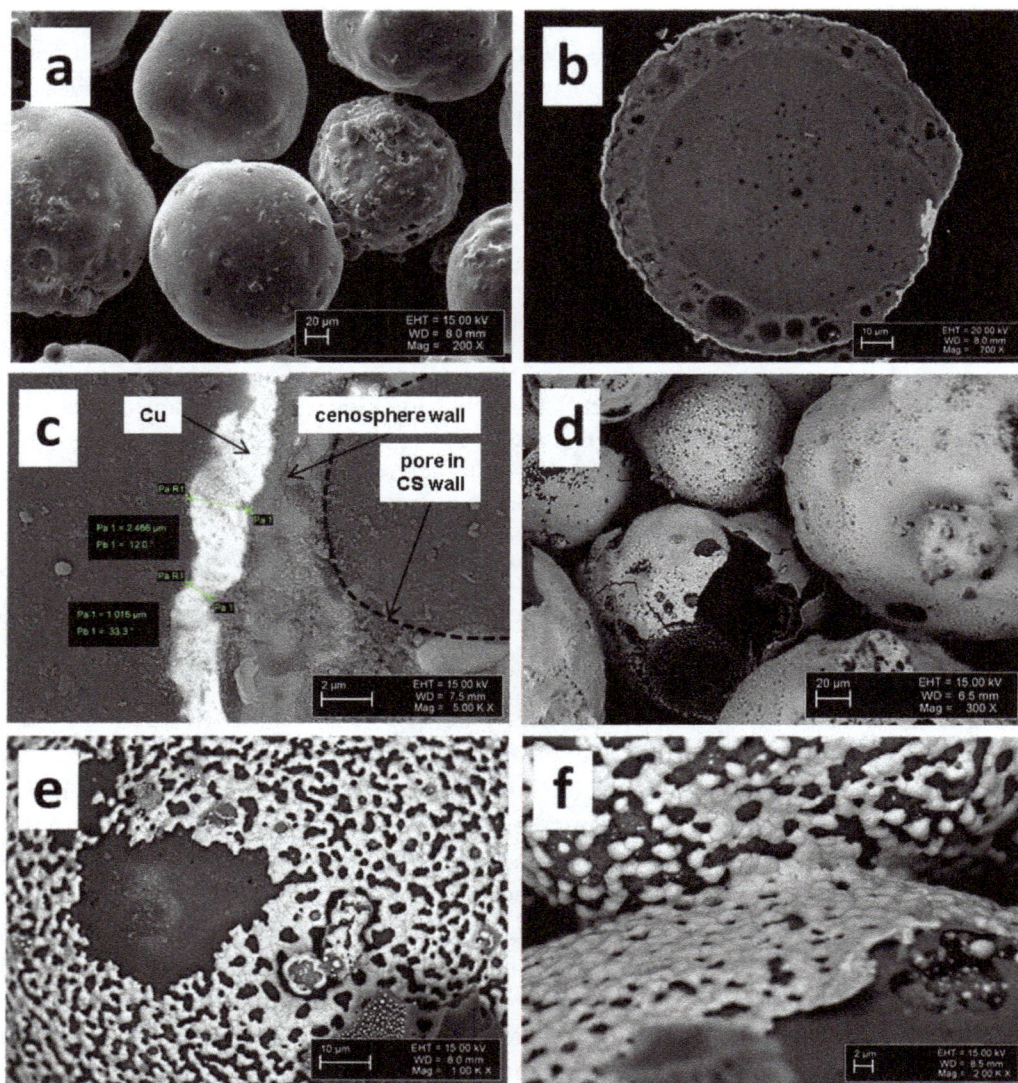

Figure 1. Various views of copper-coated cenospheres (Cu@CS materials, disc-shaped samples) before (**a–c**) and after sintering (**d–f**): (**a**) outward appearance; (**b**) cross section; (**c**) detail showing coating thickness; (**d**) sintered at 800 °C; (**e**) sintered at 900 °C; and (**f**) detail view showing inter-particle connection after sintering at 900 °C.

Due to the coating process, powder bulk density increased from an initial 0.39 ± 0.006 g/cm^3 to 0.51 ± 0.004 g/cm^3. Assuming that the basic shape and flow characteristics remained unchanged, this increase must be related to the increase in particle density caused by the Cu layer. As can be seen in Figure 1d–f, the sintering process strongly affects the coating morphology. The net-like texture observed for temperatures of 900 °C and above is associated with inferior wetting between liquid Cu and the alumino-silicate cenospheres, which causes loss of the full metal coverage originally obtained during sputtering. Occurrence of this effect below the melting point of pure copper (1094 °C) may suggest some reaction between coating and substrate during sintering, and an associated formation of lower-melting phases. This interpretation is supported by a number of studies investigating the wetting of Cu and Cu alloys on ceramic substrates, mainly focusing on alumina. The fact that no studies specifically addressing the combination of Cu and alumino-silicate as the cenosphere shell materials were found is alleviated by Chidambaram et al.'s observation that wetting in liquid metal–ceramic systems tends to be determined by the liquid metal's contact to the interface formed between the two partners, rather than the direct contact between liquid metal and ceramic [10]. An example in this respect has been reported by Espie et al., who showed that contact angles between a copper alloy on the one hand and alumina, mullite, and silicate ceramics on the other hand are almost identical [11]. Interface formation itself has been studied in detail by Kelber et al. [12], who conclude that interface-forming reactions between Cu and alumina specifically are limited to less than a full monolayer under the conditions of measurement, with metallic copper being observed at higher levels of coverage. Cu-alumina wetting is described as poor or even non-wetting under these circumstances, and the observed variation in literature values is explained based on variations in surface conditions that were not accounted for. Hydroxylation is mentioned in this context, which has been found to significantly lower the contact angle [12]. Chidambaram et al. and Meier et al. further stress the role of oxygen for improving wetting between Cu and alumina [11,13]. As contact angle value, the former report 34.4° at 1300 °C for a copper alloy containing 3 wt % of oxygen [11], while Eustathopoulos suggest between 120° and 130° both for pure copper in contact with Al_2O_3 and SiO_2 [14]. In the present case, the oxygen content of the Cu layer may be assumed to fall below the previously mentioned 3 wt %, which was achieved by oxide addition. The contact angle can thus be expected to approach the higher 120°–130° boundary rather than the lower one. In contrast, the full coverage achieved in the solid state through sputter coating would correspond to perfect wetting, and thus a contact angle approaching 0°. Given that the true contact angle value for this combination of materials is likely to be much higher, the disruption of the continuous surface layer during melting is not surprising.

Despite the observed loss of film coherence, SPS sintering still leads to good particle interconnection, as is illustrated by Figure 1f. This finding is reinforced by measurements showing varying electrical conductivities between 2.05×10^4 and 3.6×10^4 S/m for all samples, thus suggesting the existence of a continuous metallic network within the bulk of the material, even in those cases in which the aforementioned net-like structure formed.

The dependence of density on the SPS sintering parameters is represented in Figure 2 and Table 1 for cylindrical samples. Initial tests on disc-shaped samples had shown no influence of sintering parameters on density at lower processing temperatures (850, 900 °C). For this reason, the temperature range studied was limited to 950 °C and above for the cylindrical samples subsequently subjected to mechanical evaluation. The upper temperature limit of 1080 °C was motivated by the melting temperature of pure copper.

Table 1. Density and compressive strength values for Cu-coated cenospheres (Cu@CS) as measured on cylindrical samples.

Sintering Conditions		Density	Compressive Strength
T (°C)	t (min)	ϱ (g/cm^3)	σ_c (MPa)
950	0.5	0.869 [1]/0.904 [2]	8.60
	1	0.931 [1]/0.960 [2]	11.00
	2	0.999 [1]/1.025 [2]	16.60
	4	1.043 [1]/1.086 [2]	21.60
1000	0.5	0.986 [1]/1.010 [2]	17.20
	1	1.049 [1]/1.080 [2]	26.20
	2	1.003 [1]/1.016 [2]	15.70
	4	1.283 [1]/1.291 [2]	30.10
1050	1	0.973 [1]/1.085 [2]	27.70
	2	1.156 [1]/1.166 [2]	31.90
	4	1.269 [1]/1.293 [2]	36.80
1080	1	1.163 [1]/1.172 [2]	30.30
	2	1.273 [1]/1.278 [2]	37.20
	4	1.475 [1]/1.496 [2]	61.90

[1] Density determined geometrically via specimen weight and volume. [2] Density determined according to Archimedes' principle.

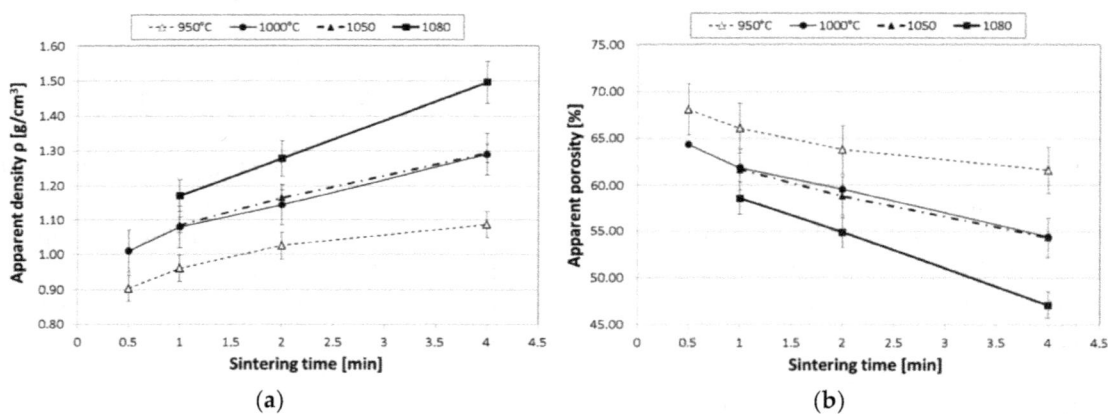

Figure 2. Dependence of (**a**) density determined according to Archimedes' principle; and (**b**) apparent porosity on spark plasma sintering (SPS) conditions for Cu@CS materials, measured on cylindrical samples.

Table 1 shows a good match between density values based on weighing and measurement of samples on the one hand and density determination via Archimedes' principle on the other, with the latter consistently leading to slightly higher results, while showing no deviation in terms of the general tendencies observed. The latter can be summarized as follows: in the temperature range studied, both increasing the sintering temperature at constant time and increasing sintering time at constant temperature will lead to an increase in density. When considering the time range from 1 to 4 min common to all temperature levels, the former approach leads to changes from 13.1% (950 °C) to 19.5% (1000 °C), 19.2% (1050 °C), and 27.6% (1080 °C). This leads to the assumption that sintering at 1000 °C already provides sufficient mobility to reach useful densities, while the added step observed at 1080 °C may be due to the copper coating changing from solid to liquid state. This transition would induce greater particle mobility and a greater likelihood of Cu surface layer interpenetration, as well as particle fracture events under the influence of the consolidation pressure. Superposition of these effects plus possible rearrangement of particles during sintering explains the difference between the coated powder's bulk density and sample densities. This interpretation, however, would imply that the network structures seen in Figure 1 must have formed in the solid state—always provided that the temperature reading at the basis of this assumption represents a homogeneous condition within the sample at all stages of the process, which is at least questionable in SPS sintering.

The corresponding perspective for density change at constant times and rising temperature levels shows increases of 22.1% (1 min), 24.7% (2 min), and 37.8% (4 min), respectively. Figure 2 represents these findings graphically, adding as further information the apparent porosity of the materials. This measure includes both the remaining open porosity between the microspheres—which typically show point-to-point extended to small area contact—and the internal porosity of the original cenospheres. It is calculated based on the measured sample density as well as the particle density of the coated cenospheres, which in turn depends on the initial density of the cenospheres, the density of the coating material, and the thickness of the coating.

4. Discussion

Besides density levels, Table 1 above reports the compressive strength values for the sintered specimens. Strength values correspond to the peak stress observed in stress–strain curves as depicted in Figure 3b. The data shown here is also depicted graphically in Figure 3a, where the relation between density and strength is evaluated. Though the density range covered is comparatively narrow, the data supports the increase of stress levels with density typical of many kinds of foams.

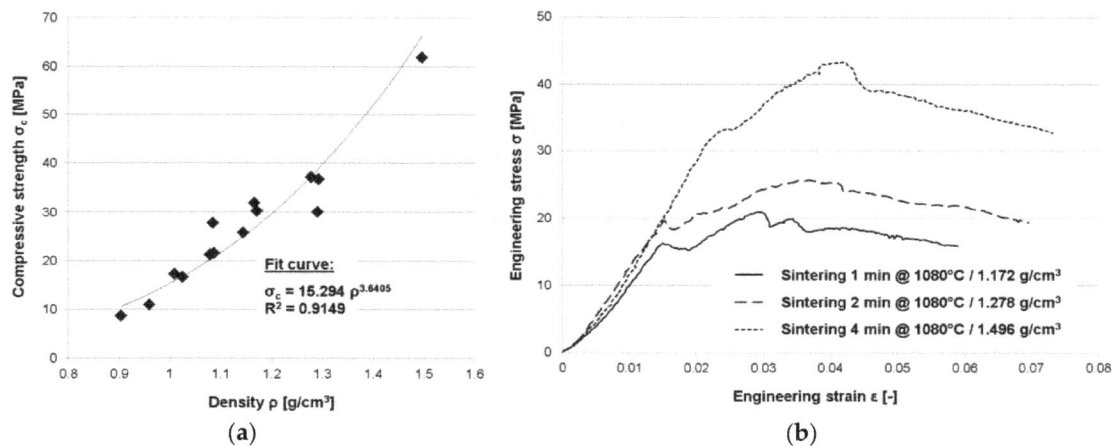

Figure 3. Mechanical performance of Cu@CS materials: (**a**) Compressive strength data plotted vs. (geometry-based, see Table 1) density values; (**b**) Examples of typical engineering stress–strain curves for different density levels, underlining the limited failure strain observed.

The general relationship between density and strength observed for the samples tested across sintering conditions is plotted in Figure 3a, which also contains a curve fit covering the relevant density range from approximately 0.90 to 1.50 g/cm^3. Once again, density values determined via Archimedes' principle have been selected as the basis of this evaluation. Considering the limited amount of data available so far, the chosen power law leads to an acceptable fit quality. However, further studies on this material will have to scrutinize this relationship both with respect to its type and actual parameter values, since it is usually associated with conventional two-phase foams rather than three phase syntactic materials [15]. Its value in the present case lies in the possibility of estimating properties at density levels not explicitly tested, and in a limited extrapolation option. Not surprisingly, coated and sintered cenospheres exhibit predominantly brittle failure as well as sample fragmentation, which together counteract the formation of a prolonged plateau region. The exemplary stress–strain curves representing different density levels in Figure 3b illustrate this general observation. Irrespective of the density and sintering conditions, none of the samples exceeds 7% of engineering strain at failure by any margin. The stress–strain curve itself has a stepped appearance, which is indicative of the observed fracture and partial disintegration of samples during compression. Apparently, the volume fraction of the (ductile) metallic phase present in the samples does not suffice to counterbalance the failure mechanism associated with the cenospheres and thus the ceramic phase.

Comparison with other types of materials in terms of strength is difficult due to the specific character of the material, which takes an intermediate position between a true metal matrix syntactic foam and a ceramic foam, though leaning towards the latter. In addition, the mixed open- vs. closed cell character comes in. In terms of density-normalized strength values, however, the properties determined are similar to those known from several other types of foams in a matching density range. An example can be drawn from comprehensive comparisons of the latter, such as those by Weise et al. [4], which have partially been taken up, focusing on the lower density range, in Figure 4 below: note, however, that the comparison is semi-quantitative at best, since strength definitions vary among the data gathered from the literature.

Considering the comparison offered in Figure 4 in more detail, and specifically the current positioning of syntactic foams based on metal-coated cenospheres, the following tendencies are apparent when it comes to structural applications:

- Increasing the thickness of the coating while maintaining the coating materials will lead to increased density and strength, but based on the unfavorable strength vs. density ratio of pure copper, most likely also to a reduction or at least stagnation of density-related specific strength.

- Replacement of Cu as coating material with either steel or aluminum alloys should increase specific strength, in the latter case ultimately exceeding the levels of conventional aluminum foams and approaching those of aluminum matrix syntactic foams.

- While the aforementioned increase of specific strength may be reached at reduced absolute density in the case of aluminum, it will be achieved at identical or slightly increased density in the case of steel. Nevertheless, based on the added open porosity, this would still position the respective material at the lowest-density end of steel matrix foams realized so far.

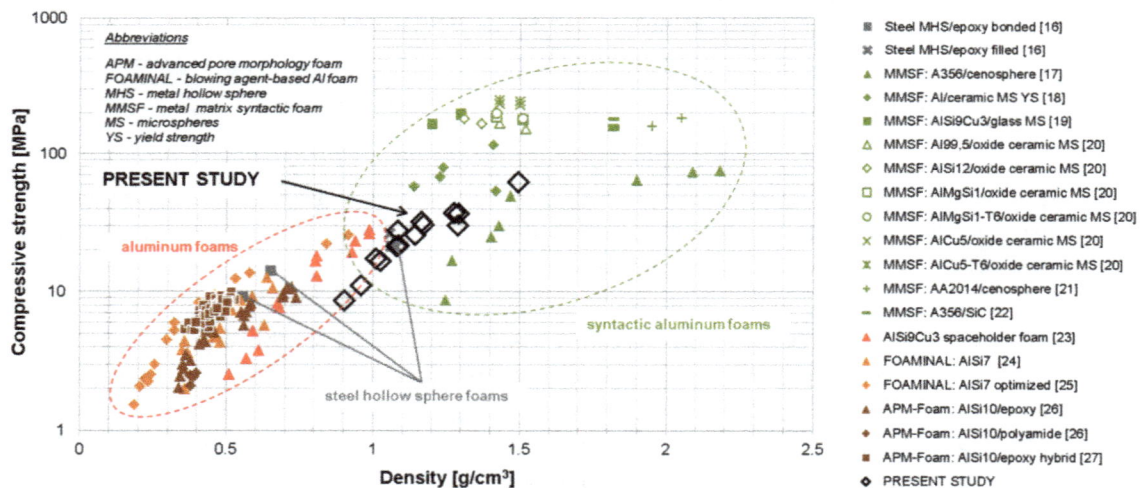

Figure 4. Comparing the mechanics of Cu@CS materials with other types of foams: Absolute compressive strength values of similar density materials are contrasted. Note that exact strength definitions may differ slightly between material variants. An explanation of abbreviations used in the diagram is provided in the upper left corner. The representation as such follows Weise et al. [4]. The data shown were collated from references [16–27].

5. Conclusions

The present investigation shows that the sputter coating of cenospheres creating thin metal layers on their surfaces is feasible and yields materials which can be consolidated directly by sintering to form performant materials with density-related compressive properties similar to those of other types of metal, ceramic, and hybrid foams, despite the fact that process optimization is still pending. Future studies will investigate the mechanical performance of this new type of material in more detail, looking at the influence of coating thickness on mechanical performance and failure mechanisms. Beyond direct consolidation of coated microspheres, their use as filler in conventional powder metallurgically produced metal matrix syntactic foams needs to be scrutinized. For this purpose, the production of steel coated microspheres will be investigated (a first sample of this type of material is depicted in Figure 5), and beyond cenospheres, the processing of high strength glass microspheres will be studied. The latter are an established filler for steel MMSF up to sintering temperatures of 1000 °C [28,29], but do not withstand the 1200 °C typical for 316 L matrices [4,30]. Benefits expected from the sputter coating of the glass spheres include the assumption that the full enclosure of the glass sphere will prevent the formation of satellite glass phases via inter-particle capillaries observed otherwise [4].

Besides this suggested switch in coating materials, it will definitely be worth evaluating Cu-coated microspheres as filler in aluminum and aluminum alloy matrices in view of the expected reaction with the matrix, starting from the formation of a Cu-enriched diffusion layer around the microspheres and up to the emergence of a transient liquid phase during sintering based on low melting eutectics in the respective systems. Besides this specific point, the fact that aluminum matrix syntactic foams are often produced from sintered filler particle preforms by pressure assisted melt infiltration [19] deserves special attention. This latter procedure will necessarily lead to a 3–3 type interpenetrating (metal and ceramic) phase composite, whereas a preform made of metal-coated filler particles could lead to a 0–3 type composite in which the brittle ceramic phase is discontinuous. Especially in terms of ductility, this configuration may be expected to bear beneficial effects.

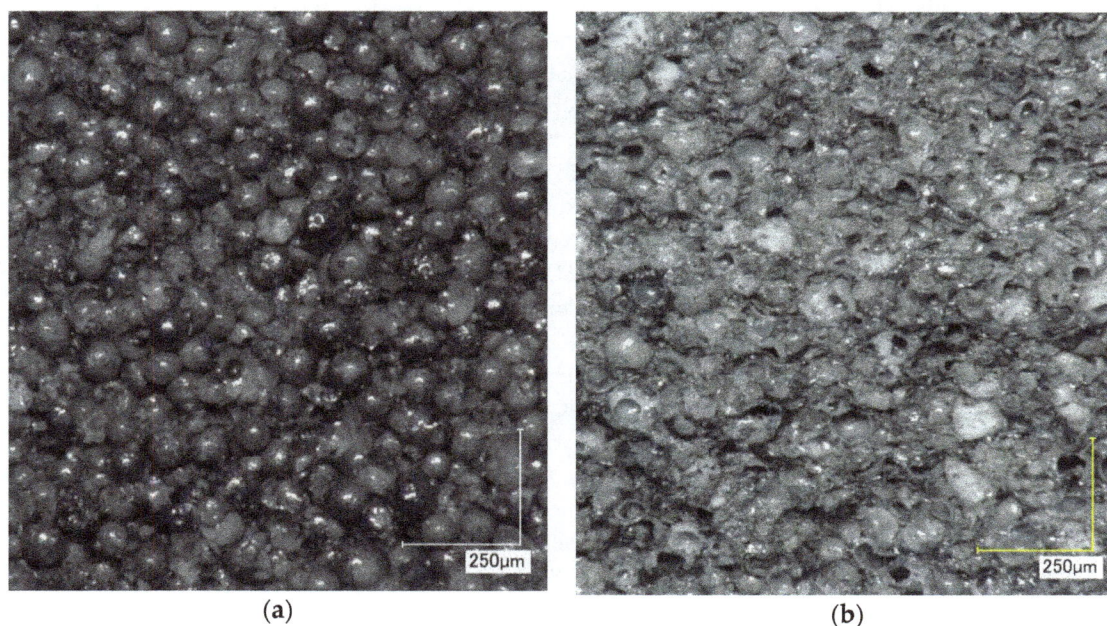

(a)   (b)

Figure 5. Early images of steel@CS materials: (**a**) Outer surface of sample sintered 4 min at 1100 °C; (**b**) Fracture surface of corresponding sample.

In the case of both the suggested steel and aluminum matrix syntactic foams, determination of mechanical properties will be an important part of evaluating their application potential. In general, properties are expected to improve with interface characteristics. With respect to processing the base materials, another notable side effect for powder-based Al matrix MMSF is the fact that Cu coating may effectively eliminate the particle density gap between Al powders on the one hand and cenospheres or glass microspheres on the other, and could thus counteract the risk of demixing and/or segregation effects during powder handling. This is of interest both for conventional powder metallurgical processes and powder-based additive manufacturing methods, and could thus widen the scope of processing options as well as manufacturable geometries for this type of material. Finally, the technique has several interesting implications for polymer matrix syntactic foams, as it can provide lightweight conductive fillers positively affecting properties like electromagnetic shielding capacity. As has been mentioned in the introduction, the effect as such has already been considered, but there is still a lot of room for performance optimization [6,7].

Acknowledgments: The authors gratefully acknowledge support by Kaspars Ozols of Institute of Technical Physics, Faculty of Material Science and Applied Chemistry, Technical Riga Technical University, Riga, Latvia, who performed electro conductivity studies on samples from the present test series.

Author Contributions: Andrei Shishkin and Irina Hussainova conceived and designed the experiments; Andrei Shishkin, Maria Drozdova and Viktor Kozlov performed the experiments; Andrei Shishkin, Irina Hussainova and Dirk Lehmhus analyzed the data; Andrei Shishkin, Maria Drozdova and Irina Hussainova contributed reagents/materials/analysis tools; Andrei Shishkin and Dirk Lehmhus wrote the paper.

Conflicts of Interest: The authors declare no conflict of interest.

References

1.	Shishkin, A.; Mironovs, V.; Lapkovskis, V.; Treijs, J.; Korjakins, A. Ferromagnetic Sorbents for Collection and Utilization of Oil Products. *Key Eng. Mater.* **2014**, *604*, 122–125. [CrossRef]

2.	Shishkin, A.; Mironovs, V.; Zemchenkov, V.; Antonov, M. Hybrid syntactic foams of metal—Fly ash cenosphere—Clay. *Key Eng. Mater.* **2016**, *674*, 35–40. [CrossRef]

3.	Gupta, N.; Rohatgi, P.K. *Metal Matrix Syntactic Foams: Processing, Microstructure, Properties and Applications*; DEStech Publications Inc.: Lancaster, PA, USA, 2014.

4.	Weise, J.; Lehmhus, D.; Baumeister, J.; Kun, R.; Bayoumi, M.; Busse, M. Production and properties of 316 L stainless steel cellular materials and syntactic foams. *Steel Res. Int.* **2014**, *85*, 486–497. [CrossRef]

5.	Peroni, L.; Scapin, M.; Lehmhus, D.; Baumeister, J.; Busse, M.; Avalle, M.; Weise, J. High Strain Rate Tensile and Compressive Testing and Performance of Mesoporous Invar (FeNi36) Matrix Syntactic Foams Produced by Feedstock Extrusion. *Adv. Eng. Mater.* **2016**, *18*. [CrossRef]

6.	Koppel, T.; Shishkin, A.; Hussainova, I.; Haldre, H.; Tint, P. Electromagnetic shielding properties of ceramic spheres coated with paramagnetic metal. *Agron. Res.* **2016**, *14*, 1015–1022.

7.	Yu, X.; Shen, Z. The electromagnetic shielding of Ni films deposited on cenosphere particles by magnetron sputtering method. *J. Magn. Magn. Mater.* **2009**, *321*, 2890–2895. [CrossRef]

8.	Meng, X.-F.; Li, D.-H.; Shen, X.-Q.; Liu, W. Preparation and magnetic properties of nano-Ni coated cenosphere composites. *Appl. Surf. Sci.* **2010**, *256*, 3753–3756. [CrossRef]

9.	Meng, X.-F.; Shen, X.-Q.; Liu, W. Synthesis and characterization of Co/cenosphere core–shell structure composites. *Appl. Surf. Sci.* **2012**, *258*, 2627–2631. [CrossRef]

10.	Chidambaram, P.R.; Meier, A.; Edwards, G.R. The nature of interfacial phenomena at copper-titanium/alumina and copper-oxygen/alumina interfaces. *Mater. Sci. Eng. A* **1996**, *206*, 249–258. [CrossRef]

11.	Espie, L.; Drevet, B.; Eustathopolous, N. Experimental study of the influence of interfacial energies and reactivity on wetting in metal/oxide systems. *Metall. Mater. Trans. A* **1994**, *25*, 599–605. [CrossRef]

12.	Kelber, J.A.; Niu, C.; Shepherd, K.; Jennison, D.R.; Bogicevic, A. Copper wetting of α-Al_2O_3(0001): Theory and experiment. *Surf. Sci.* **2000**, *446*, 76–88. [CrossRef]

13.	Meier, A.; Baldwin, M.D.; Chidambaram, P.R.; Edwards, G.R. The effect of large oxygen additions on the wettability and work of adhesion of copper-oxygen alloys on polycrystalline alumina. *Mater. Sci. Eng. A* **1995**, *196*, 111–117. [CrossRef]

14.	Eustathopoulos, N.; Nicholas, M.G.; Drevet, B. *Wettability at High Temperatures*; Pergamon Materials Series; Pergamon: Oxford, UK, 1999; Volume 3.

15.	Gibson, L.J.; Ashby, M.F. *Cellular Solids: Structure and Properties*, 2nd ed.; Cambridge University Press: Cambridge, UK, 2016.

16. Vesenjak, M.; Fiedler, T.; Ren, Z.; Öchsener, A. Behaviour of syntactic and partial hollow sphere structures under dynamic loading. *Adv. Eng. Mater.* **2008**, *10*, 185–191. [CrossRef]

17. Rohatgi, P.K.; Kim, J.K.; Gupta, N.; Alaraj, S.; Daoud, A. Compressive characteristics of A356/fly ash cenosphere composites synthesized by pressure infiltration technique. *Compos. A Appl. Sci. Manuf.* **2006**, *37*, 430–437. [CrossRef]

18. Tao, X.F.; Zhang, L.P.; Zhao, Y.Y. Al matrix syntactic foam fabricated with bimodal ceramic microspheres. *Mater. Des.* **2009**, *30*, 2732–2736. [CrossRef]

19. Weise, J.; Yezerska, O.; Busse, M.; Haesche, M.; Zanetti-Bueckmann, V.; Schmitt, M. Production and properties of micro-porous glass bubble zinc and aluminium composites. *Mater. Sci. Eng. Technol.* **2007**, *38*, 901–906.

20. Orbulov, I.N.; Ginsztler, J. Compressive characteristics of metal matrix syntactic foams. *Compos. Part A Appl. Sci. Manuf.* **2012**, *43*, 553–561. [CrossRef]

21. Dass Goel, M.; Peroni, M.; Solomos, G.; Mondal, D.P.; Matsagar, V.A.; Gupta, A.K.; Larcher, M.; Marburg, S. Dynamic compression behavior of cenosphere aluminum alloy syntactic foam. *Mater. Des.* **2012**, *42*, 418–432. [CrossRef]

22. Luong, D.D.; Strbik, O.M., III; Hammond, V.H.; Gupta, N.; Cho, K. Development of high performance lightweight aluminum alloy/SiC hollow sphere syntactic foams and compressive characterization at quasi-static and high strain rates. *J. Alloy. Compd.* **2013**, *550*, 412–422. [CrossRef]

23. Nestler, K.; Berg, A.; Rodriguez-Perez, M.A.; Busse, M. Multifunktionale Aluminiumschwämme durch Schmelzinfiltration von Platzhaltern. *Aluminium* **2006**, *82*, 688–692.

24. Avalle, M.; Lehmhus, D.; Peroni, L.; Pleteit, H.; Schmiechen, M.; Belingardi, G.; Busse, M. AlSi7 metallic foams—Aspects of material modelling for crash analysis. *Int. J. Crashworth.* **2009**, *14*, 269–285. [CrossRef]

25. Lehmhus, D.; Busse, M. Mechanical performance of structurally optimized AlSi7 aluminum foams—An experimental study. *Mater. Werkst.* **2014**, *45*, 1061–1071. [CrossRef]

26. Lehmhus, D.; Baumeister, J.; Stutz, L.; Schneider, E.; Stöbener, K.; Avalle, M.; Peroni, L.; Peroni, M. Mechanical characterization of particulate aluminum foams—Strain-rate, density and matrix alloy vs. adhesive effects. *Adv. Eng. Mater.* **2010**, *12*, 596–603. [CrossRef]

27. Weise, J.; Baumeister, J.; Hohe, J.; Böhme, W.; Beckmann, C. Epoxy Aluminum Hybrid Foam—An Innovative Sandwich Core Material with Improved Energy Absorption Characteristics. In Proceeding of the 10th International Conference on Sandwich Structures, Nantes, France, 27–29 August 2012.

28. Peroni, L.; Scapin, M.; Avalle, M.; Weise, J.; Lehmhus, D.; Baumeister, J.; Busse, M. Syntactic Iron Foams–On Deformation Mechanisms and Strain-Rate Dependence of Compressive Properties. *Adv. Eng. Mater.* **2012**, *14*, 909–918. [CrossRef]

29. Luong, D.D.; Shunmugasamy, V.S.; Gupta, N.; Lehmhus, D.; Weise, J.; Baumeister, J. Quasi-static and high strain rates compressive response of iron and Invar matrix syntactic foams. *Mater. Des.* **2015**, *66*, 516–531. [CrossRef]

30. Peroni, L.; Scapin, M.; Fichera, C.; Lehmhus, D.; Weise, J.; Baumeister, J.; Avalle, M. Investigation of the mechanical behaviour of AISI 316L stainless steel syntactic foams at different strain-rates. *Compos. Part B* **2014**, *66*, 430–442. [CrossRef]

4

Effect of Annealing Temperature on the Corrosion Protection of Hot Swaged Ti-54M Alloy in 2 M HCl Pickling Solutions

El-Sayed M. Sherif [1,2,*], Ehab A. El Danaf [3], Hany S. Abdo [1,4], Sherif Zein El Abedin [2] and Hasan Al-Khazraji [5]

[1] Deanship of Scientific Research, Advanced Manufacturing Institute (AMI), King Saud University, P.O. Box 800, Al-Riyadh 11421, Saudi Arabia; habdo@ksu.edu.sa

[2] Electrochemistry and Corrosion Laboratory, Department of Physical Chemistry, National Research Centre, El-Behoth St. 33, Dokki, 12622 Cairo, Egypt; sherifzein888@yahoo.com

[3] Mechanical Engineering Department, College of Engineering, King Saud University, P.O. Box 800, Al-Riyadh 11421, Saudi Arabia; edanaf@ksu.edu.sa

[4] Mechanical Design and Materials Department, Faculty of Energy Engineering, Aswan University, Aswan 81521, Egypt

[5] Former Ph.D. Student at Institute of Materials Science and Engineering, Clausthal University of Technology, Clausthal-Zellerfeld 38678, Germany; hamedhasan10@gmail.com

*   Correspondence: esherif@ksu.edu.sa

Academic Editor: Hugo F. Lopez

Abstract: The corrosion of Ti-54M titanium alloy processed by hot rotary swaging and post-annealed to yield different grain sizes, in 2 M HCl solutions is reported. Two annealing temperatures of 800 °C and 940 °C, followed by air cooling and furnace cooling were used to give homogeneous grain structures of 1.5 and 5 μm, respectively. It has been found that annealing the alloy at 800 °C decreased the corrosion of the alloy, with respect to the hot swaged condition, through increasing its corrosion resistance and decreasing the corrosion current and corrosion rate. Increasing the annealing temperature to 940 °C further decreased the corrosion of the alloy.

Keywords: titanium alloys; annealing; corrosion; pickling solutions; EIS; polarization

1. Introduction

Titanium and its alloys have many applications in industry for their good properties. These materials have excellent ballistic, fatigue, and corrosion resistances, low Young's modulus, high yield strength, ultimate tensile strength, and sufficient ductility [1]. These alloys have been used in automobiles, aircrafts, armors, different implant systems, in motorcycles with higher power engines, etc. [2–6]. It has been reported [7] that titanium armor provides a 15%–35% weight savings when compared to steel or aluminum armor for the same ballistic protection at areal densities of interest, which has resulted in substantial weight savings on military ground combat vehicles. Titanium has a very tenacious nascent oxide that is formed instantly on its surface upon exposure to air. The excellent corrosion resistance, low ferromagnetism, and compatibility with composites also provide significant benefits.

Titanium alloys are known to be one of the most difficult materials to machine [1]. This is due to their low thermal conductivity, which causes high cutting temperatures. Moreover, the low elastic modulus of titanium alloys leads to tool vibrations during chip formation. Several researchers have reported that the machining of titanium alloys is one of the principal challenges for their application [8–12]. Developing new kinds of titanium alloys with increased machinability has been reported [13,14]. TIMET developed a new alloy named TIMETAL54M (Ti-54M), which is an α-β alloy. Ti-54M alloy has been developed with superior machinability and strength to replace the widely-used Ti-6Al-4V alloy [10,15,16].

The corrosion of Ti-54M alloy in most known corrosive media is not available. Divi and Gruman [16] have reported the electrochemical corrosion behavior of Ti-54M in some corrosive media, including reducing, oxidizing, and chloride ones, using linear polarization resistance and potentiodynamic polarization techniques. It was claimed that the general corrosion behavior of Ti-54M and Ti-6Al-4V alloys are mostly likely similar. The authors [16] also used hydrogen uptake efficiency (HUE) technique in acidic chloride containing solutions and found that Ti-54M alloy has much lower HUE than Ti-6Al-4V. Moreover, the repassivation potential for the Ti-54M alloy was much better than Ti-6Al-4V alloy under chloride and reducing environments.

In the present study, the corrosion behavior of hot swaged Ti-54M alloy after 1.0 h and 24.0 h immersion in 2 M HCl solutions has been investigated. The effect of post swaging annealing temperatures of 800 °C and 940 °C on the protection of Ti-54M alloy against corrosion at the same conditions was also reported. Electrochemical impedance spectroscopy, cyclic potentiodynamic polarization, and chronoamperometric current-time were the corrosion test methods in this study. The work was complemented using scanning electron microscope images and energy-dispersive X-ray analysis. It is expected that the annealing of this alloy at the different temperatures, 800 °C and 940 °C, would increase its corrosion resistance in the acidic test solutions.

2. Materials and Methods

Ti-54M alloy, having (α + β) phase, was investigated in this study. This alloy was received as hot extruded bar, having a duplex microstructure and chemical composition (wt %) of 5.03% Al, 3.95% V, 0.57% Mo, 0.51% Fe, 0.11% Si, 0.10% C, 0.06% O, 0.05% N, 0.005% Zr, and the rest was Ti. A rotary swaging (RS) technique at 850 °C and a deformation degree of 3.0 was used to process this alloy. The cross-sectional area reduction, $\varepsilon_t = \ln \frac{A_0}{A}$, where A_0 is the initial cross-section and A is the final cross-section area, was used to calculate the true strain. Details of this processing technique are illustrated elsewhere [17]. Specimens, from the swaged bar were taken in the transverse direction and subjected to two heat treatment conditions. This was to obtain two differently-sized equiaxed microstructures by annealing at 800 and 940 °C for 1.0 h. The annealed samples were subjected to air cooling and furnace cooling, respectively, to yield an average grain size of 1.5 and 5 µm, respectively. The hot swaged condition exhibited an average grain size of 1.6 µm. All materials were given a final heat treatment at 500 °C for 24 h to age-harden the α phase by Ti3Al precipitates and the β phase by fine secondary α precipitates.

Hydrochloric acid (HCl, 32%) was purchased from Glassworld (Johannesburg, South Africa) and used as received. The test solution, 2 M HCl, was prepared from the stock HCl solution by dilution. All corrosion tests were carried out in a conventional three-electrode electrochemical cell accommodating for 350 mL of the test solution. An Ag/AgCl, a platinum foil, and the Ti-54M alloys were employed as the reference, counter, and working electrodes, respectively. The working electrode was prepared by welding a copper wire to one surface of the Ti-54M alloy. The sample was then mounted in an epoxy resin and one only surface was ground and exposed to the chloride acid test solution. The surface to be exposed to the solution was ground successively with metallographic emery paper of increasing fineness of up to 1000 grit.

An Autolab potentiostat-galvanostat operated by the general purpose electrochemical software (GPES, version 4.9, Amsterdam, The Netherlands) was used to perform the electrochemical measurements. The electrochemical impedance spectroscopy (EIS) measurements were performed at corrosion potentials over a change of frequency from 100 kHz to 100 mHz, with an AC wave of ±5 mV peak-to-peak overlaid on a DC bias potential, and the impedance data were collected using Powersine software at a rate of 10 points per decade change in frequency. The cyclic potentiodynamic polarization (CPP) measurements were carried out by scanning the potential from −1000 mV in the positive direction to 1800 mV at a scan rate of 3 mV/s. The potential was scanned in the backward direction at the same scan rate just after the forward scan. The change of current versus time at constant positive potential experiments were carried out by stabilizing the potential of the working electrodes at 1400 mV. All measurements were conducted after immersing our working electrodes in the hydrochloric acid solutions for 1.0 h and 24 h at room temperature. A new portion of the acid solution and a fresh surface of the alloys were employed in each experiment. SEM micrographs were obtained using a field emission scanning electron microscope (FE-SEM) Model JSM-7600F supplied by JEOL (Tokyo, Japan) after performing chronoamperometric current time experiments after 24 h immersion in 2 M HCl solutions. The applied voltage during SEM imaging was 15 kV.

3. Results and Discussion

3.1. Electrochemical Impedance Spectroscopy (EIS) Measurements

EIS method has been successfully employed to report the kinetic parameters for metals and alloys exposed to harsh environments [18–21]. The Nyquist plots obtained for (1) hot swaged; (2) annealed at 800 °C; and (3) annealed at 940 °C Ti-45M electrodes after their immersion for 1.0 h in 2 M HCl solutions are shown in Figure 1. Similar plots were also obtained for the same materials after 24 h immersion in the acid solution and depicted in Figure 2. The EIS data shown in Figures 1 and 2 were fitted to the best equivalent circuit that is shown in Figure 3. The values of the elements of the equivalent circuit (Figure 3) were obtained and listed in Table 1. The elements of this circuit can be defined as follows; R_S is the solution resistance, Q (CPEs) is the constant phase elements, R_{P1} is the polarization resistance between the surface of the alloy and a layer that may form on its surface, and R_{P2} is the polarization resistance of the interface between the surface formed layer and the hydrochloride acid solution.

It is clearly seen from Figures 1 and 2, whether after 1.0 h or 24 h immersion in the acid solution, that there is only one distorted semicircle. The diameter of the semicircle was noticed to increase with heat treating the sample at 800 °C, and further to 940 °C. It is generally agreed that the wider the diameter of the semicircle, the higher the corrosion resistance of the alloy. From this point of view, the resistance of Ti-54M increases with the increase of the annealing temperature. Prolonging the immersion time to 24 h (Figure 2) is noticed to decrease the diameter of the semicircle, which increases with the increase of annealing temperature to 800 °C, and further to 940 °C. This was further confirmed by the data listed in Table 1, where the lowest values of R_S, R_{P1}, and R_{P2} were recorded for the hot swaged alloy and increased with increasing the annealing temperature. The CPEs exponent n values vary from 0.69 to 0.77 revealing that CPEs represent a near capacitance. This is because the values of n values are known to be between 0 and 1; i.e., $0 \leq n \leq 1$, where $n = 1$ is a pure capacitance, $n = 0$ is a pure resistance and $n = 0.5$ is a Warburg element. The values of n in our study, thus, indicate that the surface of the alloys have some pores from which some dissolution occurs under the aggressiveness action of the acid solution. Moreover, the values of the CPEs decreased with the increase of annealing temperature due to the covering of the charged surfaces in order to reduce the capacitive loops. This was also confirmed by the presence of C_{dl}; the value of the C_{dl} also decreased with the annealing temperature. Increasing the exposure period of time to 24 h decreased the values of all resistances (R_S, R_{P1}, and R_{P2}) and slightly increased the values of CPEs and C_{dl} for the three alloys. This indicates that the increase of the exposure period to 24 h before measurements increases the corrosion of the alloys under investigation. However, the resistance against corrosion for these alloys whether it was immersed for 1.0 h or 24 h in the HCl solutions was found to increase according to the annealing temperature in the following order; Ti-45M annealed at 940 °C > Ti-45M annealed at 800 °C > hot swaged Ti-45M alloy.

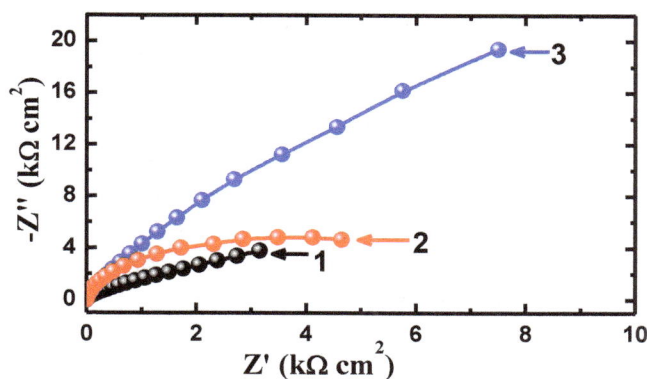

Figure 1. Nyquist plots obtained for (1) hot swaged; (2) annealed at 800 °C; and (3) annealed at 940 °C Ti-45M electrodes after their immersion for 1.0 h in 2 M HCl solutions.

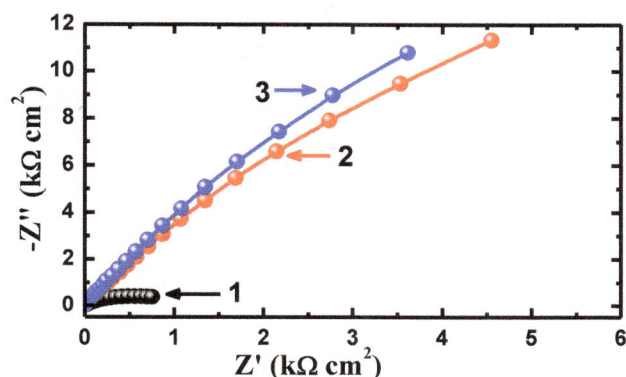

Figure 2. Nyquist plots obtained for (1) hot swaged; (2) annealed at 800 °C and (3) annealed at 940 °C Ti-45M electrodes after their immersion for 24 h in 2 M HCl solutions.

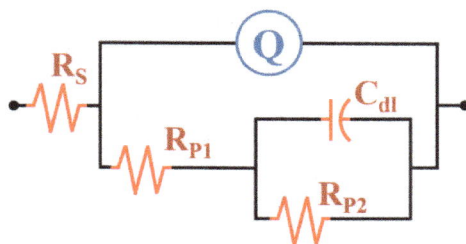

Figure 3. The equivalent circuit model used to fit EIS data shown in Figures 1 and 2.

Table 1. Electrochemical impedance spectroscopy parameters obtained for the different Ti-54M alloys in 2M HCl solutions.

Sample		EIS Parameter				
	R_S (Ω)	Q		R_{P1} (Ω)	C_{dl} (F)	R_{P2} (Ω)
		Y_{Q1} (F·cm^2)	n			
Ti-54M-HS (1 h)	1.038	0.0001378	0.73	2738	0.004712	1755
Ti-54M-800 °C (1 h)	1.739	0.0001305	0.75	3669	0.003907	4322
Ti-54M-940 °C (1 h)	1.999	0.0000385	0.77	4046	0.003146	5618
Ti-54M-HS (24 h)	1.012	0.0007654	0.70	1079	0.005547	1330
Ti-54M-800 °C (24 h)	1.357	0.0000882	0.69	2866	0.004748	3276
Ti-54M-940 °C (24 h)	1.550	0.0000791	0.72	3932	0.004316	4927

Typical Bode (a) impedance of the interface, |Z|, and (b) phase angle plots obtained for Ti-45M electrodes, (1) hot swaged; (2) annealed at 800 °C; and (3) annealed at 940 °C, after their immersion for 1.0 h in 2 M HCl solutions are shown in Figure 4. Similar Bode plots were obtained after 24 h immersion in the acid solutions and depicted in Figure 5. It is clearly seen from Figure 4 that the highest impedance value and the maximum degree of phase angle were recorded for the Ti-54M alloy that was annealed at 940 °C, followed by the alloy that was annealed at 800 °C, and the lowest was for the hot swaged alloy. Increasing the immersion time to 24 h shows almost the same behavior, but with slightly lower values for all treated alloys. This confirms that the corrosion decreases with the increase of the annealing temperature, and also that the increase of the immersion time increases the corrosion of Ti-54M alloys, whether it was hot swaged or annealed.

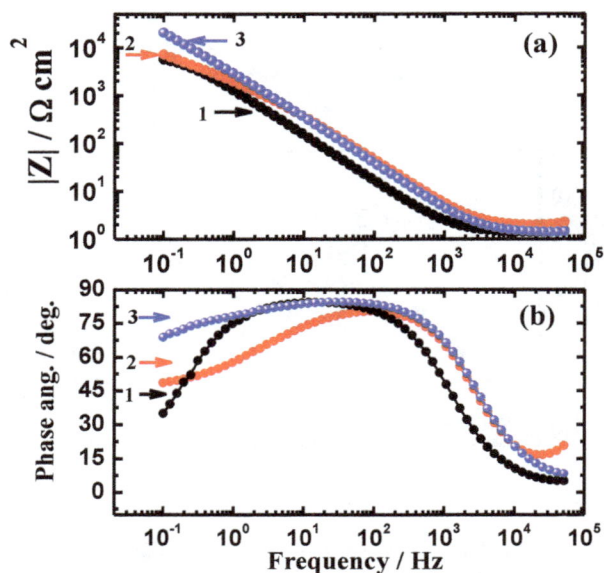

Figure 4. Typical Bode (**a**) impedance of the interface, |Z|; and (**b**) phase angle plots obtained for (1) hot swaged; (2) annealed at 800 °C; and (3) annealed at 940 °C Ti-45M electrodes after their immersion for 1.0 h in 2 M HCl solutions.

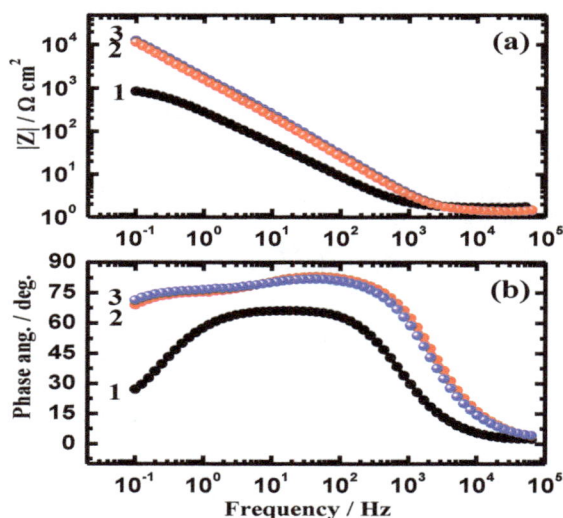

Figure 5. Typical Bode (**a**) impedance of the interface, |Z|; and (**b**) phase angle plots obtained for (1) hot swaged; (2) annealed at 800 °C; and (3) annealed at 940 °C Ti-45M electrodes after their immersion for 24 h in 2 M HCl solutions.

The increase of the corrosion resistance with the increase of the annealing temperature and its relation to the change of the grain size of the alloy can be explained here. The grain size exhibited by the sample that was hot swaged was about 1.6 μm with a high percentage of low-angle grain boundaries of 70%. This indicates the existence of high dislocation density positioned at cell boundaries or, in other words, the microstructure is characterized by a cell structure with low-angle grain boundaries. Annealing the hot swaged alloy for 800 °C, followed by air cooling, resulted in almost similar grain size with a much lower percentage of low-angle grain boundaries of 26%. This point towards reallocation of dislocations by means of the cell boundaries evolved into high-angle grain boundaries, which consumes the high dislocation density. Annealing the hot swaged alloy for 900 °C, followed by furnace cooling, resulted in a relatively higher grain size of 5 μm and an even lower percentage of low-angle grain boundaries of 18%, which points towards further reduction of dislocation density. Therefore, the increase of corrosion resistance of the hot swaged alloy after its annealing at 800 °C and the further increases at 900 °C could be due to the homogeneous distribution of dislocations at grain boundaries, which eliminates, to a great extent, the possibility of creating galvanic cells.

3.2. Cyclic Potentiodynamic Polarization

Cyclic potentiodynamic polarization (CPP) measurements have been widely used to explain the mechanism of corrosion and corrosion protection for metals and alloys in corrosive environments [18–21]. Figure 6 shows the CPP curves obtained for (a) hot swaged, (b) annealed at 800 °C, and (c) annealed at 940 °C Ti-45M electrodes after their immersion for 1.0 h in 2 M HCl solutions. It is seen from Figure 6 that the cathodic currents decreased towards the corrosion current density (j_{Corr}) with increasing the applied potential towards the less negative values. The cathodic reaction at this condition has been reported to the evolution of hydrogen as per the following equation [18,19,22],

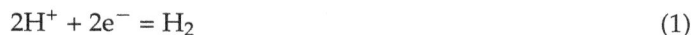

$$2H^+ + 2e^- = H_2 \tag{1}$$

The hydrogen gas produced here evolves leaving the acid solution, which, in turn, consumes more electrons and accelerates the dissolution reactions. The current rapidly increased in the anodic branch with little increase in the applied potential due to the occurrence of the anodic reaction, which, in most cases, is the dissolution of the surface and the release of electrons. The recorded current for all treated Ti-54M samples stays almost unchanged with the increase of the applied potential from −0.4 V and up to 1.8 V (Ag/AgCl). This is most probably due to the high corrosion resistance of the alloy, which allows the surface to resist the corrosive action of the concentrated 2 M HCl solution. Back scanning the applied potential produces less currents compared to the obtained currents in the forward direction, which confirms that the pitting corrosion does not occur at this condition.

In order to report the effect of prolonging the immersion time on the corrosion of the different Ti-54M electrodes in 2 M HCl solutions, CPP measurements were also performed after 24 h immersion in 2 M HCl solution and the curves are shown in Figure 7. Although prolonging the immersion time to 24 h does not change the polarization behavior of the Ti-54M in HCl solutions, it increases the corrosion of the tested alloys. The effect of 2 M HCl solutions on the corrosion of hot swaged, annealed at 800 °C, and annealed at 940 °C Ti-54M was quantified via calculating the corrosion parameters obtained from the polarization curves shown in Figures 6 and 7 as listed in Table 2. These parameters are the cathodic Tafel (β_c) slope, anodic Tafel (β_a) slope, corrosion potential (E_{Corr}), corrosion current density (j_{Corr}), polarization resistance (R_P), and corrosion rate (R_{Corr}). The values of E_{Corr} and j_{Corr} were obtained from the extrapolation of anodic and cathodic Tafel lines located next to the linearized current regions. Additionally, the values of R_P were calculated using the Stern-Geary equation as following [23,24]:

$$R_P = \frac{1}{j_{Corr}} \left(\frac{\beta_c \beta_a}{2.3 (\beta_c \beta_a)} \right) \tag{2}$$

Moreover, the values of R_{Corr} (milli-inches per year, mpy) were obtained using the following equation [25]:

$$R_{Corr} = j_{Corr}\left(\frac{k\,E_W}{d\,A}\right) \tag{3}$$

where k is a constant that defines the units for the corrosion rate ($k = 3272$), E_W is the equivalent weight in grams/equivalent of the alloy ($E_W = 11.46$ calculated), d is the density in gcm^{-3} ($d = 4.43$), and A is the area of electrode in cm^2 ($A = 1$).

It is seen from Table 2 that the value of j_{Corr} and consequently the value of R_{Corr} was the highest for the hot swaged sample and then decreased with an increase in the annealing temperature, while the values of R_P were oppositely increasing. Additionally, the increase of immersion time to 24 h led to significantly increasing the values of j_{Corr} and R_{Corr} for all tested Ti-54M electrodes, and these values decrease with the increase in the annealing temperature. This indicates that increasing the annealing temperature decreases the corrosion of Ti-54M alloy, while the increase of the time of the immersion periods increases its corrosion in 2 M HCl solutions, and that is in good agreement with the data obtained by EIS experiments.

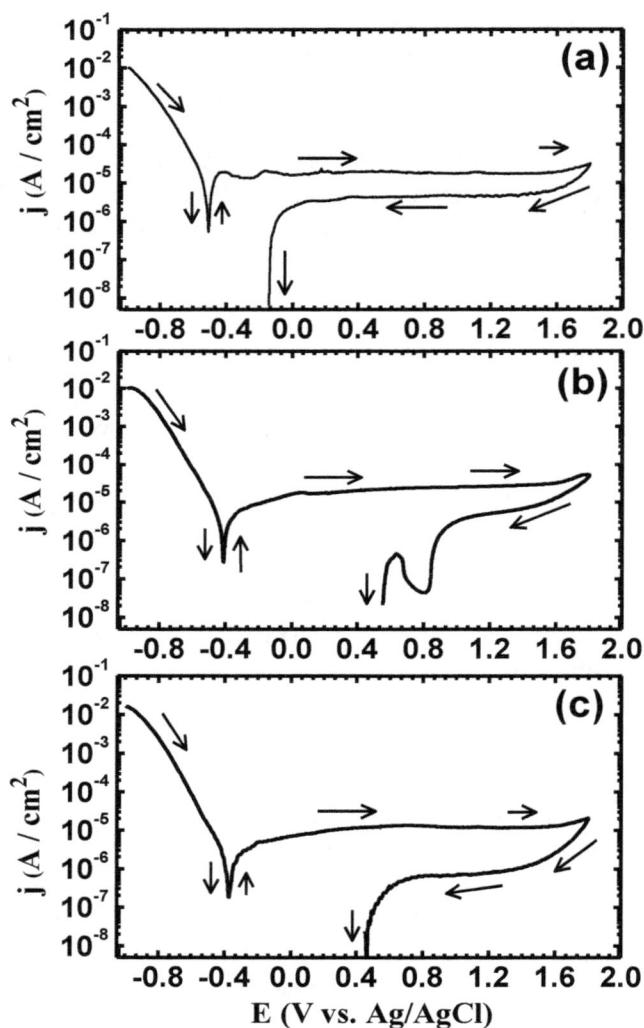

Figure 6. Cyclic potentiodynamic polarization curves obtained for (**a**) hot swaged; (**b**) annealed at 800 °C; and (**c**) annealed at 940 °C Ti-45M electrodes after their immersion for 1.0 h in 2 M HCl solutions.

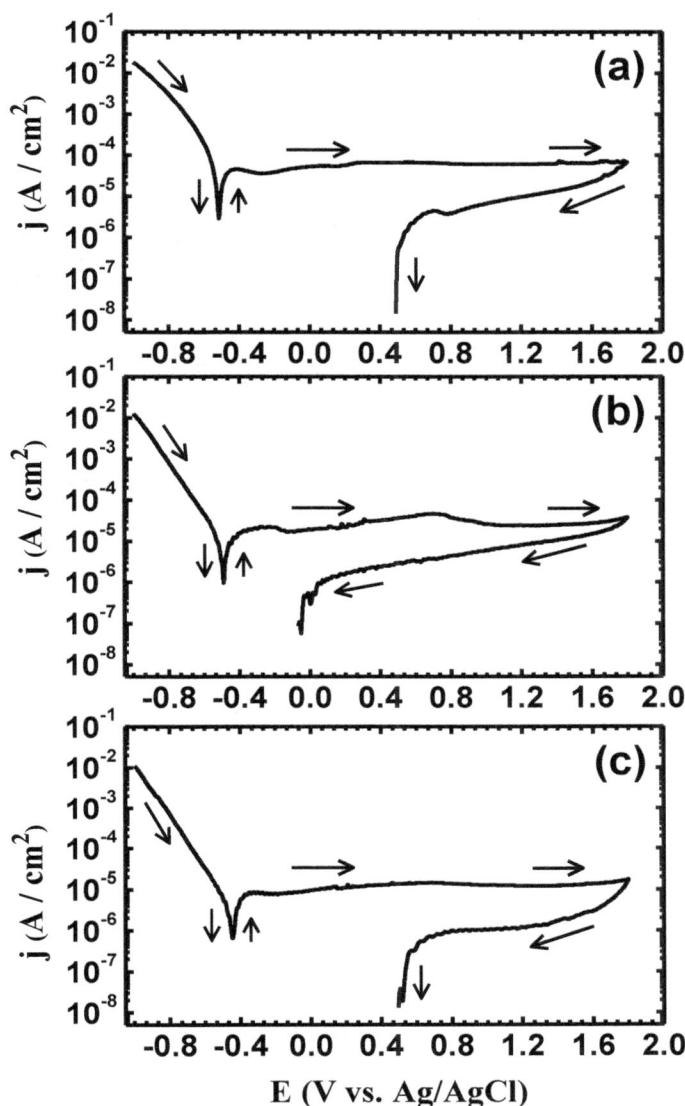

Figure 7. Cyclic potentiodynamic polarization curves obtained for (**a**) hot swaged; (**b**) annealed at 800 °C; and (**c**) annealed at 940 °C Ti-45M electrodes after their immersion for 24 h in 2 M HCl solutions.

Table 2. Parameters obtained from the polarization curves for the different Ti-54M alloys in 2 M HCl solutions.

Sample	Corrosion Parameter					
	β_c (V/dec^{-1})	E_{Corr} (V)	β_a (V/dec^{-1})	j_{Corr} (μA·cm^{-2})	R_p (Ω)	R_{Corr} (mpy)
Ti-54M-HS (1 h)	0.150	−0.515	0.400	12.0	3953	0.1016
Ti-54M-800 °C (1 h)	0.145	−0.390	0.400	4.6	10,059	0.0389
Ti-54M-940 °C (1 h)	0.135	−0.367	0.395	2.2	19,884	0.0186
Ti-54M-HS (24 h)	0.145	−0.472	0.295	38.0	1112	0.3216
Ti-54M-800 °C (24 h)	0.155	−0.470	0.320	9.0	5044	0.0762
Ti-54M-940 °C (24 h)	0.155	−0.460	0.325	4.5	10,140	0.0381

3.3. Chronoamperometric Current-Time Measurements

We have been using chronoamperometric current-time at constant anodic potential experiments to shed more light into whether pitting corrosion occurs for metals and alloys in aggressive media [18–22]. The applied potential value is usually chosen from the anodic polarization branch. Figure 8 shows the chronoamperometric curves obtained for Ti-45M electrodes, (1) hot swaged; (2) annealed at 800 °C; and (3) annealed at 940 °C, after their immersions in 2 M HCl for 1 h, respectively, followed by stepping the potential to +1.4 V vs. Ag/AgCl for 1.0 h. The same measurements were carried out for the Ti-54M alloys after 24 h immersion in the acid solutions as shown in Figure 9. This current-time technique was employed to confirm whether pitting corrosion takes place for the tested alloy in the acid solutions and also to report the effect of increasing immersion time before measurements on the intensity of both uniform and pitting corrosion. It is seen from Figure 8 that the current increased for Ti-54M hot swaged alloy after its immersion for 1.0 h (curve 1), then the current rapidly decreased, as it did for all heat-treated alloys, from the first moment of applying potential until the end of the run. It is also seen that the increase of the annealing temperature led to a decrease in the absolute values of current with time, where the lowest current values were obtained for the Ti-54M alloy that was annealed at 940 °C (curve 3).

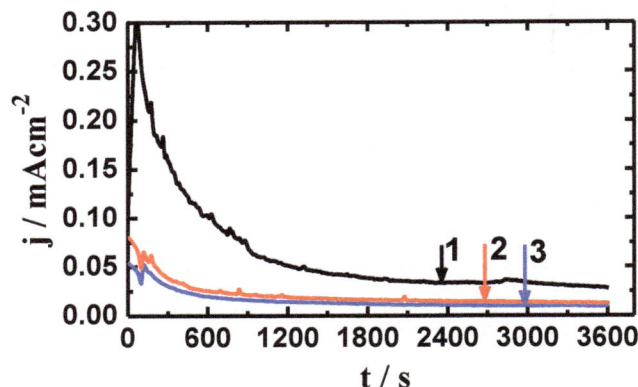

Figure 8. Chronoamperometric curves obtained for (1) hot swaged; (2) annealed at 800 °C; and (3) annealed at 940 °C Ti-45M electrodes after their immersions in 2 M HCl for 1 h, respectively, followed by stepping the potential to +1.4 V vs. Ag/AgCl for 1.0 h.

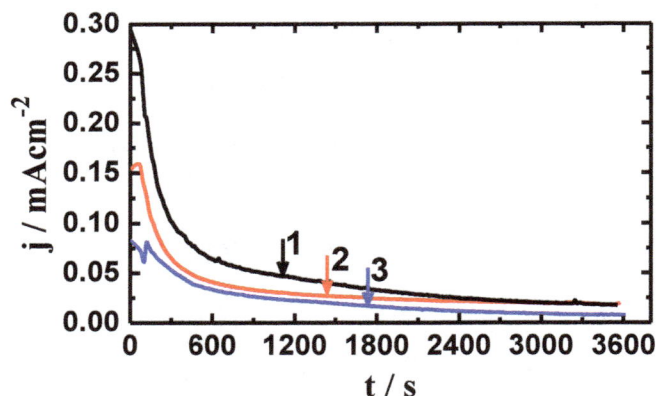

Figure 9. Chronoamperometric curves obtained for (1) hot swaged; (2) annealed at 800 °C; and (3) annealed at 940 °C Ti-45M electrodes after their immersions in 2 M HCl for 24 h, respectively, followed by stepping the potential to +1.4 V vs. Ag/AgCl for 1.0 h.

Prolonging the immersion time to 24 h showed almost the same current-time behavior with higher absolute current values for all investigated alloys. This indicates that increasing the immersion time from 1.0 h to 24 h increases the severity of the uniform corrosion via increasing the absolute current values. Moreover, and although the applied potential was very high value (+1.4 V), this current-time behavior does not give any indications on the occurrence of pitting corrosion of any of the heat-treated

Ti-54M alloys. Where the current values did not increase with time, there are no fluctuations that appear on the values of currents. The chronoamperometric current-time measurements, thus, are in good agreement with the cyclic potentiodynamic polarization data for all Ti-54M alloys at the same conditions.

3.4. Scanning Electron Microscope (SEM) and X-ray Energy Dispersive Spectroscopy (EDX)

In order to report the effect of 2 M HCl solutions on the morphology of the Ti-54M alloys, SEM micrographs and EDX spectra were carried out after immersing Ti-54M alloy in the three conditions mentioned earlier for five days in the acid solutions. Figure 10a,b shows SEM micrographs at different magnifications for the surface of the hot swaged Ti-54M alloy, Figure 10c shows the EDX spectra for the dark areas depicted on the SEM image shown in Figure 10b, and Figure 10d shows the EDX spectra for the white areas of SEM image shown in Figure 10b. Figures 11 and 12 show the SEM images and EDX spectra obtained for Ti-54M alloys annealed at 800 °C and 940 °C, respectively. It is seen from the image shown in Figure 10a that the morphology of the hot swaged alloy is smooth and homogeneous, while the image in Figure 10b declares that the surface developed layers of corrosion products (white areas). The weight percentage of the elements obtained from the EDX spectra for the dark areas (Figure 10c) recorded 93.12% Ti, 4.37% Al, and 2.52% V. On the other hand, the elements obtained for the white areas (Figure 10d) were 78.44% Ti, 2.70% Al, 11.98% V, 3.88% Mo, 2.31% Fe, 0.29% Cl, 0.23% Ca, and 0.18% Pt. The obtained weight percentages thus indicate that the surface of the dark areas contains the main elements of the alloy. At the same time, the low percentages of Ti and Al and the high percent of V, in addition to the presence of Cl recorded for the white areas, indicate that these areas represent thin films of corrosion products and confirm also that the hot swaged alloy does not suffer severe corrosion in the high concentration of the 2M HCl solution.

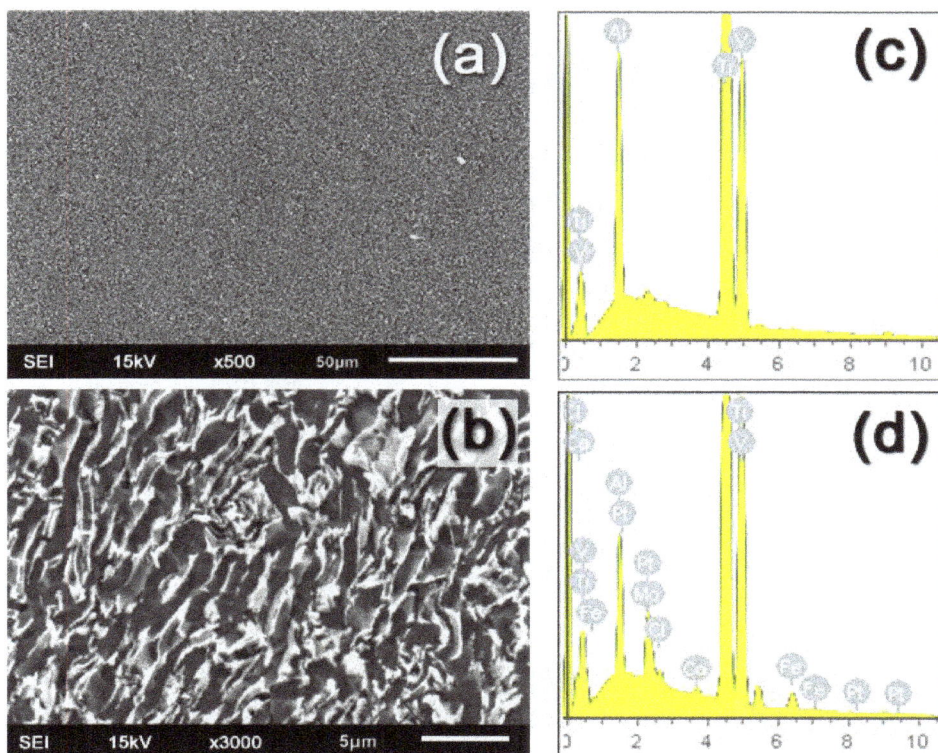

Figure 10. SEM/EDX for the as hot swaged Ti-54M sample after five days of immersion in 2 M HCl solutions, (**a,b**) are the SEM images; (**c,d**) are the EDX profiles for the area shown in (**a,b**), respectively.

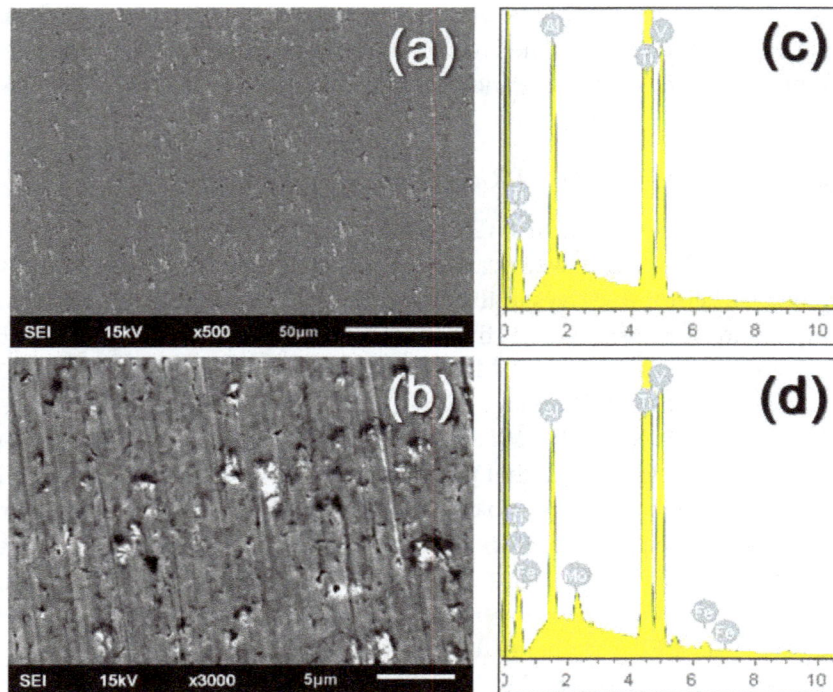

Figure 11. SEM/EDX for the as-annealed Ti-54M sample at 800 °C and after five days of immersion in 2 M HCl solutions, (**a**,**b**) are the SEM images; (**c**,**d**) are the EDX profiles for the area shown in (**a**,**b**), respectively.

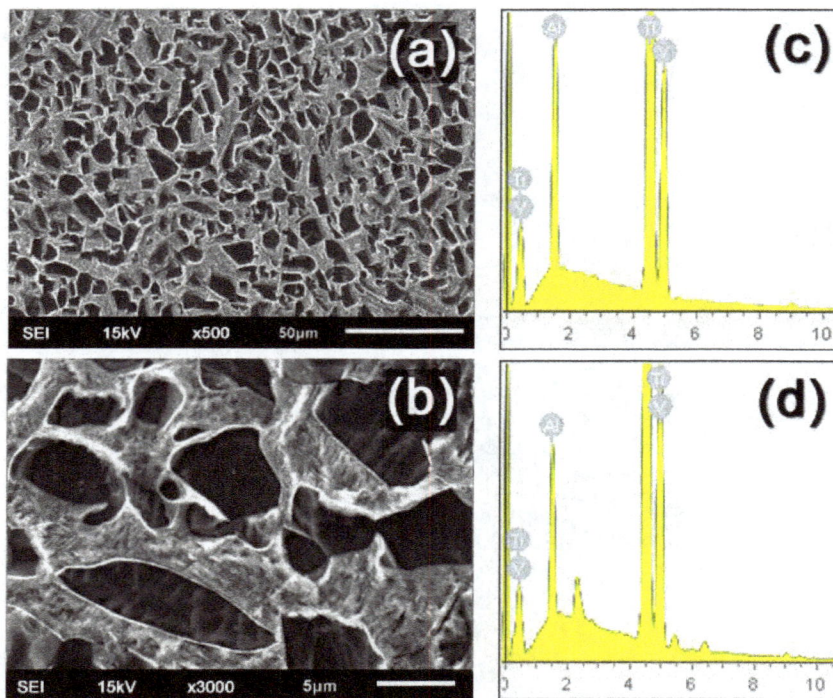

Figure 12. SEM/EDX for the as-annealed Ti-54M sample at 940 °C and after five days of immersion in 2 M HCl solutions, (**a**,**b**) are the SEM images; (**c**,**d**) are the EDX profiles for the area shown in (**a**,**b**), respectively.

The SEM images obtained for the Ti-54M alloy that was heat treated at 800 °C (Figure 11) showed that the surface has, unexpectedly, many pits, which were most probably due to a harsh localized attack towards that surface of the alloy via the chloride ions present in the acid molecules. The EDX spectra detected that the weight percentages of the elements found outside the pits (most of the surface) has 92.93% Ti, 4.55% Al, and 2.51% V, which represents the main elements of the Ti-54M alloy. On the other hand, the weight percentages for the elements found inside the pits (white areas) were 87.50% Ti, 3.69% Al, 5.97% V, 1.74% Mo, and 1.11% Fe. The elements found inside the pits indicate that the pits were filled with corrosion products, which were deposited after the dissolution of the main elements of the alloy.

The SEM micrographs obtained for Ti-54M alloy that was heat treated at 940 °C (Figure 12) showed clearly that the surface has two distinguished areas; dark and bright ones. The dark regions consist mainly of the original percentages of the elements of the alloy, where the weight percent of Ti was 92.87, Al was 5.75, and V was 1.37. The bright regions were similar to a net on the surface of the alloy and composed of 89.56% Ti, 3.07% Al, and 7.37% V in weight.

4. Conclusions

The corrosion behavior of hot swaged Ti-54M alloy in 2 M HCl solutions was reported. The effect of annealing temperature, namely, at 800 °C and 940 °C on the corrosion of the alloy in the acid solution, was also investigated. It has been found that the corrosion resistance of Ti-54M alloy is high against the harsh effect of the concentrated acid solutions. Annealing the alloy at 800 °C further increases the resistance against uniform corrosion, but allows the alloy to corrode via pitting attack. Increasing the annealing temperature to 940 °C remarkably increases the corrosion resistance of the Ti-54M alloy towards both uniform and pitting corrosion. Results together were in good agreement with each other and indicated that increasing the annealing temperature decreases the corrosion current and corrosion rate, and increases the polarization resistance. This effect was also found to decrease the absolute values of currents recorded for the alloy at a more active potential, 1.4 V (Ag/AgCl).

Acknowledgments: This project was supported by King Saud University, Deanship of Scientific Research, College of Engineering Research Center. The authors would also like to thank Twasol Research Excellence Program (TRE Program), King Saud University, Riyadh, Saudi Arabia for support.

Author Contributions: El-Sayed M. Sherif contributed reagents/materials/analysis tools, designed the experimental work and edited the final version of the paper; Ehab A. Danaf delivered the samples and helped in writing the draft version of the manuscript, Hany S. Abdo did the experimental work to corrosion and surface analysis; Sherif Zein El Abedin analyzed the data and wrote the draft version of the manuscript; Hasan Al-Khazraji delivered the samples and helped in writing the draft version of the manuscript.

Conflicts of Interest: The authors declare no conflict of interest.

References

1. Geetha, M.; Singh, A.K.; Asokamani, R.; Gogia, C. Ti based biomaterials, the ultimate choice for orthopaedic implants—A review. *Prog. Mater. Sci.* **2009**, *54*, 397–425. [CrossRef]
2. AlOtaibi, A.; El-Sayed, M.S.; Zinelis, S.; Al Jabbari, Y. Corrosion Behavior of Two cp Titanium Dental Implants Connected by Cobalt Chromium Metal Superstructure in Artificial Saliva and the Influence of Immersion Time. *Int. J. Electrochem. Sci.* **2016**, *11*, 5877–5890. [CrossRef]
3. Veiga, C.; Davim, J.P.; Loureiro, A.J.R. Properties and Applications of Titanium Alloys—A brief Review. *Adv. Mater. Sci.* **2012**, *32*, 133–148.
4. Boyer, R.-R. Titanium for Aerospace: Rationale and applications. *Adv. Perform. Mater.* **1995**, *2*, 349–368. [CrossRef]
5. Seikh, A.H.; Mohammad, A.; El-Sayed, M.S.; Al-Ahmari, A. Corrosion Behavior in 3.5% NaCl Solutions of γ-TiAl Processed by Electron Beam Melting Process. *Metals* **2015**, *5*, 2289–2302. [CrossRef]
6. Gopi, D.; El-Sayed, M.S.; Rajeswari, D.; Kavitha, L.; Pramod, R.; Dwivedi, J.; Polaki, S.R. Evaluation of the mechanical and corrosion protection performance of electrodeposited hydroxyapatite on the high energy electron beam treated titanium alloy. *J. Alloy. Compd.* **2014**, *616*, 498–504. [CrossRef]

7. The Mechanical and Ballistic Properties of an Electron Beam Single Melt of Ti-6A1-4V Plate. Available online: www.arl.army.mil/arlreports/2001/ARL-MR-515.pdf (accessed on 12 January 2017).

8. Machado, A.R.; Wallbank, J. Machining of titanium and its alloys—A review. *Proc. Inst. Mech. Eng. Part B J. Eng. Manuf.* **1990**, *204*, 53–60. [CrossRef]

9. Rahman Rashid, R.A.; Sun, S.; Wang, G.; Dargusch, M.S. Machinability of a near beta titanium alloy. *Proc. Inst. Mech. Eng. Part B J. Eng. Manuf.* **2011**, *225*, 2151–2162. [CrossRef]

10. Armendia, M.; Garay, A.; Iriarte, L.M.; Arrazola, P.-J. Comparison of the machinabilities of Ti6Al4V and TIMETAL® 54M using uncoated WC-Co tools. *J. Mater. Process. Technol.* **2010**, *210*, 197–203. [CrossRef]

11. Khanna, N.; Garay, A.; Iriarte, L.M.; Soler, D.; Sangwan, K.S.; Arrazola, P.J. Effect of heat treatment conditions on the machinability of Ti64 and Ti54M alloys. *Proced. CIRP* **2012**, *1*, 477–482. [CrossRef]

12. Khanna, N.; Sangwan, K.S. Cutting tool performance in machining of Ti555.3, Timetal®54M, Ti 6-2-4-6 and Ti 6-4 alloys: A review and analysis. In Proceedings of the 2nd CIRP International Conference on Process Machine Interactions, Vancouver, BC, Canada, 10–11 June 2010.

13. Khanna, N.; Sangwan, K.S. Comparative machinability study on Ti54M titanium alloy in different heat treatment conditions. *Proc. Inst. Mech. Eng. Part B J. Eng. Manuf.* **2013**, *227*, 96–101. [CrossRef]

14. Gabriela, S.B.; Dille, J.; Rezende, M.C.; Mei, P.; de Luiz, H.A.; Baldan, R.; Nunes, C.A. Mechanical Characterization of Ti–12Mo–13Nb Alloy for Biomedical Application Hot Swaged and Aged. *Mater. Res.* **2015**, *18*, 8–12. [CrossRef]

15. Al-Khazraji, H.; El-Danaf, E.; Wollmann, M.; Wagner, L. Microstructure, Mechanical, and Fatigue Strength of Ti-54M Processed by Rotary Swaging. *J. Mater. Eng. Perform.* **2015**, *24*, 2074–2084. [CrossRef]

16. Divi, S.; Grauman, J. Electrochemical Corrosion Properties of TIMETAL®54M. In Proceedings of the 13th World Conference on Titanium, San Diego, CA, USA, 16 August 2015.

17. ALkhazraji, H.; El-Danaf, E.; Wollmann, M.; Wagner, L. Enhanced Fatigue Strength of Commercially Pure Ti Processed by Rotary Swaging. *Adv. Mater. Sci. Eng.* **2015**, *2015*, 301837. [CrossRef]

18. Sherif, E.-S.M.; Potgieter, J.H.; Comins, J.D.; Cornish, L.; Olubambi, P.A.; Machio, C.N. Effects of minor additions of ruthenium on the passivation of duplex stainless steel corrosion in concentrated hydrochloric acid solutions. *J. Appl. Electrochem.* **2009**, *39*, 1385–1392. [CrossRef]

19. Latief, F.H.; El-Sayed, M.S. Effects of sintering temperature and graphite addition on the mechanical properties of aluminum. *J. Ind. Eng. Chem.* **2012**, *18*, 2129–2134. [CrossRef]

20. Sherif, E.-S.M. Electrochemical investigations on the corrosion inhibition of aluminum by 3-amino-1,2,4-triazole-5-thiol in naturally aerated stagnant seawater. *J. Ind. Eng. Chem.* **2013**, *19*, 1884–1889. [CrossRef]

21. Sherif, E.-S.M.; Ammar, H.R.; Khalil, K.A. A comparative study on the electrochemical corrosion behavior of microcrystalline and nanocrystalline aluminum in natural seawater. *Appl. Surf. Sci.* **2014**, *301*, 775–785.

22. Potgieter, J.H.; Olubambi, P.A.; Cornish, L.; Machio, C.N.; Sherif, K.A. Influence of nickel additions on the corrosion behaviour of low nitrogen 22% Cr series duplex stainless steels. *Corros. Sci.* **2008**, *50*, 2572–2579. [CrossRef]

23. Roberge, P.R. *Handbook of Corrosion Engineering*; McGrow-Hill: New York, NY, USA, 2000.

24. Sherif, E.M.; Park, S.-M. Inhibition of copper corrosion in acidic pickling solutions by *N*-phenyl-1,4-phenylenediamine. *Electrochim. Acta* **2006**, *51*, 4665–4673. [CrossRef]

25. Scully, J.R. Polarization resistance method for determination of instantaneous corrosion rates. *Corrosion* **2000**, *56*, 199–218. [CrossRef]

5

Improved Dehydrogenation Properties of 2LiNH$_2$-MgH$_2$ by Doping with Li$_3$AlH$_6$

Shujun Qiu [1], Xingyu Ma [1], Errui Wang [1], Hailiang Chu [1,2,*], Yongjin Zou [1], Cuili Xiang [1], Fen Xu [1] and Lixian Sun [1,*]

[1] Guangxi Key Laboratory of Information Materials, Guangxi Collaborative Innovation Center of Structure and Property for New Energy and Materials, School of Materials Science and Engineering, Guilin University of Electronic Technology, Guilin 541004, China; qiushujun@guet.edu.cn (S.Q.); maxingyuma@163.com (X.M.); Errui1990@163.com (E.W.); zouy@guet.edu.cn (Y.Z.); xiangcuili@guet.edu.cn (C.X.); xufen@guet.edu.cn (F.X.)

[2] Key Laboratory of Advanced Energy Materials Chemistry (Ministry of Education), Nankai University, Tianjin 300071, China

* Correspondence: chuhailiang@guet.edu.cn (H.C.); sunlx@guet.edu.cn (L.S.)

Academic Editor: Jacques Huot

Abstract: Doping with additives in a Li-Mg-N-H system has been regarded as one of the most effective methods of improving hydrogen storage properties. In this paper, we prepared Li$_3$AlH$_6$ and evaluated its effect on the dehydrogenation properties of 2LiNH$_2$-MgH$_2$. Our studies show that doping with Li$_3$AlH$_6$ could effectively lower the dehydrogenation temperatures and increase the hydrogen content of 2LiNH$_2$-MgH$_2$. For example, 2LiNH$_2$-MgH$_2$-0.1Li$_3$AlH$_6$ can desorb 6.43 wt % of hydrogen upon heating to 300 °C, with the onset dehydrogenation temperature at 78 °C. Isothermal dehydrogenation testing indicated that 2LiNH$_2$-MgH$_2$-0.1Li$_3$AlH$_6$ had superior dehydrogenation kinetics at low temperature. Moreover, the release of byproduct NH$_3$ was successfully suppressed. Measurement of the thermal diffusivity suggests that the enhanced dehydrogenation properties may be ascribed to the fact that doping with Li$_3$AlH$_6$ could improve the heat transfer for solid–solid reaction.

Keywords: hydrogen storage; Li-Mg-N-H system; Li$_3$AlH$_6$; thermal diffusivity; ball milling; phase transformation

1. Introduction

Safe and efficient storage of hydrogen is one of the major technological challenges associated with the use of hydrogen as an energy carrier, which is a vitally important process for the subsequent transition to the so called "hydrogen economy" [1]. Hydrogen can be stored as a compressed gas or as a cryogenic liquid; however, solid-state materials have the great potential to provide significantly high hydrogen storage densities, which draws a significant amount of research interest [2]. A long list of materials with a much higher hydrogen density have been synthesized and investigated during the past few decades. Several types of such materials mainly include microporous media that can physically adsorb hydrogen molecules at low temperatures [3], intermetallic hydrides that absorb atomic hydrogen as an interstitial, and complex hydrides that chemically absorb/desorb hydrogen [4,5]. Owing to the high hydrogen content, lightweight complex hydrides mostly containing Li, B, Na, Mg, and Al, such as alanates [AlH$_4$]$^-$, amides [NH$_2$]$^-$, amidoboranes [NH$_2$BH$_3$]$^-$, and borohydrides [BH$_4$]$^-$, are considered to be particularly promising as hydrogen storage materials [6–17]. The extensive studies of metal-N-H systems in recent years were initially prompted by Chen and coworkers, who reported the absorption and desorption of hydrogen gas by lithium nitride (Li_3N) at high temperatures (195–255 °C) [18] according to Equation (1).

$$Li_3N + H_2 \leftrightarrow Li_2NH + LiH + H_2 \leftrightarrow LiNH_2 + 2LiH \tag{1}$$

The second step in Equation (1) with about 6.5 wt % of hydrogen was given much attention due to a fairly good reversibility. However, $LiNH_2$-LiH suffers from high operating temperatures and an emission of ammonia. To address these problems, the substitution of LiH by MgH_2 to form a $2LiNH_2$-MgH_2 system has a remarkable destabilization effect [19]. Complete dehydrogenation of the mixture of $2LiNH_2$-MgH_2 produces a new ternary imide of $Li_2Mg(NH)_2$, and the following rehydrogenation of $Li_2Mg(NH)_2$ is converted to a mixture of $2LiH$-$Mg(NH_2)_2$ due to the thermodynamic stability. The whole reaction path can be expressed by Equation (2):

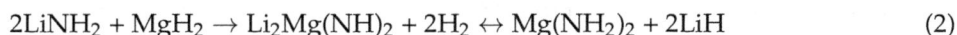

$$2LiNH_2 + MgH_2 \rightarrow Li_2Mg(NH)_2 + 2H_2 \leftrightarrow Mg(NH_2)_2 + 2LiH \tag{2}$$

Unfortunately, the dehydrogenation temperature of the $2LiNH_2$-MgH_2 system discussed above is still too high for practical applications (<100 °C) due to its high kinetic barriers. Many studies have been reported to focus on further altering the thermodynamics/kinetics of the $2LiNH_2$-MgH_2 system [20–23]. Doping with high-performance additives exhibits an excellent effect on reducing the temperature for hydrogen uptake and release. Lithium aluminum hydride ($LiAlH_4$) has a high hydrogen storage capacity (10.5 wt % H_2) and an excellent performance of hydrogen desorption at low temperature; thus, it has received significant attention from researchers. $LiAlH_4$ decomposes through a two-step process into Al, LiH, and H_2 at $T < 250$ °C through the intermediate Li_3AlH_6, according to reaction scheme (3) [24–29].

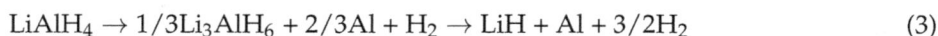

$$LiAlH_4 \rightarrow 1/3Li_3AlH_6 + 2/3Al + H_2 \rightarrow LiH + Al + 3/2H_2 \tag{3}$$

In particular, hexahydride of lithium alanate (Li_3AlH_6) releases hydrogen according to reaction (4) [13].

$$Li_3AlH_6 \rightarrow 3LiH + Al + 3/2H_2 \tag{4}$$

Hydrogen release from the $LiAlH_4$-$LiNH_2$ system was first reported by Xiong and coworkers [30]. It was found that $LiNH_2$ could effectively destabilize $LiAlH_4$ during the dehydrogenation process. The overall reaction of hydrogen release from this mixture was proposed as given by reaction (5).

$$LiAlH_4 + 2LiNH_2 \rightarrow Li_3AlN_2 + 4H_2 \tag{5}$$

Lu et al. [31] found that the reversible storage capacity of the Li_3AlH_6-$3LiNH_2$ system is increased to 7.0 wt % of hydrogen under 300 °C, according to the following reaction (6):

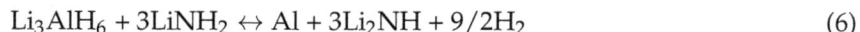

$$Li_3AlH_6 + 3LiNH_2 \leftrightarrow Al + 3Li_2NH + 9/2H_2 \tag{6}$$

The aforementioned reactions between lithium aluminum hydrides and lithium amide demonstrate the great potentials for the approach of destabilizing alanate materials with amides. In this study, the additives are focused on the catalytic enhancement of the dehydrogenation of $2LiNH_2$-MgH_2. We prepared Li_3AlH_6 and examined its effect on the hydrogen storage properties of $2LiNH_2$-MgH_2. The dehydrogenation properties and the thermal diffusivity of the combined system are discussed.

2. Materials and Methods

2.1. Sample Preparation

Li_3AlH_6 sample was prepared by mechanically milling LiH (95% purity, Sigma-Aldrich, St. Louis, MO, USA) and $LiAlH_4$ (95% purity, Sigma-Aldrich) in a molar ratio of 2:1 on a Retsch PM400 planetary mill (Haan, Germany) at 200 rpm under 0.1 MPa of an argon atmosphere. The ball-to-powder weight ratio was set to about 30:1. $2LiNH_2$-MgH_2-XLi_3AlH_6 (X = 0, 0.05, 0.10, 0.15, and 0.20) composites were prepared by mechanically milling $LiNH_2$ (95% purity, Sigma-Aldrich) and MgH_2 (98% purity, Sigma-Aldrich) with and without Li_3AlH_6 additive. The ball-to-powder weight ratio was set to be about 60:1. To minimize the temperature increment of the samples during the ball milling process, there was a 30 s pause for each 2 min of milling. The total milling time was 20 h. All the sample handling was performed in an Ar-filled glove box, in which the typical H_2O/O_2 levels were below 1 ppm.

2.2. Structural Characterization and Property Evaluation

Temperature-programmed desorption (TPD) properties were measured on an automated chemisorption analyzer (ChemBet Pulsar TPD, Quantachrome, Boynton Beach, FL, USA). The sample was heated up to 300 °C at a rate of 5 °C/min in a reactor in flowing Ar gas. The temperature-dependence of hydrogen desorption was performed on a thermogravimetric apparatus (TG, SETSYS Evolution, SETARAM Instrumentation, Lyon, France)-mass spectrometer (MS, GAM 200, InProcess Instruments, Bremen, Germany) combined system to analyze the evolved gas composition. The sample was heated up to 300 °C at a rate of 5 °C/min in flowing Ar gas. The dehydrogenation capacity based on volumetric release was measured on a HyEnergy PCTPRO-2000 Sieverts-type apparatus (SETARAM Instrumentation, Lyon, France). Approximately 150 mg of sample powder was loaded into the sample holder and heated at 2 °C/min from room temperature to 300 °C initially under dynamic vacuum.

Structural identification of the phases in the samples at different stages was performed on a Bruker D8 Advance diffractometer (Cu Kα radiation, 40 kV and 40 mA, Karlsruhe, Germany). The thermal diffusivity of the samples was measured on a LINSEIS XFA 500 instrument (Linseis Messgeräte GmbH, Selb, Germany) under dynamic vacuum at different temperatures of 30, 60, 90, and 120 °C.

3. Results and Discussion

Figure 1A presents the XRD patterns for the as-prepared Li_3AlH_6 sample. It can be observed that the majority of peaks can be ascribed to Li_3AlH_6, accompanied by a few peaks from impurities of metallic Al. Moreover, the thermal gas desorption properties of the Li_3AlH_6 sample were determined by PCTPRO-2000 and are shown in Figure 1B. A total of 4.63 wt % of hydrogen was liberated from Li_3AlH_6 sample when heated up to 300 °C, which is consistent with the previous studies [7,13]. These results illustrate that Li_3AlH_6 was successfully prepared through ball milling the mixture of $2LiH$/$LiAlH_4$.

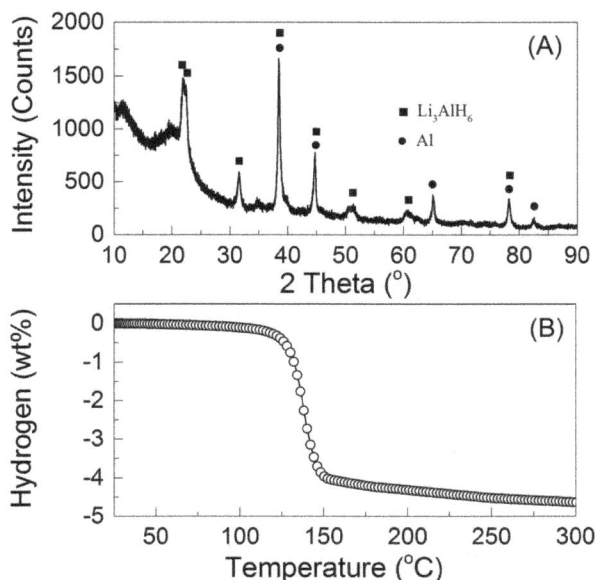

Figure 1. (**A**) XRD pattern and (**B**) Non-isothermal dehydrogenation of the as-prepared Li_3AlH_6 sample.

As shown in Figure 2A, the hydrogen desorption performance of the as-prepared $2LiNH_2$-MgH_2-XLi_3AlH_6 was first evaluated by means of TPD and MS. The operating temperatures for the dehydrogenation of the $2LiNH_2$-MgH_2 system were significantly reduced through the addition of Li_3AlH_6. Interestingly, the dehydrogenation process of the samples with $X = 0.05$–0.20 exhibited three peaks, which is different from the pristine sample, with only one desorption peak at 184 °C. For the $2LiNH_2$-MgH_2-$0.1Li_3AlH_6$ sample, three dehydrogenation peaks were seen at temperatures of 96, 128, and 180 °C, respectively. A reduction of 52 °C in the first dehydrogenation peak was achieved as compared to the pristine sample of $2LiNH_2$-MgH_2 [32]. MS examination shows that $2LiNH_2$-MgH_2 generated gaseous products including hydrogen and ammonia in a wide heating process. After the addition of Li_3AlH_6, the ammonia emission in the heating process was dramatically suppressed and almost completely inhibited for the sample of $2LiNH_2$-MgH_2-$0.1Li_3AlH_6$.

The hydrogen desorption performance of the as-prepared $2LiNH_2$-MgH_2 samples doped with different amounts of Li_3AlH_6 is shown in Figure 2B. Obviously, the operating temperatures for dehydrogenation were significantly decreased, and the amount of hydrogen released was found to be increased after the addition of Li_3AlH_6. A total of 5.15 wt % of hydrogen was liberated from pristine $2LiNH_2$-MgH_2 when heated up to 300 °C, while 6.47 wt % of hydrogen was released from $2LiNH_2$-MgH_2-$0.05 Li_3AlH_6$ with an onset temperature of about 102 °C. It is worth noting that the increase of hydrogen capacity for $2LiNH_2$-MgH_2-XLi_3AlH_6 is not proportional to the quantity of Li_3AlH_6 added, implying that Li_3AlH_6 may participate in the dehydrogenation reaction of $2LiNH_2$-MgH_2 during ball milling or heating processes. The onset desorption temperature was found to decrease gradually with an increasing amount of the doped Li_3AlH_6. Considering the hydrogen capacity and the operating temperature, the sample of $2LiNH_2$-MgH_2-$0.1Li_3AlH_6$ exhibited an optimal overall performance in the present study, since it could release 6.43 wt % of hydrogen. Therefore, subsequent investigation of the relationship between hydrogen storage properties and thermal diffusivity was focused on the $2LiNH_2$-MgH_2-$0.1Li_3AlH_6$ sample.

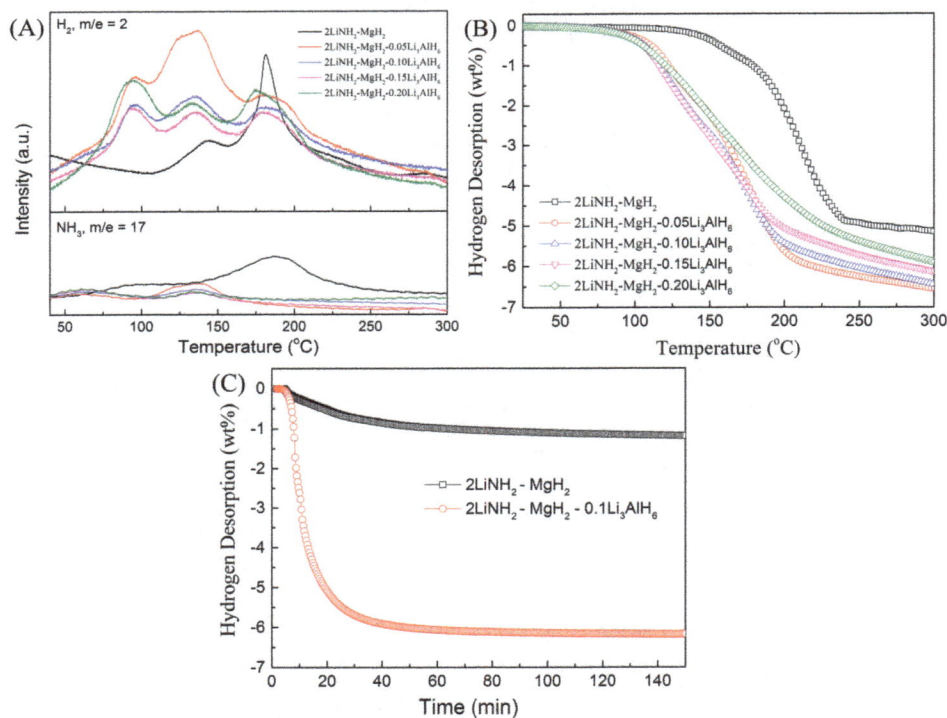

Figure 2. (A) Temperature-dependent gas (hydrogen (top) and ammonia (bottom)) released and (B) non-isothermal dehydrogenation curves of $2LiNH_2$-MgH_2-XLi_3AlH_6 samples; (C) Isothermal dehydrogenation curves of $2LiNH_2$-MgH_2 and $2LiNH_2$-MgH_2-$0.1Li_3AlH_6$ at 160 °C.

The isothermal dehydrogenation curves shown in Figure 2C indicate that the dehydrogenation rate of $2LiNH_2$-MgH_2 was remarkably enhanced by the addition of $0.1Li_3AlH_6$. At 160 °C, about 5.82 wt % of hydrogen was desorbed from $2LiNH_2$-MgH_2-$0.1Li_3AlH_6$ within 30 min, whereas only 0.57 wt % of hydrogen desorbed from $2LiNH_2$-MgH_2. When the dehydrogenation period was extended to 150 min, the amount of hydrogen desorbed from $2LiNH_2$-MgH_2-$0.1Li_3AlH_6$ increased to 6.11 wt %, which is very close to the total hydrogen capacity of 6.43 wt % heated up to 300 °C (Figure 2B). That is to say, the dehydrogenation kinetics was enhanced through the addition of $0.1Li_3AlH_6$, even at low temperature. We performed reversibility tests of the $2LiNH_2$-MgH_2-$0.1Li_3AlH_6$ sample (i.e., rehydrogenation at 200 °C and 50 bar hydrogen pressure). The initial rate for isothermal hydrogen absorption was so quick that the pressure-composition-temperature (PCT) could not accurately record the data. So, the absorption capacity was much lower than the theoretical value, and the hydrogen absorption capacity decreased with the increase of running cycles. Despite all this, it is worth noting that the doped sample had a much better reabsorption property than that of pristine sample.

The study of the dehydrogenation mechanism of the Li-Mg-N-H system shows that poor mass and/or heat transfer for solid–solid reaction is one of the critical issues for altering the thermodynamic and kinetic performance for hydrogen storage [33]. Figure 3 shows the thermal diffusivity of studied samples measured under the same conditions. Obviously, the thermal diffusivity increased after the addition of Li_3AlH_6. Doping with $0.1Li_3AlH_6$ or more gave rise to a significant increase of the thermal diffusivity. The thermal diffusivity of $2LiNH_2$-MgH_2-$0.1Li_3AlH_6$ is about 0.0035 cm^2/s, almost two times higher than that of pristine $2LiNH_2$-MgH_2. With the increase of temperature, the thermal diffusivity of both samples remained roughly constant. As discussed above in this study, the hydrogen storage properties were been significantly improved (i.e., lower dehydrogenation temperature and suppression of the NH_3 evolution after the addition of Li_3AlH_6). It can be concluded that these improvements could be ascribed to the significant increase of thermal diffusivity, helpful to improve the performance of heat transfer for solid–solid reaction, eventually resulting in an enhancement of hydrogen desorption performance.

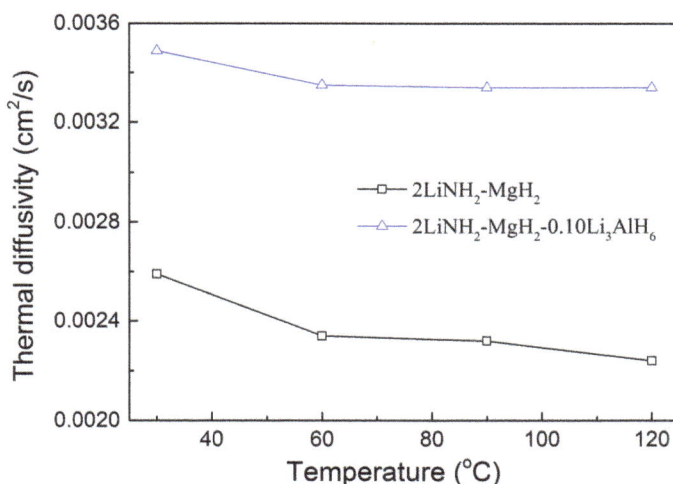

Figure 3. Thermal diffusivity for $2LiNH_2$-MgH_2 and $2LiNH_2$-MgH_2-$0.1Li_3AlH_6$ samples at different temperatures.

On the basis of the results discussed above, it can be deduced that the added Li_3AlH_6 should participate in the dehydrogenation reaction. Therefore, the phase evolution of $2LiNH_2$-MgH_2-$0.1Li_3AlH_6$ during heating process was studied in detail with the XRD patterns shown in Figure 4. It can be observed that in the initial stage, $LiNH_2$ and MgH_2 diffraction peaks were observed for sample after ball milling without detectable Li_3AlH_6, which means that Li_3AlH_6 may transform to an amorphous structure in the process of ball milling. After desorption at 100 °C, about 0.40 wt % of hydrogen was desorbed from $2LiNH_2$-MgH_2-$0.1Li_3AlH_6$, which gives a similar XRD pattern. When heated to 160 °C, LiNH$_2$ and MgH$_2$ diffraction peaks almost disappeared. Meanwhile, α-phase Li$_2$Mg(NH)$_2$ with an orthorhombic structure was formed. Upon further increasing the temperature to 220 °C with about 6.31 wt % of hydrogen released, α-phase Li$_2$Mg(NH)$_2$ was still the main product. Note that after complete dehydrogenation at 300 °C, β-phase Li$_2$Mg(NH)$_2$ with a primitive cubic structure was observed, indicating that a solid phase transition of Li$_2$Mg(NH)$_2$ occurred. The structural transition from an orthorhombic phase to a primitive cubic phase was reported to always occur at an elevated temperature of 400 °C or under a treatment of 36 h of high-energetic ball milling [34]. It should be highlighted that this phase transition occurred below 300 °C in our case, which may be related to the addition of Li$_3$AlH$_6$. The study of the underlying mechanism is underway.

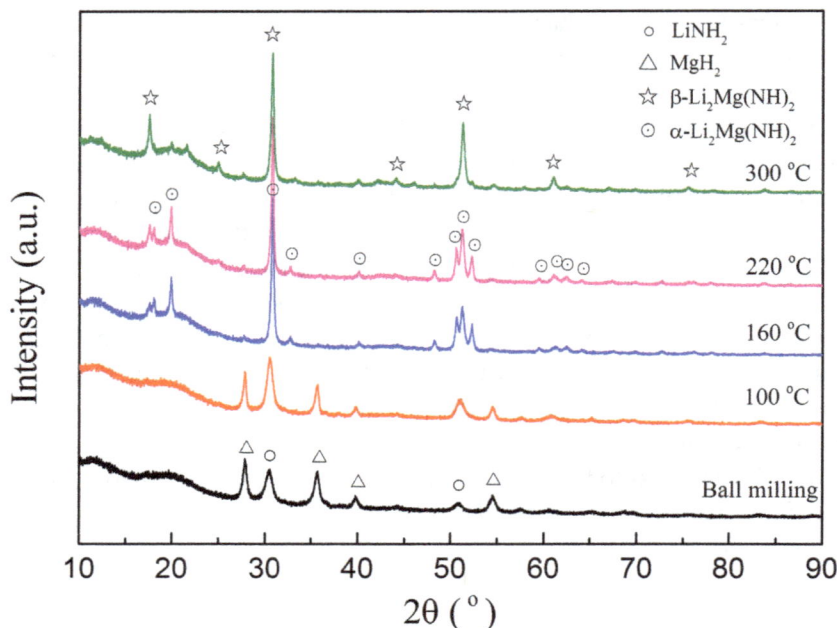

Figure 4. XRD patterns for 2LiNH$_2$-MgH$_2$-0.1Li$_3$AlH$_6$ sample at different stages.

4. Conclusions

Li$_3$AlH$_6$ was prepared by ball milling the mixture of 2LiH/LiAlH$_4$. Then, it was doped into 2LiNH$_2$-MgH$_2$, which resulted in an improvement of the dehydrogenation properties. The addition of Li$_3$AlH$_6$ not only reduced the dehydrogenation temperatures and increased the amount of hydrogen released from the 2LiNH$_2$-MgH$_2$ system, but also inhibited the release of ammonia as byproduct. 2LiNH$_2$-MgH$_2$-0.1Li$_3$AlH$_6$ had a reduced onset dehydrogenation temperature of 78 °C without detectable ammonia emission during the whole heating process. Moreover, 2LiNH$_2$-MgH$_2$-0.1Li$_3$AlH$_6$ had excellent low temperature hydrogen releasing performance (i.e., 6.11 wt % of hydrogen released at 160 °C in 150 min). Moreover, the kinetics for hydrogen reabsorption of 2LiNH$_2$-MgH$_2$-0.1Li$_3$AlH$_6$ was much better than that of pristine sample, which needs to be confirmed by non-isothermal absorption tests in the future. Doping with 0.1Li$_3$AlH$_6$ gave rise to a high thermal diffusivity, almost two times higher than that of 2LiNH$_2$-MgH$_2$, probably contributing to the improved hydrogen storage properties.

Acknowledgments: This research was financially supported by NSFC (51401059, 51361006, 51461010, 51361005, 51371060, U1501242, and 51461011), the Innovation Project of GUET Graduate Education (2016YJCX22) and GXNSF (2014GXNSFAA118043 and 2014GXNSFAA118333).

Author Contributions: H.C., F.X. and L.S. conceived and designed the experiments; S.Q., X.M. and E.W. performed the experiments; S.Q., Y.Z. and H.C. analyzed the data; S.Q., X.M. and C.X. contributed reagents/materials/analysis tools; S.Q., X.M. and H.C. wrote the paper.

Conflicts of Interest: The authors declare no conflict of interest.

References

1. Meeks, N.D.; Baxley, S. Fuel cells and the hydrogen economy. *Chem. Eng. Prog.* **2016**, *112*, 34–38.
2. Graetz, J. New approaches to hydrogen storage. *Chem. Soc. Rev.* **2009**, *38*, 73–82. [CrossRef] [PubMed]
3. Thomas, K.M. Hydrogen adsorption and storage on porous materials. *Catal. Today* **2007**, *120*, 389–398. [CrossRef]
4. Schuth, F.; Bogdanovic, B.; Felderhoff, M. Light metal hydrides and complex hydrides for hydrogen storage. *Chem. Commun.* **2004**, *20*, 2249–2258. [CrossRef] [PubMed]
5. Orimo, S.I.; Nakamori, Y.; Eliseo, J.R.; Zuttel, A.; Jensen, C.M. Complex hydrides for hydrogen storage. *Chem. Rev.* **2007**, *107*, 4111–4132. [CrossRef] [PubMed]
6. Xiong, Z.T.; Wu, G.T.; Hu, J.J.; Chen, P. Ternary imides for hydrogen strogen. *Adv. Mater.* **2004**, *16*, 1522–1525. [CrossRef]
7. Xiong, Z.T.; Wu, G.T.; Hu, J.J.; Chen, P. Investigation on chemical reaction between $LiAlH_4$ and $LiNH_2$. *J. Power Sources* **2006**, *159*, 167–170. [CrossRef]
8. Wang, H.; Cao, H.J.; Wu, G.T.; He, T.; Chen, P. The improved hydrogen storage performances of the multi-component composite: $2Mg(NH_2)_2–3LiH-LiBH_4$. *Energies* **2015**, *8*, 6898–6909. [CrossRef]
9. Lu, J.; Fang, Z.Z. Dehydrogenation of a Combined $LiAlH_4/LiNH_2$ System. *J. Phys. Chem. B* **2005**, *109*, 20830–20834. [CrossRef] [PubMed]
10. Wei, J.; Leng, H.Y.; Li, Q.; Chou, K.C. Improved hydrogen storage properties of $LiBH_4$ doped Li-N-H system. *Int. J. Hydrog. Energy* **2014**, *39*, 13609–13615. [CrossRef]
11. Cao, H.J.; Chua, Y.S.; Zhang, Y.; Xiong, Z.T.; Wu, G.T.; Qiu, J.S.; Chen, P. Releasing 9.6 wt % of H_2 from $Mg(NH_2)_2–3LiH-NH_3BH_3$ through mechanochemical reaction. *Int. J. Hydrog. Energy* **2013**, *38*, 10446–10452. [CrossRef]
12. Kwak, Y.J.; Kwon, S.N.; Song, M.Y. Preparation of $Zn(BH_4)_2$ and diborane and hydrogen release properties of $Zn(BH_4)_2 + xMgH_2$ (x = 1, 5, 10, and 15). *Met. Mater. Int.* **2015**, *21*, 971–976. [CrossRef]
13. Chen, J.; Kuriyama, N.; Xu, Q.; Takeshita, H.T.; Sakai, T. Reversible hydrogen storage via titanium-catalyzed $LiAlH_4$ and Li_3AlH_6. *J. Phys. Chem. B* **2001**, *105*, 11214–11220. [CrossRef]
14. Chu, H.L.; Xiong, Z.T.; Wu, G.T.; He, T.; Wu, C.Z.; Chen, P. Hydrogen storage properties of Li–Ca–N–H system with different molar ratios of $LiNH_2/CaH_2$. *Int. J. Hydrog. Energy* **2010**, *35*, 8317–8321. [CrossRef]
15. Chua, Y.S.; Chen, P.; Wu, G.T.; Xiong, Z.T. Development of amidoboranes for hydrogen storage. *Chem. Commun.* **2011**, *47*, 5116–5129. [CrossRef] [PubMed]
16. David, W.I.F. Effective hydrogen storage: A strategic chemistry challenge. *Faraday Discuss.* **2011**, *151*, 399–414. [CrossRef] [PubMed]
17. Chen, P.; Zhu, M. Recent progress in hydrogen storage. *Mater. Today* **2008**, *11*, 36–43. [CrossRef]
18. Chen, P.; Xiong, Z.T.; Luo, J.Z.; Lin, J.Y.; Tan, K.L. Interaction of hydrogen with metal nitrides and imides. *Nature* **2002**, *420*, 302–304. [CrossRef] [PubMed]
19. Luo, W.F. ($LiNH_2-MgH_2$): A viable hydrogen storage system. *J. Alloys Compd.* **2004**, *381*, 284–287. [CrossRef]
20. Amica, G.; Larochette, P.A.; Gennari, F.C. Hydrogen storage properties of $LiNH_2$-LiH system with MgH_2, CaH_2 and TiH_2 added. *Int. J. Hydrog. Energy* **2015**, *40*, 9335–9346. [CrossRef]
21. Li, Y.T.; Ding, X.L.; Wu, F.L.; Sun, D.L.; Zhang, Q.A.; Fang, F. Enhancement of hydrogen storage in destabilized $LiNH_2$ with $KMgH_3$ by quick conveyance of N-containing species. *J. Phys. Chem. C* **2016**, *120*, 1415–1420. [CrossRef]
22. Shukla, V.; Bhatnagar, A.; Pandey, S.K.; Shahi, R.R.; Yadav, T.P.; Shaz, M.A.; Srivastava, O.N. On the synthesis, characterization and hydrogen storage behavior of $ZrFe_2$ catalyzed Li-Mg-N-H hydrogen storage material. *Int. J. Hydrog. Energy* **2015**, *40*, 12294–12302. [CrossRef]
23. Principi, G.; Agresti, F.; Maddalena, A.; Russo, S. The problem of solid state hydrogen storage. *Energy* **2009**, *34*, 2087–2091. [CrossRef]
24. Nakamori, Y.; Ninomiya, A.; Kitahara, G.; Aoki, M.; Noritake, T.; Miwa, K.; Kojima, Y.; Orimo, S. Dehydriding reactions of mixed complex hydrides. *J. Power Sources* **2006**, *155*, 447–455. [CrossRef]
25. Ono, T.; Shimoda, K.; Tsubota, M.; Kohara, S.; Ichikawa, T.; Kojima, K.; Tansho, M.; Shimizu, T.; Kojima, Y. Ammonia desorption property and structural changes of $LiAl(NH_2)_4$ on thermal decomposition. *J. Phys. Chem. C* **2011**, *115*, 10284–10291. [CrossRef]

26. Siangsai, A.; Suttisawat, Y.; Sridechprasat, P.; Rangsunvigit, P.; Kitiyanan, B.; Kulprathipanja, S. Effect of Ti compounds on hydrogen desorption/absorption of LiNH$_2$/LiAlH$_2$/MgH$_2$. *J. Chem. Eng. Jpn.* **2008**, *43*, 95–98. [CrossRef]

27. Andreasen, A.; Vegge, T.; Pedersen, A.S. Dehydrogenation kinetics of as-received and ball-milled LiAlH$_4$. *J. Solid State Chem.* **2005**, *178*, 3672–3678. [CrossRef]

28. Andreasen, A. Effect of Ti-doping on the dehydrogenation kinetic parameters of lithium aluminum hydride. *J. Alloys Compd.* **2006**, *419*, 40–44. [CrossRef]

29. Lu, J.; Fang, Z.Z.; Sohn, H.Y.; Bowman, R.C.; Hwang, S.J. Potential and reaction mechanism of Li-Mg-Al-N-H system for reversible hydrogen storage. *J. Phys. Chem. C* **2007**, *111*, 16686–16692. [CrossRef]

30. Xiong, Z.T.; Wu, G.T.; Hu, J.J.; Liu, Y.F.; Chen, P.; Luo, W.F.; Wang, J. Reversible hydrogen storage by a Li-Al-N-H complex. *Adv. Funct. Mater.* **2007**, *17*, 1137–1142. [CrossRef]

31. Lu, J.; Fang, Z.Z.; Sohn, H.Y. A new Li-Al-N-H system for reversible hydrogen storage. *J. Phys. Chem. B* **2006**, *110*, 14236–14239. [CrossRef] [PubMed]

32. Wang, J.H.; Liu, T.; Wu, G.T.; Li, W.; Liu, Y.F.; Araujo, C.M.; Scheicher, R.H.; Blomqvist, A.; Ahuja, R.; Xiong, Z.T.; et al. Potassium-modified Mg(NH$_2$)$_2$/2LiH system for hydrogen storage. *Angew. Chem. Int. Ed.* **2009**, *48*, 5828–5832. [CrossRef] [PubMed]

33. Chen, P.; Xiong, Z.T.; Yang, L.F.; Wu, G.T.; Luo, W.F. Mechanistic investigations on the heterogeneous solid-state reaction of magnesium amides and lithium hydrides. *J. Phys. Chem. B* **2006**, *110*, 14221–14225. [CrossRef] [PubMed]

34. Liang, C.; Gao, M.X.; Pan, H.G.; Liu, Y.F. Structural transitions of ternary imide Li$_2$Mg(NH)$_2$ for hydrogen storage. *Appl. Phys. Lett.* **2014**, *105*, 083909. [CrossRef]

6

The Morphologies of Different Types of Fe$_2$SiO$_4$–FeO in Si-Containing Steel

Mingxing Zhou, Guang Xu *, Haijiang Hu, Qing Yuan and Junyu Tian

The State Key Laboratory of Refractories and Metallurgy, Key Laboratory for Ferrous Metallurgy and Resources Utilization of Ministry of Education, Wuhan University of Science and Technology, Wuhan 430081, China; kdmingxing@163.com (M.Z.); hhjsunny@sina.com (H.H.); 15994235997@163.com (Q.Y.); 13164178028@163.com (J.T.)
* Correspondence: xuguang@wust.edu.cn

Academic Editor: Hugo F. Lopez

Abstract: Red scale defect is known to be mainly caused by net-like Fe$_2$SiO$_4$–FeO. In the present study, the morphology of Fe$_2$SiO$_4$–FeO in a Si-containing steel was investigated by simultaneous thermal analysis, high-temperature laser scanning confocal microscopy, scanning electron microscopy, and energy dispersive spectroscopy. Only liquid Fe$_2$SiO$_4$–FeO can form a net-like morphology. Liquid Fe$_2$SiO$_4$–FeO is classified into two types in this work. Type-1 liquid Fe$_2$SiO$_4$–FeO is formed by melting pre-existing solid Fe$_2$SiO$_4$–FeO that already exists before the melting point of Fe$_2$SiO$_4$–FeO. Type-2 liquid Fe$_2$SiO$_4$–FeO is formed at a temperature higher than the melting point of Fe$_2$SiO$_4$–FeO. The results show that type-1 liquid Fe$_2$SiO$_4$–FeO is more likely to form a net-like morphology than is type-2 liquid Fe$_2$SiO$_4$–FeO. The penetration depth of type-1 liquid Fe$_2$SiO$_4$–FeO is also larger at the same oxidation degree. Therefore, type-1 liquid Fe$_2$SiO$_4$–FeO should be avoided in order to eliminate red scale defect. Net-like Fe$_2$SiO$_4$–FeO may be alleviated by two methods: decreasing the oxygen concentration in the heating furnace before the melting point of Fe$_2$SiO$_4$–FeO is reached and increasing the reheating rate before the melting point. In addition, FeO is distributed with a punctiform or lamellar morphology on Fe$_2$SiO$_4$.

Keywords: Fe$_2$SiO$_4$–FeO; morphology; Si-containing steel; mechanism

1. Introduction

Silicon (Si) is a common alloying element in advanced high strength steels [1–3], such as dual phase (DP) steel and transformation induced plasticity (TRIP) steel. However, the addition of Si often leads to red scale (mainly consisting of Fe$_2$O$_3$), a surface defect of hot rolled steels [4]. Some research has investigated the formation of red scale [5–10], commonly regarded as directly related to the presence of Fe$_2$SiO$_4$–FeO eutectic, which is formed by the combination of SiO$_2$ and FeO [5,6]. The theoretical eutectic temperature (melting point) of Fe$_2$SiO$_4$–FeO is recognized as 1173 °C [7]. When the reheating temperature of slabs is above 1173 °C, the liquid Fe$_2$SiO$_4$–FeO penetrates into the external scale along the grain boundary of the scale and forms a net-like distribution [8,9]. If the subsequent descaling temperature is below 1173 °C, the liquid net-like Fe$_2$SiO$_4$ solidifies and firmly bonds the steel substrate and iron scale, making it difficult to completely remove the FeO layer during descaling. The remaining FeO scale is oxidized into red Fe$_2$O$_3$ (red scale defect) during the subsequent cooling and rolling processes [5,10].

Due to the close relationship between red scale and Fe$_2$SiO$_4$–FeO, some studies on Fe$_2$SiO$_4$–FeO in Si-containing steels have been carried out [11–14]. Yuan et al. [11] reported that the net-like morphology of Fe$_2$SiO$_4$–FeO is not obvious when the Si content is low. Mouayd et al. [12] and Suarez et al. [13] found that the amount and penetrative depth of Fe$_2$SiO$_4$–FeO increases with the Si content. In addition,

He et al. [14] reported that the morphology of Fe_2SiO_4–FeO is blocky when the reheating temperature is below 1173 °C, because solid Fe_2SiO_4–FeO cannot penetrate into the external scale. However, when the reheating temperature is above 1173 °C, the morphology of Fe_2SiO_4–FeO is net-like. Net-like Fe_2SiO_4–FeO is well known to more easily lead to red scale compared with blocky Fe_2SiO_4–FeO.

In summary, red scale is mainly caused by net-like Fe_2SiO_4–FeO, and only liquid Fe_2SiO_4–FeO can form a net-like morphology. Thus, more attention should be given to the liquid Fe_2SiO_4–FeO. During the industrial reheating process, solid Fe_2SiO_4–FeO forms first before 1173 °C is reached and then melts into liquid at temperatures above 1173 °C. Besides, new liquid Fe_2SiO_4–FeO is gradually formed by the combination of SiO_2 and FeO at temperatures above 1173 °C. Therefore, liquid Fe_2SiO_4–FeO can be classified into two types when the reheating temperature is above 1173 °C. One forms by the melting of pre-existing solid Fe_2SiO_4–FeO, which has already formed below 1173 °C. The other appears above 1173 °C which is liquid once it forms. The former is termed as type-1 liquid Fe_2SiO_4–FeO and the latter is termed as type-2 liquid Fe_2SiO_4–FeO. The biggest difference between two types of liquid Fe_2SiO_4–FeO is that type-1 liquid Fe_2SiO_4–FeO is solid before 1173 °C is reached, whereas type-2 liquid Fe_2SiO_4–FeO is liquid from the time it forms. The distributions and morphologies of both types of liquid Fe_2SiO_4–FeO may be different. It is necessary to study the difference in their morphologies due to the close relationship between red scale and liquid Fe_2SiO_4–FeO. Thus far, research on this subject has been rarely reported. The present study investigates the morphologies of different types of Fe_2SiO_4–FeO and provides a theoretical reference toward preventing red scale defect in Si-containing steels.

2. Materials and Methods

The chemical composition of the experimental steel is Fe-0.06C-1.21Si-1.4Mn-0.035Al-0.01P-0.001S (wt. %). The steel was obtained from a hot strip plant (WISCO, Wuhan, China). The oxidation tests were carried out on a Setaram Setsys Evo simultaneous thermal analyzer (STA, Setaram, Lyon, France). The dimensions of the samples were 15 mm × 10 mm × 3 mm. A hole with a diameter of 4 mm was drilled near the edge center of each sample for suspension in the oxidation chamber. The surfaces of all samples were polished to remove the scale before the tests. As shown in Figure 1, two types of experimental routes were designed. For Routes 1–3, a binary gas mixture of oxygen and nitrogen with an oxygen concentration of 4.0 vol % was introduced into the STA chamber at the beginning of the experiments to obtain a certain amount of solid Fe_2SiO_4–FeO. Then, the binary gas mixture was replaced with 100 vol % nitrogen at the end of isothermal holding. Route 1 was set to observe the morphology of solid Fe_2SiO_4–FeO. For Routes 2 and 3, only type-1 liquid Fe_2SiO_4–FeO forms at temperatures higher than the melting point of Fe_2SiO_4–FeO. For Routes 4–6, a binary gas mixture of oxygen and nitrogen with an oxygen concentration of 4.0 vol % was not introduced into the STA chamber until the isothermal holding at 1260 °C. Thus, only type-2 liquid Fe_2SiO_4–FeO forms. Different isothermal holding time was set to investigate the effect of holding time on the morphology of Fe_2SiO_4–FeO. In short, the morphology of solid Fe_2SiO_4–FeO can be observed in Route 1 and the morphology of type-1 liquid Fe_2SiO_4–FeO can be observed in Routes 2 and 3. The morphology of type-2 liquid Fe_2SiO_4–FeO can be observed in Routes 4–6. There is no standard procedure of heating and oxidation routes. The experimental procedures are set to observe the separate morphology of different types of Fe_2SiO_4–FeO. The oxidizing atmosphere (4.0 vol % O_2-96.0 vol % N_2) is similar to that in the industrial reheating furnace. The accuracy of temperature measurement is ±0.5 °C. The mass gain of the samples and the temperature were digitally recorded during the whole oxidation processes.

After the oxidation tests, the samples were molded in resins at room temperature to protect the integrity of the oxide scale. Cross sections of the mounted samples were ground and polished. The microstructures of the oxide scale were observed by using a Nova 400 Nano scanning electron microscope (SEM, FEI, Hillsboro, OR, USA) operated at an accelerating voltage of 20 kV. The components of the oxide scale were analyzed with an energy dispersive spectrometer (EDS, OIMS, Oxford, UK). In addition, high-temperature laser scanning confocal microscopy (LSCM) was used for in situ observation of the melting process of Fe_2SiO_4–FeO. Samples for LSCM were selected from oxidized specimens and machined into a cylinder 6 mm in diameter and 4 mm in height. The investigations were conducted on a VL2000DXSVF17SP LSCM (lasertec, Yokohama, Japan). The specimen chamber was initially evacuated to 6×10^{-3} Pa before heating and argon was used to protect the specimens from surface oxidation. The sample was heated to 1260 °C at 20 °C/min and held for 10 min, followed by cooling to room temperature at 50 °C/min. Fifteen photographs per second were taken during the LSCM experiments.

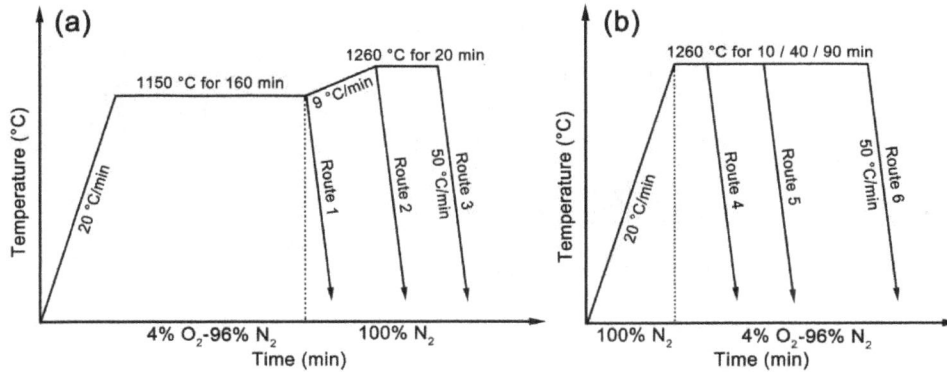

Figure 1. Oxidation experiment routes: (**a**) Routes 1–3; (**b**) Routes 4–6.

3. Results and Discussions

3.1. In Situ Observation

Route 1 and Routes 2 and 3 are designed for observing the morphologies of solid Fe_2SiO_4–FeO and type-1 liquid Fe_2SiO_4–FeO, respectively; thus, it is necessary to ensure that Fe_2SiO_4–FeO does not melt at 1150 °C. The theoretical melting point of Fe_2SiO_4–FeO is 1173 °C. However, the real value is influenced by the composition of the steel. The melting process of Fe_2SiO_4–FeO was observed in Figure 2. Fe_2SiO_4–FeO is solid at 1000 °C (Figure 2a). Solid Fe_2SiO_4–FeO begins to melt at 1170 °C (Figure 2b), so that the real melting point of Fe_2SiO_4–FeO is 1170 °C for the tested steel. Figure 2c indicates that Fe_2SiO_4–FeO completely melts at 1190 °C. Therefore, Fe_2SiO_4–FeO is always solid in Route 1 and only type-1 liquid Fe_2SiO_4–FeO forms after 1170 °C in Routes 2 and 3 (type-2 liquid Fe_2SiO_4–FeO does not form due to nitrogen protection).

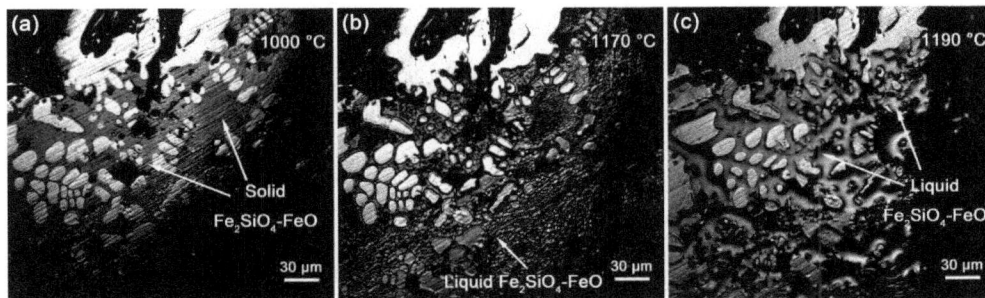

Figure 2. In situ observation of the melting process of Fe_2SiO_4–FeO: (**a**) 1000 °C, before melting; (**b**) 1170 °C, at the start of melting; (**c**) 1190 °C, after melting.

3.2. SEM Observations

Previous studies have confirmed that the iron scale in this steel contains Fe_2O_3, Fe_3O_4, FeO, and Fe_2SiO_4 [11,14,15]. The intimal scale consists of Fe_2SiO_4 and FeO. Figure 3 shows the typical morphology of Fe_2SiO_4–FeO in Routes 1–3. The Fe_2SiO_4–FeO layers are marked by red lines. When the heating temperature is 1150 °C (Route 1), Fe_2SiO_4–FeO is blocky and dispersively distributed

(Figure 3a); thus, solid Fe_2SiO_4–FeO is not net-like even when the holding time is as long as 160 min. Figure 3b shows that a large amount of net-like Fe_2SiO_4–FeO appears during Route 2, in which the melting time of the preexisting solid Fe_2SiO_4–FeO is 10 min; this indicates that the morphology of Fe_2SiO_4–FeO changes quickly and significantly after melting. Therefore, type-1 liquid Fe_2SiO_4–FeO can quickly form a net-like morphology within 10 min. When the melting time increases to 30 min, Fe_2SiO_4–FeO penetrates into a deeper area (Figure 3c).

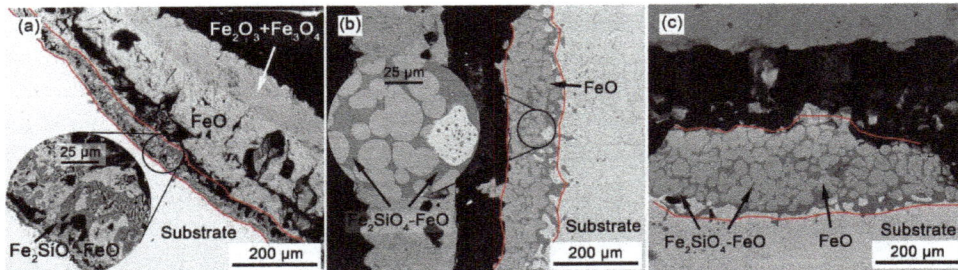

Figure 3. The typical morphology of Fe_2SiO_4–FeO in Routes 1-3: (**a**) Route 1, without melting; (**b**) Route 2, melting for 10 min; (**c**) Route 3, melting for 30 min.

The distribution of FeO in Fe_2SiO_4–FeO has been rarely reported. Figure 4 shows the typical distribution of Fe_2SiO_4–FeO. According to EDS results (Figure 4c,d), the darker scale is Fe_2SiO_4 and the lighter one is FeO. Fe_2SiO_4 surrounds FeO. Interestingly, FeO is distributed on Fe_2SiO_4 with a punctiform (Figure 4a) or lamellar morphology (Figure 4b), which is similar to the morphology of pearlite in steel. A possible mechanism for the structure of Fe_2SiO_4–FeO may be that, during the cooling process, FeO separates out from Fe_2SiO_4–FeO in the form of a lamella or sphere due to the diffusion of Fe, O, and Si elements. Lamellar FeO has a larger surface area and interfacial energy compared with punctiform FeO. The nonuniform concentrations of Si, O, and Fe lead to two different morphologies of FeO.

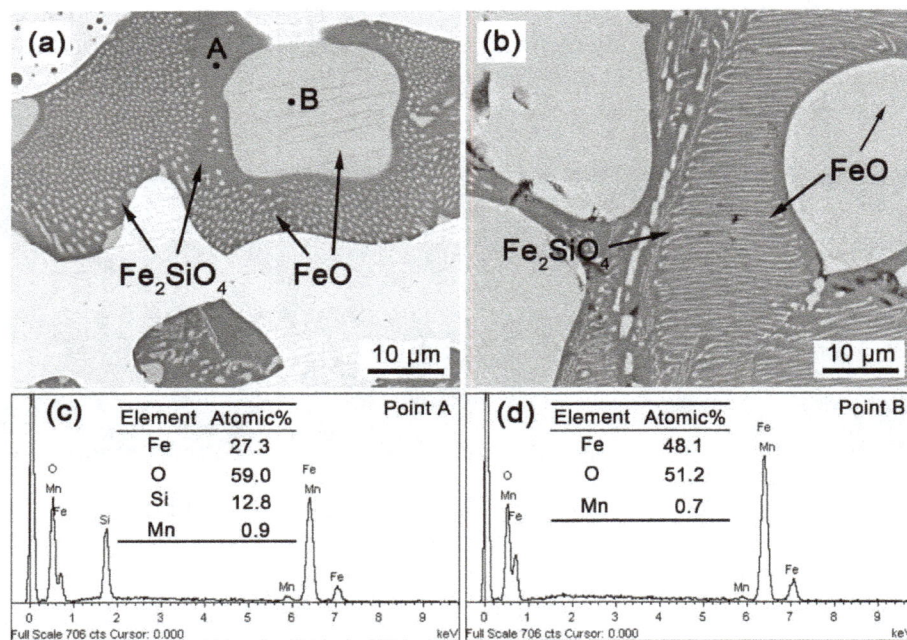

Figure 4. (**a,b**) The typical distribution of Fe_2SiO_4–FeO; (**c**) The energy spectra of Point A (Fe_2SiO_4); (**d**) The energy spectra of Point B (FeO).

Figure 5 shows the typical morphology of Fe_2SiO_4–FeO for Routes 4–6, in which only type-2 liquid Fe_2SiO_4–FeO forms during isothermal holding at 1260 °C. Figure 5a shows that type-2 liquid Fe_2SiO_4–FeO is blocky after 10 min of oxidation. As the time increases to 40 min, the penetration depth increases, but Fe_2SiO_4–FeO is still blocky (Figure 5b). When the oxidation time increases to 90 min, the net-like morphology of Fe_2SiO_4–FeO appears. Therefore, liquid Fe_2SiO_4–FeO is not necessarily net-like, and type-2 liquid Fe_2SiO_4–FeO does not form a net-like morphology before 40 min.

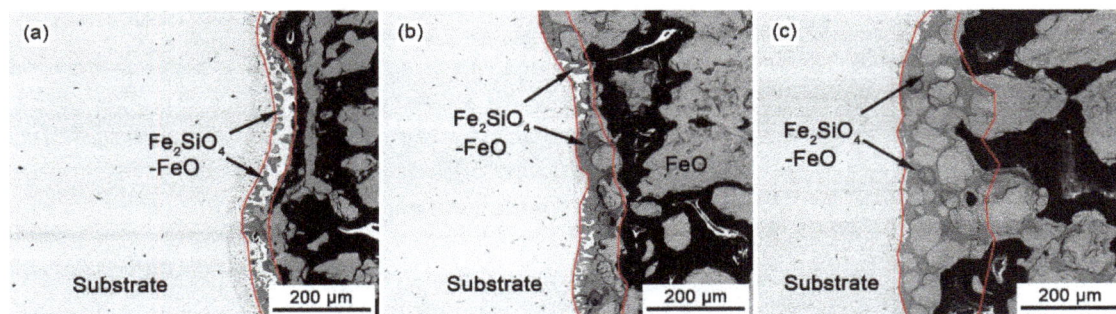

Figure 5. The typical morphology of Fe_2SiO_4–FeO for Routes 4–6: (**a**) Route 4, 1260 °C for 10 min; (**b**) Route 5, 1260 °C for 40 min; (**c**) Route 6, 1260 °C for 90 min.

The total mass gain recorded in the oxidation experiments represents the oxidation degree. Figure 6a shows the total mass gains for Routes 1–6. The oxidation degrees are almost the same for Routes 1–3 because of the same oxidation temperature and time (Note that the samples were not oxidized after isothermal holding at 1150 °C in Routes 1–3). The oxidation degrees for Routes 2, 3, and 5 are similar. However, the morphology of Fe_2SiO_4–FeO is significantly different. With similar oxidation degree, type-1 liquid Fe_2SiO_4–FeO (Routes 2 and 3; Figure 3b,c) is obviously net-like, whereas type-2 Fe_2SiO_4–FeO is blocky (Route 5; Figure 5b). Note that type-2 Fe_2SiO_4–FeO can also form a net-like morphology (Figure 5c); however, it requires a much higher oxidation degree compared with type-1 Fe_2SiO_4–FeO. In addition, the oxidation degree in Route 6 is larger than that in Route 3, whereas the penetration depth of Fe_2SiO_4–FeO in Route 6 (Figure 5c) is smaller than that in Route 3 (Figure 3c), indicating that type-1 liquid Fe_2SiO_4–FeO is more likely to form a net-like morphology. The penetration depth of Fe_2SiO_4–FeO is measured by using the software Image-Pro Plus 6.0 and then normalized by dividing the total mass gain (normalized penetration depth = real penetration depth/total mass gain), as is shown in Figure 6b. The normalized penetration depth of type-1 liquid Fe_2SiO_4–FeO is larger than that of type-2 liquid Fe_2SiO_4–FeO, indicating that type-1 liquid Fe_2SiO_4 penetrates more easily into the external scale. Moreover, the penetration depth of solid Fe_2SiO_4–FeO is much smaller than that of liquid Fe_2SiO_4–FeO.

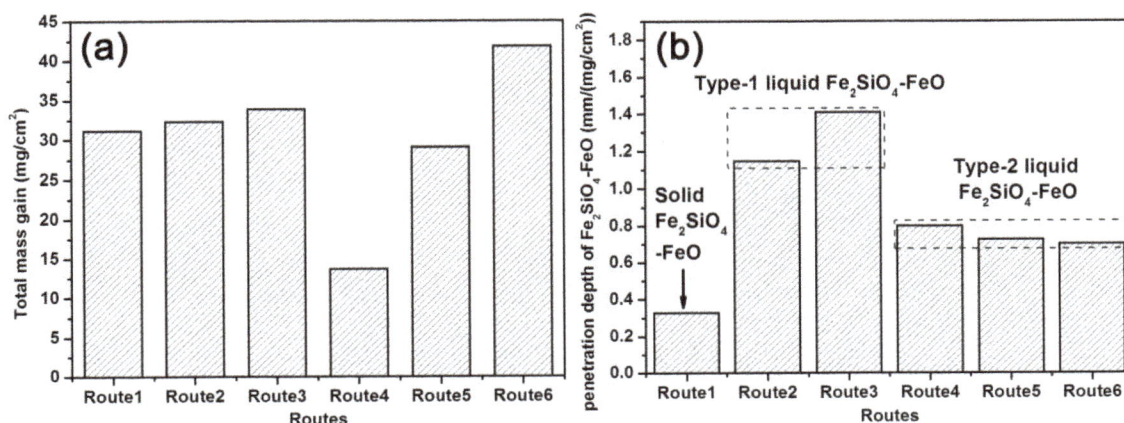

Figure 6. (**a**) The total mass gain for Routes 1–6. (**b**) The normalized penetration depth of Fe_2SiO_4–FeO.

The above results can be interpreted as follows. The Pilling–Bedworth ratio (PBR) is the ratio of the oxide volume to the consumed metal volume [16]. The PBR of Fe oxide or Si oxide is larger than 1 because the volume of the oxide is larger than that of the consumed metal, leading to a compressive stress in the oxide [8]. Liquid Fe_2SiO_4–FeO can penetrate into outer scale under the effect of this compressive stress, so that net-like morphology of Fe_2SiO_4–FeO forms after a certain time. The scale adjacent to Fe_2SiO_4–FeO (whether solid or liquid) is solid. The compressive stress between two solids should be larger than that between a solid and a liquid; thus, solid Fe_2SiO_4–FeO should be subjected to a larger stress compared with liquid Fe_2SiO_4–FeO. When preexisting solid Fe_2SiO_4–FeO melts into type-1 liquid Fe_2SiO_4–FeO, it can quickly penetrate into an outer place under a larger compressive stress. On the other hand, type-2 liquid Fe_2SiO_4–FeO is subjected to a smaller stress because it is liquid when formed; thus, its penetration rate is smaller. In addition, a large amount of Fe_2SiO_4–FeO has accumulated before the penetration of type-1 liquid Fe_2SiO_4–FeO, whereas the penetration and formation of type-2 liquid Fe_2SiO_4–FeO take place simultaneously. Therefore, it is easier for type-1 liquid Fe_2SiO_4–FeO to penetrate into the scale.

Red scale defect is well known to be caused by the net-like morphology of Fe_2SiO_4–FeO [5,10]. Type-1 liquid Fe_2SiO_4–FeO is more likely to form a net-like morphology, so that it should be avoided in order to eliminate red scale defect. The amount of type-1 liquid Fe_2SiO_4–FeO can be decreased by hindering the formation of solid Fe_2SiO_4–FeO. One way to do this is by decreasing the oxygen concentration in the heating furnace before the melting point of Fe_2SiO_4–FeO (1170 °C in the present study) is reached [17,18]. In addition, increasing the reheating rate before 1170 °C can also decrease the amount of solid Fe_2SiO_4–FeO due to a shorter oxidation time.

4. Conclusions

Liquid Fe_2SiO_4–FeO is classified into two types. The present study investigates the difference in morphology between these two types of liquid Fe_2SiO_4–FeO. The results show that, compared with type-2 liquid Fe_2SiO_4–FeO, type-1 liquid Fe_2SiO_4–FeO is more likely to form a net-like morphology. The penetration depth of type-1 liquid Fe_2SiO_4–FeO is also larger at the same oxidation degree. Red scale defect is known to be caused by the net-like Fe_2SiO_4–FeO. Therefore, type-1 liquid Fe_2SiO_4–FeO should be avoided in order to eliminate red scale defect. Net-like Fe_2SiO_4–FeO may be alleviated by two methods: decreasing the oxygen concentration in the heating furnace and increasing the reheating rate before the melting point of Fe_2SiO_4–FeO is reached. In addition, FeO is distributed with a punctiform or lamellar morphology on Fe_2SiO_4.

Acknowledgments: The authors gratefully acknowledge the financial supports from National Natural Science Foundation of China (NSFC) (No. 51274154) and the National High Technology Research and Development Program of China (No. 2012AA03A504).

Author Contributions: Guang Xu and Mingxing Zhou conceived and designed the experiments; Mingxing Zhou performed the experiments; Mingxing Zhou, Haijiang Hu, and Qing Yuan analyzed the data; Junyu Tian contributed materials tools; Mingxing Zhou wrote the paper.

Conflicts of Interest: The authors declare no conflict of interest. The founding sponsors had no role in the design of the study; in the collection, analyses, or interpretation of data; in the writing of the manuscript; or in the decision to publish the results.

Abbreviations

The following abbreviations are used in this manuscript:

STA	simultaneous thermal analyzer
LSCM	high temperature laser scanning confocal microscopy
SEM	scanning electron microscope

References

1.   Hu, H.J.; Xu, G.; Wang, L.; Xue, Z.L.; Zhang, Y.; Liu, G. The effects of Nb and Mo addition on transformation and properties in low carbon bainitic steels. *Mater. Des.* **2015**, *84*, 95–99. [CrossRef]

2. Hu, H.J.; Xu, G.; Zhou, M.X.; Yuan, Q. Effect of Mo Content on Microstructure and Property of Low-Carbon Bainitic Steels. *Metals* **2016**, *6*, 173–182. [CrossRef]
3. Zhou, M.X.; Xu, G.; Wang, L.; Yuan, Q. The Varying Effects of Uniaxial Compressive Stress on the Bainitic Transformation under Different Austenitization Temperatures. *Metals* **2016**, *6*, 119–130. [CrossRef]
4. Takeda, M.; Onishi, T. Oxidation behavior and scale properties on the Si containing steels. *Mater. Sci. Forum.* **2006**, *522*, 477–488. [CrossRef]
5. Okada, H.; Fukagawa, T.; Ishihara, H. Prevention of red scale formation during hot rolling of steels. *ISIJ Int.* **1995**, *35*, 886–891. [CrossRef]
6. Yang, Y.L.; Yang, C.H.; Lin, S.N.; Chen, C.H.; Tsai, W.T. Effect of Si and its content on the scale formation on hot-rolled steel strips. *Mater. Chem. Phys.* **2008**, *112*, 566–571. [CrossRef]
7. Liu, X.J.; Cao, G.M.; He, Y.Q.; Jia, T.; Liu, Z.Y. Effect of temperature on scale morphology of Fe-1.5Si Alloy. *J. Iron Steel Res. Int.* **2013**, *20*, 73–78. [CrossRef]
8. Garnaud, G.; Rapp, R.A. Thickness of the oxide layers formed during the oxidation of iron. *Oxid. Met.* **1977**, *11*, 193–198. [CrossRef]
9. Liu, X.J.; Cao, G.M.; Nie, D.M.; Liu, Z.Y. Mechanism of black strips generated on surface of CSP hot-rolled silicon steel. *J. Iron. Steel Res. Int.* **2013**, *20*, 54–59. [CrossRef]
10. Fukagawa, T.; Okada, H.; Maeharara, Y. Mechanical of red scale defect formation in Si-added hot-rolled steels. *ISIJ Int.* **1994**, *34*, 906–911. [CrossRef]
11. Yuan, Q.; Xu, G.; Zhou, M.X.; He, B. The effect of the Si content on the morphology and amount of Fe_2SiO_4 in low carbon steels. *Metals* **2016**, *6*, 94–103. [CrossRef]
12. Mouayd, A.A.; Koltsov, A.; Sutter, E.; Tribollet, B. Effect of silicon content in steel and oxidation temperature on scale growth and morphology. *Mater. Chem. Phys.* **2014**, *143*, 996–1004. [CrossRef]
13. Suarez, L.; Schneider, J.; Houbaert, Y. High-Temperature oxidation of Fe-Si alloys in the temperature range 900–1250 °C. *Defect. Diffus. Forum.* **2008**, *273–276*, 661–666. [CrossRef]
14. He, B.; Xu, G.; Zhou, M.X.; Yuan, Q. Effect of Oxidation Temperature on the Oxidation Process of Silicon-Containing Steel. *Metals* **2016**, *6*, 137–145. [CrossRef]
15. Yuan, Q.; Xu, G.; Zhou, M.X.; He, B. New insights into the effects of silicon content on the oxidation process in silicon-containing steels. *Int. J. Min. Met. Mater.* **2016**, *23*, 1–8. [CrossRef]
16. Staettle, R.W.; Fontana, M.G. *Advances in Corrosion Science and Technology*; Springer: New York, NY, USA, 1974; pp. 239–356.
17. Abuluwefa, H.; Guthrie, R.I.L.; Ajersch, F. The Effect of Oxygen Concentration on the Oxidation of Low-Carbon Steel in the Temperature Range 1000 to 1250 °C. *Oxid. Met.* **1996**, *46*, 423–440. [CrossRef]
18. Chen, R.Y.; Yuen, W.Y.D. Review of the High-Temperature Oxidation of Iron and Carbon Steels in Air or Oxygen. *Oxid. Met.* **2003**, *59*, 433–468. [CrossRef]

7

Effects of EMS Induced Flow on Solidification and Solute Transport in Bloom Mold

Qing Fang, Hongwei Ni *, Bao Wang, Hua Zhang and Fei Ye

The State Key Laboratory of Refractories and Metallurgy, Wuhan University of Science and Technology, Wuhan 430081, China; qfang525@sina.com (Q.F.); wangbao1983@wust.edu.cn (B.W.); huazhang@wust.edu.cn (H.Z.); yefeishangnan@163.com (F.Y.)
* Correspondence: nihongwei@wust.edu.cn

Academic Editor: Mohsen Asle Zaeem

Abstract: The flow, temperature, solidification, and solute concentration field in a continuous casting bloom mold were solved simultaneously by a multiphysics numerical model by considering the effect of in-mold electromagnetic stirring (M-EMS). The mold metallurgical differences between cases with and without EMS are discussed first, and then the solute transport model verified. Moreover, the effects of EMS current intensity on the metallurgical behavior in the bloom mold were also investigated. The simulated solute distributions were basically consistent with the test results. The simulations showed that M-EMS can apparently homogenize the initial solidified shell, liquid steel temperature, and solute element in the EMS effective zone. Meanwhile, the impingement effect of jet flow and molten steel superheat can be reduced, and the degree of negative segregation in the solidified shell at the mold corner alleviated from 0.74 to 0.78. However, the level fluctuation and segregation degree in the shell around the center of the wide and narrow sides were aggravated from 4.5 mm to 6.2 mm and from 0.84 to 0.738, respectively. With the rise of current intensity the bloom surface temperature, level fluctuation, stirring intensity, uniformity of molten steel temperature, and solute distribution also increased, while the growth velocity of the solidifying shell in the EMS effective zone declined and the solute mass fraction at the center of the computational outlet ($z = 1.5$ m) decreased. M-EMS with a current intensity of 600 A is more suitable for big bloom castings.

Keywords: continuous casting mold; flow pattern; solidification; solute transport; electromagnetic stirring

1. Introduction

Most defects affecting steel quality in the continuous casting (CC) process are associated with metallurgical behavior in the mold including flow pattern, heat transfer, initial solidification, and solute transport etc. At present, in-mold electromagnetic stirring (M-EMS) used to optimize molten steel flow is a widely accepted technique in big bloom casting, which has been proved to be beneficial to enhance the columnar-to-equiaxed transition during solidification and reduce the surface and subsurface defects [1–3]. Geng et al. [4] and Yu et al. [5] indicated that the EMS parameters affects the metallurgical behavior and the steel quality significantly, an optimum EMS parameter for a certain bloom should be proposed for a higher quality. Recently, numerous mathematical and experimental studies on the CC process involving M-EMS were conducted to investigate the flow field distribution of molten steel [6–9], and the heat transfer and initial solidification behaviors influenced by EMS induced fluid flow were also studied in past years [10–13]. Yang et al. [10] and Ren et al. [13] recently discussed the effects of EMS parameters on the flow pattern and initial solidification in the big bloom mold fed by a normal nozzle with a single outlet, and stated that M-EMS can promote superheat dissipation and enhance the percentage of the equiaxed zone, while the mutual effect between the EMS induced flow and the growth of the solidifying shell lacked discussion. Sun et al. [14] who designed a swirling flow nozzle for round bloom, concluded that a horizontal swirling flow in the mold can eliminate the superheat and depress local shell thinning. However, the solute transport phenomena associated with the EMS induced turbulent flow during the initial solidification process in the bloom mold has been rarely reported, although this deeply affects the macrosegregation profile of the final products. Aboutalebi et al. [15] and Yang et al. [16] developed a fully coupled flow, heat and solute transport model which predicted a turbulent flow induced solidification and segregation profile for CC round bloom and slab, respectively, while the relation of flow and segregation profile was not clearly explained and the effect of M-EMS was ignored. Lei et al. [17] and KwanGu et al. [18] preliminarily investigated the flow, solidification, and solute transport in CC slab mold with an electromagnetic brake, but without detailed explanation of the effect of the electromagnetic field on the solidification process and solute transport, while the verification of the solute transport model was neglected. Sun et al. [19] studied the effect of a swirling flow nozzle (SFN) on the flow, solidification, and macrosegregation behavior of casting blooms, and concluded that the formation of positive and negative segregation in the initial solidified shell can contribute to the flotation of solute-richer molten steel and the "solute washing effect" [20], which could offer the reader useful enlightenment for analyzing the swirling flow induced macrosegregation in the CC mold. However, the effects of M-EMS parameters on the macro-transport phenomena in CC bloom mold, especially the solute transport behavior, need detailed investigation.

In order to gain a deep insight into the M-EMS effects, the flow, temperature field, solidification behavior, and solute concentration field in the CC bloom mold of U71Mn steel were solved simultaneously by a three-dimensional multiphysics numerical model by considering the electromagnetic stirring force. In this paper, the metallurgical effects with EMS on the flow, level fluctuation, heat transfer, initial solidification, and solute transport phenomena in the CC mold were discussed first, and compared to that without EMS. The simulated segregation profiles of solute element carbon (C) in the region of initial solidified shell with a thickness of 30 mm at both the wide and the narrow sides were examined by the infrared carbon-sulfur determinator. Furthermore, the influences of EMS current intensity on the mold metallurgical behavior were also investigated and an optimized current intensity was suggested for big bloom casting.

2. Model Descriptions

2.1. Basic Assumptions

A three dimensional coupled model was established by combining electromagnetism, fluid flow, heat, and solute transport based on the following assumptions:

(1)   The transport phenomena in the mold are assumed to be at steady state and the influence of inclusions on the fluid flow, heat transfer, and species transport is neglected to simplify the simulation.
(2)   The impact of fluid flow on the internal heat transfer of molten steel is ignored in this model, and the liquid steel is assumed to be an incompressible Newtonian fluid. The influence of mold oscillation and mold taper on the fluid flow is also ignored.
(3)   The effect of thermal contraction on the fluid flow and temperature field in the bloom is neglected.
(4)   The mold arc is neglected, and the computational zone is assumed to be a vertical model.
(5)   The effect of the melt flow on the electromagnetic field is ignored due to the small magnetic Reynolds number (about 0.01) in the stirring process.

2.2. Governing Equations

2.2.1. Turbulent Flow

The melt flow pattern in the mold is numerically simulated by solving the continuity and momentum equations under turbulent conditions:

$$\frac{\partial(\rho u_i)}{\partial x_i} = 0 \tag{1}$$

$$\rho \frac{\partial u_i u_j}{\partial x_j} = \frac{\partial}{\partial x_j}\left[\mu_{\text{eff}}\left(\frac{\partial u_i}{\partial x_j} + \frac{\partial u_j}{\partial x_i}\right)\right] - \frac{\partial P}{\partial x_i} + \rho g_i + F_B + F_E + S_P \tag{2}$$

where μ_{eff} is the effective viscosity coefficient, which is the sum of the laminar (μ_l) and turbulent (μ_t) viscosity. μ_t can be defined as:

$$\mu_t = \rho \times f_\mu \times C_\mu \times \frac{K^2}{\varepsilon} \tag{3}$$

where K and ε are the turbulent kinetic energy and dissipation rate of this energy, respectively. The low-Reynolds number K-ε turbulent model [21,22] is applied to describe the turbulence flow and solidification phenomena in the computational domain. The empirical constants are set as $C_\mu = 0.09$ and $f_\mu = \exp\left[\frac{-3.4}{\left(1 + \frac{\rho K^2}{50\varepsilon}\right)^2}\right]$, respectively.

The last term in Equation(2) accounts for the phase interaction force within the mushy zone, and it can be described by Darcy's law, the permeability K_p is presented as follows [23]:

Permeability parallel to primary dendrite arms:

$$K_p^{\text{par}} = \begin{cases} f_l < 0.7, & [4.53 \times 10^{-4} + 4.02 \times 10^{-6}(f_l + 0.1)^{-5} \cdot \frac{\lambda_2^2 \cdot f_l^3}{1 - f_l} \\ f_l \geq 0.7, & 0.07425 \cdot \lambda_1^2[-\ln(1 - f_l) - 1.487 + 2(1 - f_l) - 0.5(1 - f_l)^2] \end{cases} \tag{4}$$

Permeability perpendicular to primary dendrite arms:

$$K_p^{\text{per}} = \begin{cases} f_l < 0.7, \, 0.00173 \times \left(\frac{\lambda_1}{\lambda_2}\right)^{1.09} \times \frac{\lambda_2^2 \cdot f_l^3}{(1 - f_l)^{0.749}} \\ f_l \geq 0.7, \, 0.03978 \cdot \lambda_1^2[-\ln f_s - 1.476 + 2f_s - 1.774f_s^2 + 4.076f_s^3] \end{cases} \tag{5}$$

$$S_p = \frac{\mu_l}{K_p}(u_i - u_{s,\,i}) \tag{6}$$

where λ_1, λ_2 are the primary arm space and secondary arm space respectively; $u_{s,\,i}$ is the velocity of solid in the direction of i; f_l is the local liquid fraction.

The source term F_B in the momentum equation represents the full buoyancy force which includes the solutal buoyancy and the thermal buoyancy effect. The detailed information is described by the following equation.

$$F_B = \rho \times g \times \beta_t \times (T - T_l) + \sum_i \rho \times g \times \beta_{C_i} \times (C_{l,i} - C_{l,\,0}^i) \tag{7}$$

where T_l is liquidus temperature, $C_{l,i}$ is locally averaged concentration of solute element i in liquid phase, and $C_{l,\,0}^i$ is solute concentration at the liquidus temperature. The density ρ is assumed to be a constant except in the buoyancy terms according to the Boussinesq approximation [24].

2.2.2. Heat Transfer Model

To obtain a precise prediction of the temperature field and solidification behavior, the enthalpy equation is employed.

$$\rho u_i \frac{\partial H}{\partial x_i} = -\frac{\partial}{\partial x_i}\left[\left(k_l + \frac{\mu_t}{Pr_t}\right)\frac{\partial H}{\partial x_i}\right] \tag{8}$$

where u_i represents fluid flow velocity in direction of i, the turbulent Prandtl number Pr_t is set to the be 0.9 [25], the total enthalpy H can be obtained by the following equation:

$$H = h_{\text{ref}} + \int_{T_{\text{ref}}}^{T} c_p \, dT + f_l L \tag{9}$$

where c_p is specific heat, L is latent heat, and liquid fraction f_l is updated by:

$$f_l = \frac{T - T_s}{T_l - T_s}(T_s < T < T_l) \tag{10}$$

$$T_l = T_{\text{pure}} - \sum_i m_i \times C_{l,\,i}, \; T_s = T_{\text{pure}} - \sum_i m_i \times C_{l,\,i}/k_i \tag{11}$$

$$k_i = C^*_{s,\,i}/C^*_{l,\,i} \tag{12}$$

where T_l and T_s are the liquidus and solidus temperature, respectively; T_{pure} is the fusion temperature of pure iron, m_i, k_i are the slope of liquidus line and equilibrium partition coefficient of solute element i, respectively. $C^*_{s,\,i}$ and $C^*_{l,\,i}$ are the interface concentration of solute element i in the solid and liquid phase, respectively.

2.2.3. Solute Transport Model

The conservation equation for multicomponents in the CC system is expressed as:

$$\frac{\partial(\rho u_i c)}{\partial x_i} = \frac{\partial}{\partial x_i}\left[\left(\frac{\mu_l}{S_{c_l}} + \frac{\mu_t}{S_{c_t}}\right)\frac{\partial c}{\partial x_i}\right] + S_c \tag{13}$$

where $S_{c_l} = \frac{\mu_l}{\rho D_l}$ and $S_{c_t} = 1$ [15] represent the laminar and turbulent Schmidt numbers, respectively. D_l is the liquid diffusive coefficient. S_c is the source term:

$$S_c = \frac{\partial}{\partial x_i}\left[\rho f_s D_s \frac{\partial(c_s - c)}{\partial x_i}\right] + \frac{\partial}{\partial x_i}\left[f_l\left(\frac{\mu_l}{S_{c_l}} + \frac{\mu_t}{S_{c_t}}\right)\right]\frac{\partial(c_l - c)}{\partial x_i} - \frac{\partial}{\partial x_i}\rho(c_l - c)(u_i - u_{i,\,s}) \tag{14}$$

where D_s is the solid diffusive coefficient, c_s and c_l are the species mass fraction in the solid and liquid steel, respectively.

2.2.4. Electromagnetism Model

For the source term in Equation (2), F_E represents the electromagnetic force, which contains radial force ($\overline{F_r}$) and tangential force ($\overline{F_\theta}$) [26]:

$$\overline{F_r} = -\frac{1}{8}B_0^2\sigma^2(2\pi f - \frac{V_\theta}{r})^2\mu_0 r^3 \tag{15}$$

$$\overline{F_\theta} = \frac{1}{2}B_0^2\sigma(2\pi f - \frac{V_\theta}{r})r \tag{16}$$

where B_0 is the magnetic induction density at the boundary of molten steel; σ is the electrical conductivity of liquid steel; f is the current frequency of the EMS; V_θ is the tangential velocity; μ_0 is the magnetoconductivity; r is the value of radial displacement. The radial force ($\overline{F_r}$) is ignored for $\overline{F_r}$ is much smaller than $\overline{F_\theta}$ in a rotating magnetic field. The electromagnetic force is added into FLUENT by user defined functions as a momentum source. The distribution of electromagnetic intensity in the mold at the axis under different current intensity is measured by the Gauss meter, the result is presented in Figure 1.

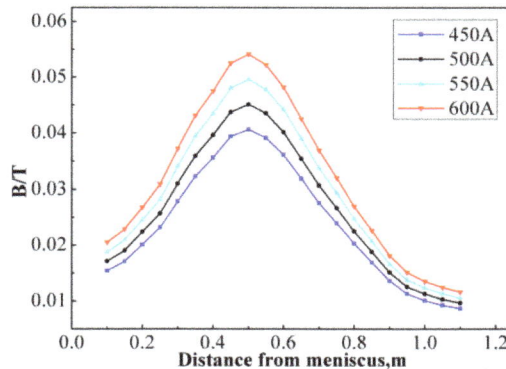

Figure 1. Measured distribution of electromagnetic intensity in the mold under different current intensity.

2.2.5. VOF Model

The steel/slag interface fluctuation in the mold is tracked by the volume fraction of fluid (VOF) method, which can track the shape of the fluctuating steel/slag interface. The detailed information and equations on the VOF model in representing the movement of the mold slag layer can be found elsewhere [27].

3. Simulation Procedure and Verification

The fluid flow, heat, initial solidification behavior, and solute transport in the CC bloom mold were calculated using FLUENT 15.0. The segregation profiles of C in the region of the initial solidification shell at both the wide and the narrow face were measured by an infrared carbon-sulfur determinator and compared with the calculation results.

3.1. Operating Condition and Parameters

The target chemical composition of U71Mn steel in the plant is listed in Table 1. The construction and installation method of submerged entry nozzle (SEN) for bloom casting is presented in Figure 2, where the immersion depth is 210 mm and the port downward angle is 15°, respectively. SEN has been proved to be a practical design in optimizing the fluid and heat transfer in the mold while producing U71Mn steel [28]. Table 2 lists the industrial conditions for producing the steel, and the thermo-solutal and physical parameters of U71Mn steel are presented in Table 3.

Table 1. Target chemical composition of U71Mn steel.

Chemical Composition	C	Si	Mn	P	S	Cr	Mo	Ni	Cu
Mass%	0.73	0.25	1.2	≤0.02	≤0.02	≤0.15	≤0.02	≤0.10	≤0.15

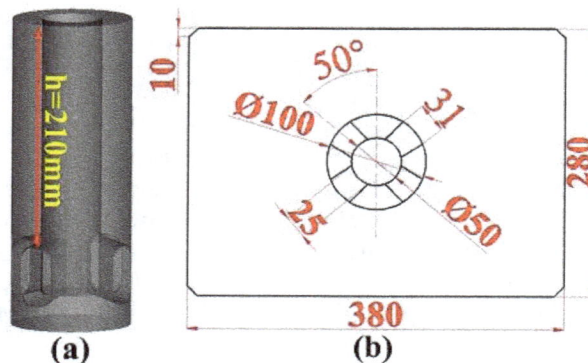

Figure 2. The three-dimensional view (**a**) and installation method (**b**) of submerged entry nozzle (SEN).

Table 2. Industrial conductions for U71Mn steel.

Parameters	Value
Cross section of bloom, mm^2	380 × 280
Casting speed, m/min	0.63
Casting temperature, K	1765
Nozzle adaption	Figure 2
Calculation length, mm	1500
Running current of M-EMS, A	450–600
Running frequency of M-EMS, Hz	2.0
EMS center(distance from meniscus), mm	420
Height of EMS, mm	480
Water quantity in mold, L/min	2600

Table 3. Thermo-solutal and physical parameters of U71Mn steel.

Parameters	Value
Operation density, kg/m^3	7020
Latent heat, J/kg	272,000
Specific heat of liquid, J/(kg·K)	810
Specific heat of solid, J/(kg·K)	682
Electric conductivity, S/m	7.14×10^5
Viscosity, Pa·s	Figure 3a
Thermal conductivity, W/(m·K)	Figure 3b
Diffusion coefficient of liquid, cm^2/s	$0.0052 \exp\left(\frac{-11700}{8.314 \cdot T}\right)$
Diffusion coefficient of solid, cm^2/s	$0.0761 \exp\left(\frac{-134557}{8.314 \cdot T}\right)$
Equilibrium partition coefficient	0.4
Slope of liquidus line	78

Viscosity and thermal conductivity are very important thermo-physical parameters for predicting the flow and solidification behavior of the steel, especially the mushy zone. The variations of viscosity and thermal conductivity with local temperature for U71Mn steel shown in Figure 3, which are calculated by the code JMatPro7.0, were applied in the calculation process.

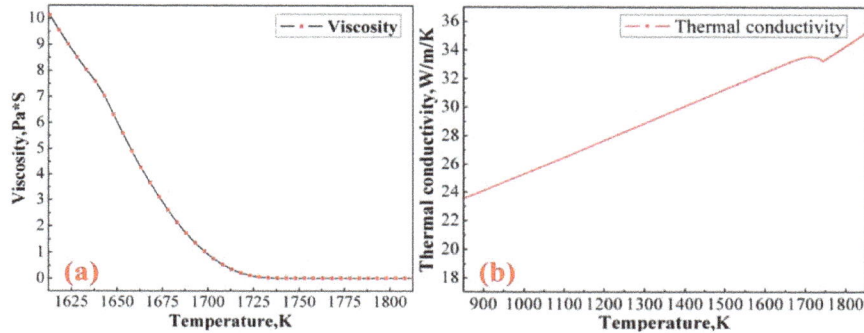

Figure 3. Variations of viscosity (**a**) and thermal conductivity (**b**) with local temperature.

3.2. Model Building

A quarter of the three-dimensional (3D) schematic sketch of the top part of a real mold is given in Figure 4a. To simulate the behavior of the initial solidified shell more accurately, the technology of local grid refinement is applied, the meshes of FLUENT computational domain include non-uniform grids with cells about 3,100,000. The meshed geometrical model for the computational domain is presented in Figure 4b. The meshed computational model has the same geometrical parameters as the real caster.

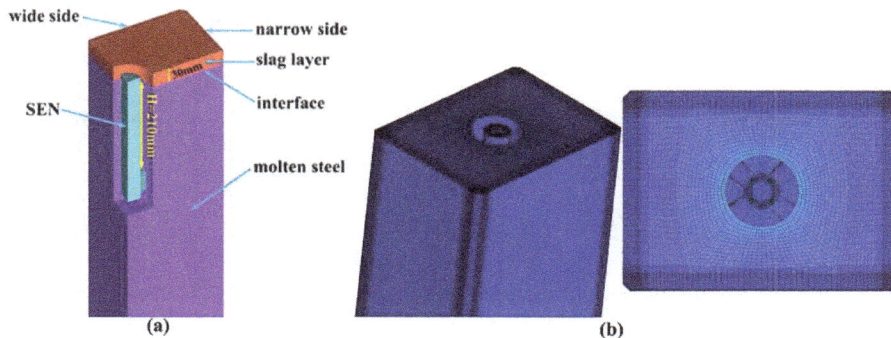

Figure 4. The schematic sketch of mold and meshed model of computational domain. (**a**) mold top part; (**b**) meshed model.

3.3. Initial and Boundary Conditions

The initial and boundary condition for velocity, temperature, turbulent kinetic energy, dissipation rate, and species concentration are set as follows:

For computational inlet, $v_z = v_{in} = v_{cast} \times \frac{S_{mold}}{S_{SEN}} = 0.5679$ m/s, $v_x = v_y = 0$, $k_{in} = 0.01v_{in}^2$, $\varepsilon = k_{in}^{1.5}/D$, $T_{in} = 1765.15$ K, and $C_{in} = C_0 = 0.7246\%$. k_{in} and ε are calculated by the semiempirical equations [29]. For outlet, the fully developed flow condition is adopted, where the normal gradients of all variables are set to zero. For the free surface, the normal derivative of all variables is set to be zero, and the adiabatic condition is adopted for the free surface.

For bloom wall, the detailed boundary conditions for heat transfer at the bloom wall are presented as:

In the mold region, the average heat flux is applied to describe the heat exchange effect in the actual domain [30]:

$$\bar{q} = \frac{\rho_w \times c_w \times W \times \Delta T}{S} \tag{17}$$

where $\bar{q}$ is the average heat flux density, W/m^2; ρ_w is the water density, kg/m^3; c_w is the specific heat capacity of the cooling water, J/(kg·K); W is the water flow rate, m^3/s; ΔT is the water temperature difference between in and out, K; S is the effective contact area of the steel and the mold wall, m^2. A correction factor (α) is necessary while calculating the solidification behavior at the bloom mold corner.

$$\bar{q}_g = \alpha \times \bar{q} \tag{18}$$

The heat transfer coefficient in the secondary cooling zone I is calculated by the following equation, in which water cooling is applied.

$$h = \beta \times 581 w^{0.541} (1 - 0.0075 T_w) \tag{19}$$

where h is the heat transfer coefficient, W/(m^2·K); w is the water density, L/(m^2·s); and T_w represents the ambient temperature; β is the correction factor.

3.4. Verification of Solute Transport

To validate the mathematical coupled model, the mass fraction of the chemical element C in the initial solidified shell thickness of 30 mm is evaluated by an infrared carbon-sulfur determinator. The specific locations of the drilling samples are presented in Figure 5. Where the samples are drilled with a 4.2 mm diameter drill up to a depth about 20 mm and the weight is around 2 g. The test and simulated results for the verification are without EMS.

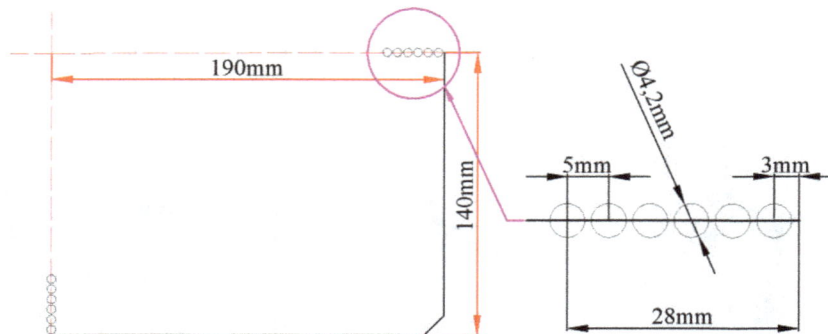

Figure 5. Distributions of drilling samples in the initial solidified shell.

Figure 6 shows the comparison between the simulation results and the result from the infrared carbon-sulfur determinator. A basically consistent variation tendency can be observed between the simulation results and testing results along both width direction and thickness direction.

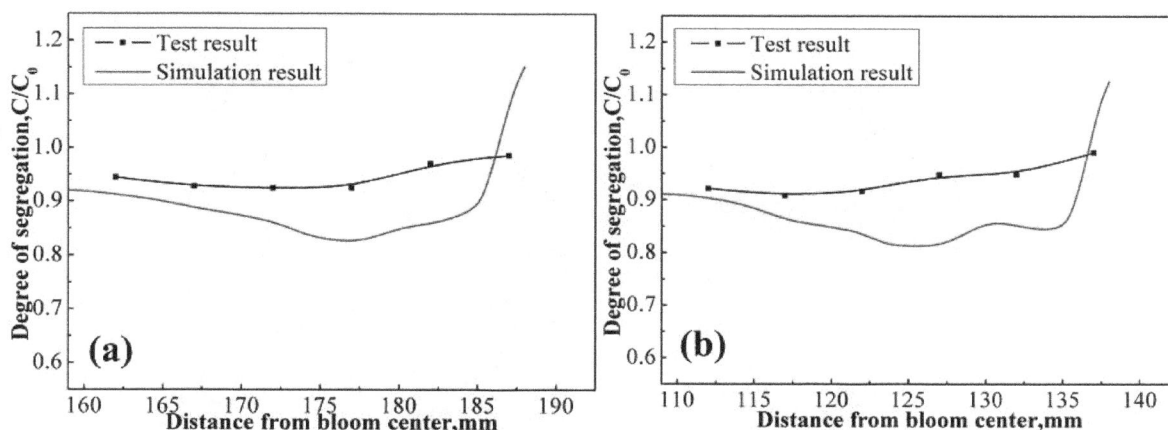

Figure 6. Comparison between measured and simulated segregation profiles of C in the region of initial solidification shell. (**a**) width direction; (**b**) thickness direction.

4. Results and Discussion

4.1. Metallurgical Effects of M-EMS

Two simulation cases in this paper are discussed first and compared to find out the exact effect of mold EMS on flow, heat transfer, initial solidification, and solute transport in the mold region. One is the case without EMS, the other is the case loaded with EMS, in which the current intensity is 600 A and the electromagnetic frequency is 2.0 Hz.

4.1.1. Flow Field

Figure 7 shows the contour and vector of flow patterns in the upper part of the bloom without EMS (a) and with EMS (b). Also, the velocity contour of several key cross sections in the mold zone for the two cases is presented in Figure 7, which are $z = 0.234$ m for the center of the SEN outlets, $z = 0.42$ m for the EMS center and $z = 0.7$ m for the mold outlet, respectively. In the bloom mold without EMS, the bulk flow leaving the SEN outlets impinges on the mold corner wall and is split into two opposing directions. Upper circulation is confined by the walls and meniscus and the lower circulation is formed by the mold corner and the return flow near the mold center. The center of the upper circulation is located at around 0.09 m bloom from the meniscus, and the lower circulation center is locating at approximately 0.45 m. When the mold is loaded with EMS, four obvious circulations are formed upon and under the EMS center, the two lower circulations can improve the heat transfer mechanism of molten steel in the EMS effect area, eliminate steel superheat, and promote uniform growth of the solidifying shell. The location of the upper circulation center in the diagonal plane changes to 0.21 m, with a higher velocity than the case without EMS, but the circulation under the nozzle outlets is eliminated and the flow is changed to the casting direction because of the increasing stirring effect of the EMS. In the center cross section of the SEN outlets, the four jet flows impinge on the mold corner, rebound to the bloom center and form four horizontal circulations near the narrow sides with a high velocity around the mold corner without EMS. The horizontal circulations remain for a while, then are eliminated at the mold outlet. The horizontal circulations created by the mold wall and molten steel in the bloom center can be changed into a strong horizontal swirling flow gradually by EMS, and the strongest flow occurs in the cross section of the EMS center. In the horizontal swirling flow, the largest tangential velocity locates at the solidification front near the center of the wide and narrow sides. The strong tangential velocity can affect the solidification behavior in the mold, effectively prevent the growth of columnar crystal, promote the growth of equiaxed grain [31], and be beneficial to the homogeneity of temperature and solute element in the molten steel.

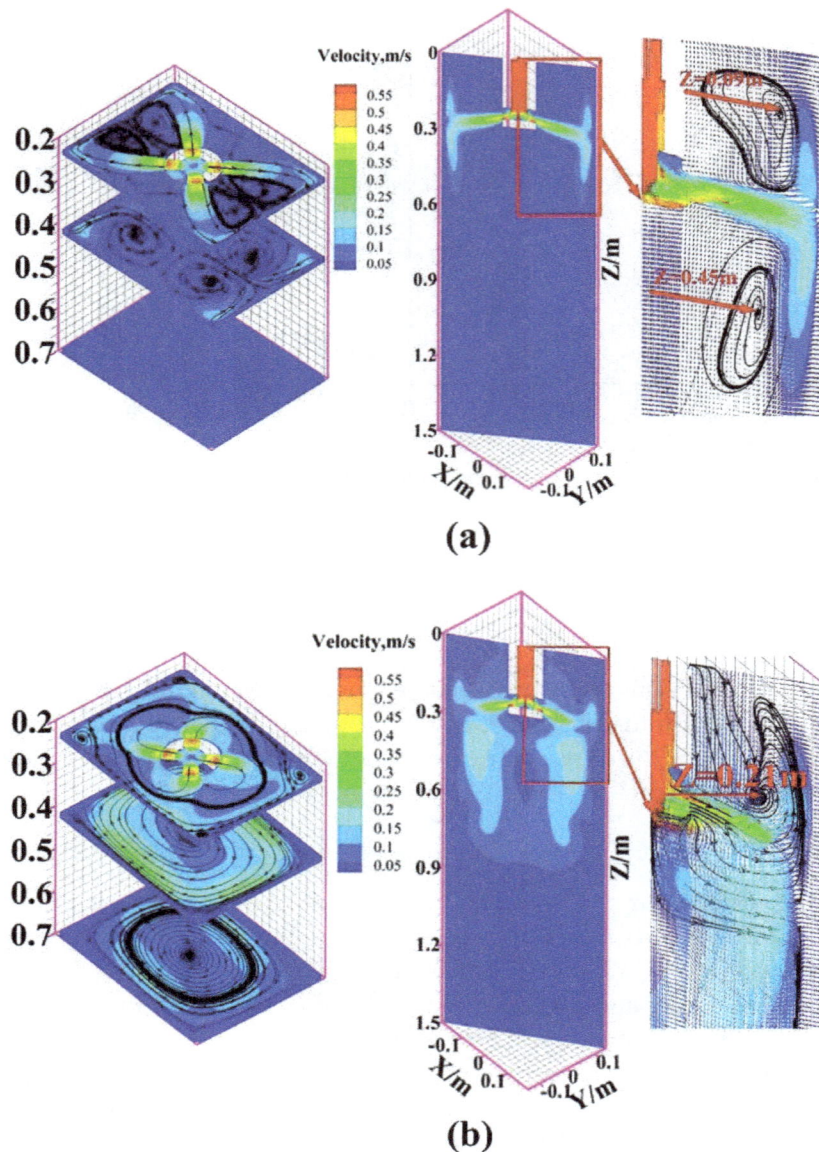

Figure 7. Velocity contour and vector in the upper part of the mold for the cases without (**a**) and with (**b**) electromagnetic stirring (EMS).

The comparison of the three-dimensional time-averaged level fluctuation with and without EMS is presented in Figure 8. The height difference between the zenith and the nadir of the free surface level is defined as the largest level fluctuation value of the mold. The largest level fluctuation value without EMS is about 4.5 mm, which happens at very few positions, for most of the other places the level fluctuation is less than 1 mm, the free surface is relatively stable. After the EMS is loaded, the level fluctuation at the center and edge of the free surface in the mold moves up and down respectively, which is affected by the electromagnetic force at the upper mold. The largest value of level fluctuation is 6.2 mm in the case with EMS. For bloom castings, the value of level fluctuation should be controlled as low as possible while the M-EMS can aggravate the movement of mold flux, which will increase the chance of slag entrapment. A favorable electromagnetic stirring effect along with a relatively stable free surface should be obtained to guarantee proper metallurgical behavior.

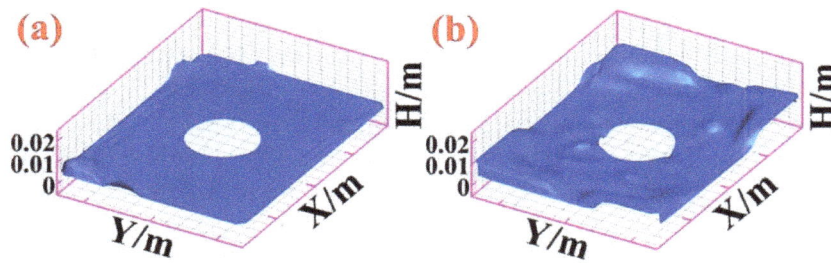

Figure 8. Comparison of three dimensional level fluctuations in the mold without (**a**) and with (**b**) EMS.

4.1.2. Heat Transfer and Solidification

Figure 9 presents the temperature contour at the diagonal plane ((a) without EMS, (b) 600 A, 2.0 Hz) and the temperature distribution of molten steel at the computational centerline for the two cases (c). The temperature at the molten steel core without EMS rapidly decreases from 1765 K to 1735.7 K at 0.3 m below the meniscus, and with little change in the rest of the computational zone, at which the centerline temperature decreases from 1735.7 K at 0.3 m below the meniscus to 1728 K at the computational outlet. While the temperature at the liquid core at 0.3 m below the meniscus is 1737 K after the EMS is loaded, the average temperature in the molten steel between 0.3 m and 0.48 m below the meniscus is higher than the case without EMS, and the temperature decreases from 1735.2 K at 0.48 m below the meniscus to 1725.4 K at the computational outlet. The temperature of molten steel with EMS is uniformly distributed and remains at around 1725 K from 0.8 m to 1.5 m below the meniscus, while the molten steel temperature without EMS is non-uniformly distributed and remains in remarkable decline at about 2.6 K higher at the center of the computational outlet. Therefore, the mold EMS can effectively reduce and even eliminate the superheat of the molten steel by retaining the molten steel at a high temperature in the upper zone of the mold.

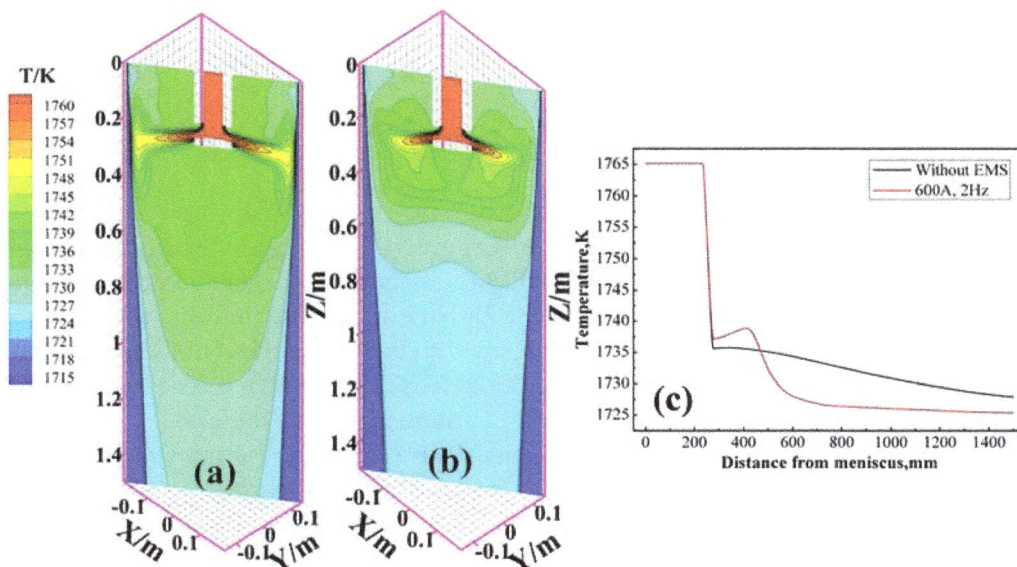

Figure 9. Temperature contour at the diagonal planes ((**a**) without EMS; (**b**) 600 A, 2.0 Hz) and molten steel temperature at computational centerline (**c**) for the two cases.

Figure 10 shows the temperature variation at the centerline of the bloom wide face and chamfered corner in the two cases, in which the solid and dashed lines represent the corner and surface temperature, respectively. The variation rule of temperature at both the bloom surface and the corner with and without can be easily understood by combining the flow results mentioned above.

The surface and corner temperature at 0–0.2 m below the meniscus is almost the same in the two cases because of the small EMS effect in this region. In the EMS effect area, the descent velocity of the mold corner temperature in the case loaded with EMS is larger than the other one, because the electromagnetic force can apparently lighten the scouring effect of the jet flow on the corner wall and accelerate the heat transfer process of the corner. The corner temperatures at the mold outlet in these two cases are 1286 K and 1180 K, respectively, and the temperature difference decreases gradually as the casting process continues, which is changed to 1241 K and 1195 K at the computational outlet, respectively. The corner temperature changes about ±30 K till the bloom reaches to the straightening area, which effectively avoids the high-temperature brittle zone of the steel and prevents the chance of transverse cracks. The surface temperature in the EMS effective zone is higher than the case without EMS, because the horizontal swirling flow generated by EMS can strongly move the solidification front into the bloom center, then the solidification front is replaced by high temperature molten steel. The bloom surface temperature increases with the change of the cooling condition, which increases from 1272 K at the mold outlet to 1411 K at the computational outlet without EMS and from 1355 K to 1440 K with EMS, respectively. The values of temperature rise are 136 K without EMS and 85 K with EMS, which indicates that EMS in the bloom mold can reduce the possibility of surface cracks.

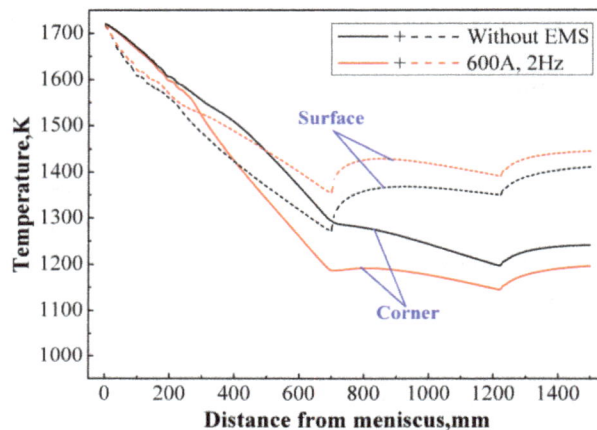

Figure 10. Temperature variations at the centerline of the bloom wide face and chamfered corner.

Figure 11 shows the liquid fraction distributions at the diagonal plane and cross section of the EMS center ($z = 0.42$ m) for the two different cases. The growth velocity of the solidifying shell at the mold corner in the impact area in the case without EMS is lower than the other because of a remarkable impingement of the jet flow on the corner wall, and the solidified shell thickness at the corner of the computational outlet is apparently thinner. Compared with the case without EMS, the solidified shell in the cross section $z = 0.42$ m in the case loaded with EMS is thicker at the corner, but thinner at both the wide and narrow sides, and the solidified shell distribution is more uniform.

Figure 12 presents the shell thickness and liquid/solid distributions at the y–z plane ($x = 0$ m) along the casting direction for the two cases, where the solid line refers to the case without EMS, while the dashed line represents the case loaded with EMS. The growth rhythm of the solidifying shell in the width direction is associated with the results shown in Figure 11, the growth velocity of the solidifying shell at the narrow side in the case loaded with EMS is slower in the EMS effective zone and the shell thickness at the mold outlet ($z = 0.7$ m) is 14.2 mm, which is about 3 mm thinner than the case without EMS. With the decrease of electromagnetic force, the shell thickness difference between the two cases decreases as well, the solidified shell at the computational outlet for the two cases is 26.86 mm with EMS and 28.3 mm without EMS, respectively. The difference in mushy zone length variation between the case with and without EMS can also be observed in Figure 12, in which the length of the mushy zone under the EMS effective zone is significantly smaller, especially in the EMS center, where the mushy zone lengths are 3.25 mm and 5.4 mm respectively. The main reason for the difference in mushy zone length is the remarkable tangential velocity at the solidification front of the narrow side, which is generated by the electromagnetic force. M-EMS can reduce the growth velocity of the solidifying shell and the length of the mushy zone in the EMS effective zone, which makes the shell thickness at the computational outlet 1.44 mm thinner than the case without EMS.

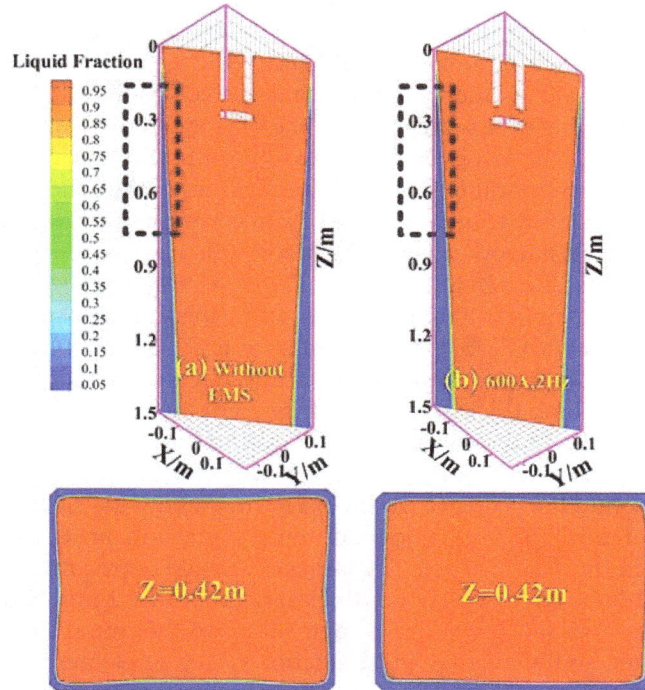

Figure 11. Liquid fraction distributions at the diagonal plane and the EMS center for different cases.

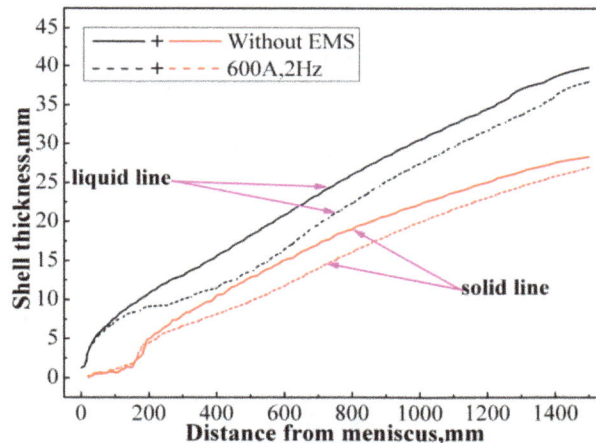

Figure 12. Liquid and solid distributions at the y–z plane along the casting direction for different cases.

4.1.3. Solute Transport

Figure 13 presents the distributions of solute element C on the diagonal plane ((a) the case without, (b) the case loaded with EMS) and at the diagonal line of the computational outlet(c) in the two different cases. Figure 14 shows the distributions of solute element C on the y–z plane, various cross sections for the bloom ((a) without EMS, (b) 600 A, 2.0 Hz) and at the centerline of the computational outlets (c) in the two different casting cases, in which the cross sections of z = 0.234 m (center cross section of SEN outlets), z = 0.42 m(EMS center), z = 0.7 m (mold outlet) and z = 1 m are chosen to display more intuitively the effect of EMS on the distribution of solute element C at the bloom cross section. For the case without EMS, the original liquid steel with initial mass fraction of solute element C (0.725%) is sprayed into the mold through the four outlets of SEN which are pointed directly at the mold corners, the primitive steel channels are formed by the effect of the inertia force, and the liquid steel with a high solute content floats upward to the meniscus dead zone because of the density change of liquid steel with temperature, which leads to the appearance of positive segregation in the 4 mm initial solidified shell. When arriving at the flow impact area, the impingement of liquid steel on the mold corner can dispel the enrichment of the solute at the solidification front and as there is not enough time to supply a rich solute content while the solidification continues, a negative position will be developed under this condition, and the maximum negative segregation in the diagonal plane will occur at the impact point. The high level of negative segregation is gradually reduced after the bloom is pulled out of the EMS effective zone, where the enriched solute steel will again gather at the solidification front. Compared with the result of solute element distribution at the bloom corner, a much weaker impingement effect happens around the wide and narrow sides of the bloom, the degree of negative segregation in the solidified shell of the narrow side is obviously lower. The maximum negative segregation is 0.74 when the corner solidified shell thickness is 10 mm and 0.84 when the shell thickness at the width direction is 15 mm. For the case loaded with EMS, the jet trajectory is gradually changed by the electromagnetic force, the impact pressure is lightened, and the maximum negative segregation at the bloom corner improves to 0.78 where the shell thickness is 12.5 mm compared with the case without EMS. The degree of negative segregation in the solidified shell around the wide and narrow sides in the EMS effective zone is more serious than the case without EMS, because the high tangential velocity near the solidification front around the center of the bloom narrow and wide sides can wash out the solute enriched mushy steel and move it into the bloom center, thereby enhancing the average mass fraction of the solute element C in the molten steel. The largest negative segregation at the width centerline of the computational outlet is 0.738, where the thickness is about 9.2 mm. Compared with the case without EMS, the degree and area of negative segregation in the initial solidified shell in the case loaded with EMS is larger, so the average mass fraction of solute element C in the bloom center is relatively higher due to the mass conservation law. The mass fractions of element C at the center of the computational outlet with and without EMS are 0.7904% and 0.7743%, respectively.

For big bloom castings, the main functions of M-EMS are in eliminating the superheat of the molten steel, making the temperature and solute in the molten steel uniform, improving the non-metallic inclusion floatation, promoting the ability of fluxes for adsorption of inclusion, and improving the solidification structure of the bloom. There is no obvious improvement of M-EMS on centerline segregation of big bloom casting, the effective technologies to reduce the centerline segregation for big bloom casting are soft reduction and final electromagnetic stirring [32]. In many steel plants, the combination of M-EMS with soft reduction or final EMS is widely applied to improve the internal quality of the casting bloom.

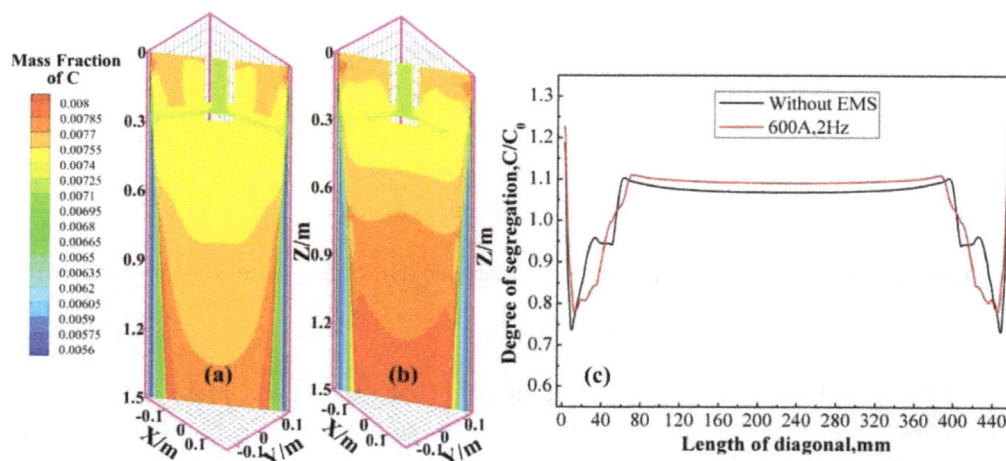

Figure 13. Distributions of solute element C on the diagonal plane ((**a**) without EMS; (**b**) 600 A, 2 Hz) and at the diagonal line of the computational outlets (**c**).

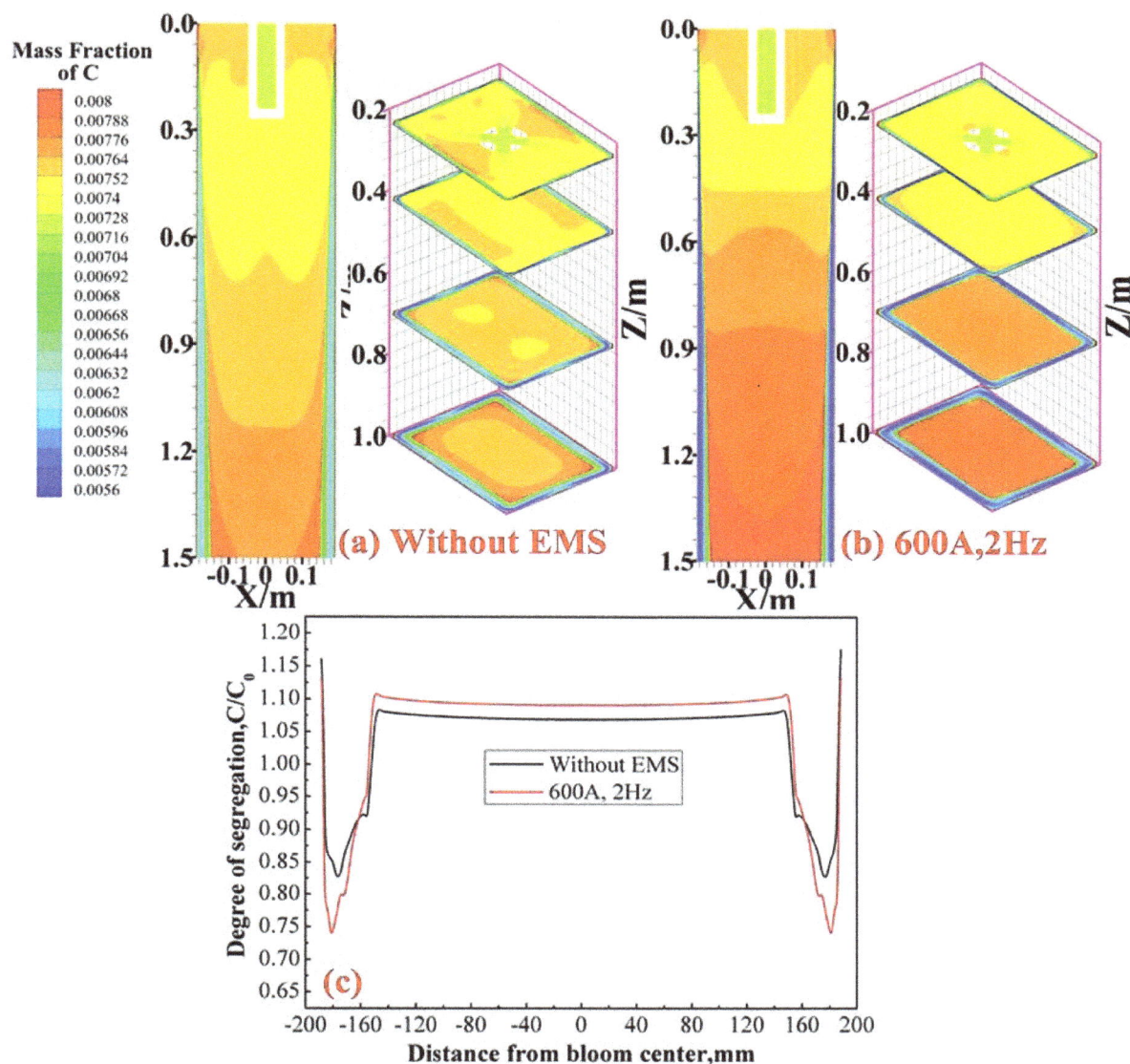

Figure 14. Distributions of solute element C on the y–z plane, at various cross sections of the bloom (**a,b**), and the centerline of the computational outlets (**c**).

4.2. Effect of Current Intensity

After analyzing and comparing the results of the flow pattern, heat transfer, initial solidification, and solute transport in the CC bloom mold with and without EMS (600 A, 2.0 Hz), the exact effect of EMS current intensity on the metallurgical behavior in the mold will also now be discussed in this paper. The EMS parameters for the four cases are 450 A, 500 A, 550 A, and 600 A respectively, of which the frequencies are fixed at 2.0 Hz. To avoid repeating, the influence of different EMS frequencies is ignored because of the similar changing rule to current intensity.

Figure 15 presents the three dimensional level fluctuation in the mold under different current intensities. The fluctuation range of the liquid level in the mold rises with the increase of current intensity and the maximum site of the level fluctuation occurs near the four corners of the bloom, and the largest fluctuating values are 5.3 mm, 5.7 mm, 5.9 mm, and 6.2 mm, respectively. The fluctuation degree increases with the rise of current intensity. For big bloom castings, the value of level fluctuation should be controlled within around ±3 mm, and the maximum fluctuation of the four current intensities are close to this range.

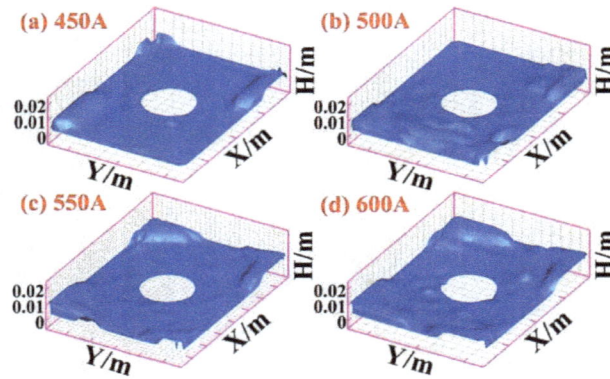

Figure 15. Comparison of three dimensional level fluctuations under different current intensity.

The velocity vector (a) and distribution of element C (b) at the cross section of EMS center ($z = 0.42$ m) are presented in Figure 16. With the increase of current intensity, the tangential velocity at the solidification front is obviously increased, especially at the places near the center of the wide and narrow sides; the largest velocity at this moment for these four cases is 0.197 m/s, 0.219 m/s, 0.24 m/s, and 0.271 m/s respectively. The stirring effect is not satisfactory when the current intensity is less than 500 A, at which the distribution of solute element C in the cross section of the EMS center still remains part of the initial injection trajectory of the molten steel, the stirring force is not strong enough to improve the uniformity of solute and temperature in the molten steel, and the ability of inclusion flotation is not significantly enhanced. When the current intensity is larger than 500 A, the stirring effect increases with the rise of current intensity, and the distribution of solute is more and more uniform in the bloom cross section of the EMS effective zone.

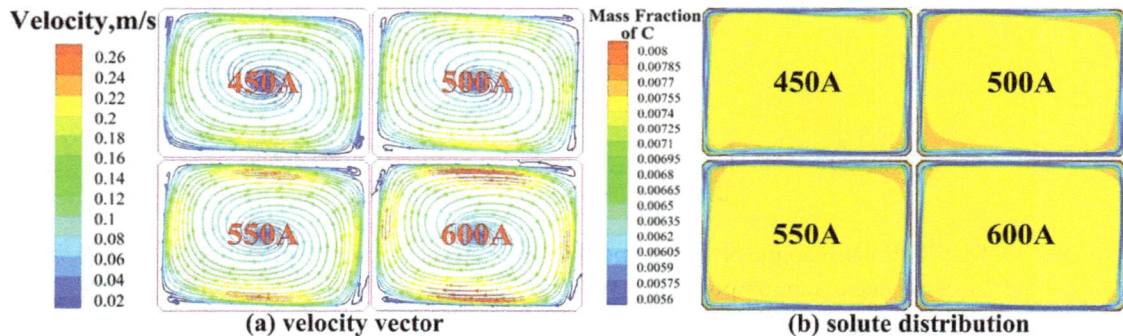

Figure 16. Velocity vector (**a**) and distribution of element C (**b**) at EMS center.

Figure 17 shows the distribution of solute element C at computational outlets (a) and the centerline along the casting direction (b) under different current intensities. It can be seen that the distribution law of the solute element in the solidified shell under different EMS current intensities is very similar, the maximum mass fraction of solute element C is presented at the solidification front, where the positive segregation will be formed in the mushy zone about to solidify. Compared with the other three cases, the area of solute enriched steel at the solidification front is smaller and the mass fraction of the solute element C at the bloom center is lower in the case with 600 A EMS. The mass fractions of C at the center of the computational outlet are 0.79275% for 450 A, 0.79389% for 500 A, 0.79214% for 550 A, and 0.79042% for 600 A, respectively.

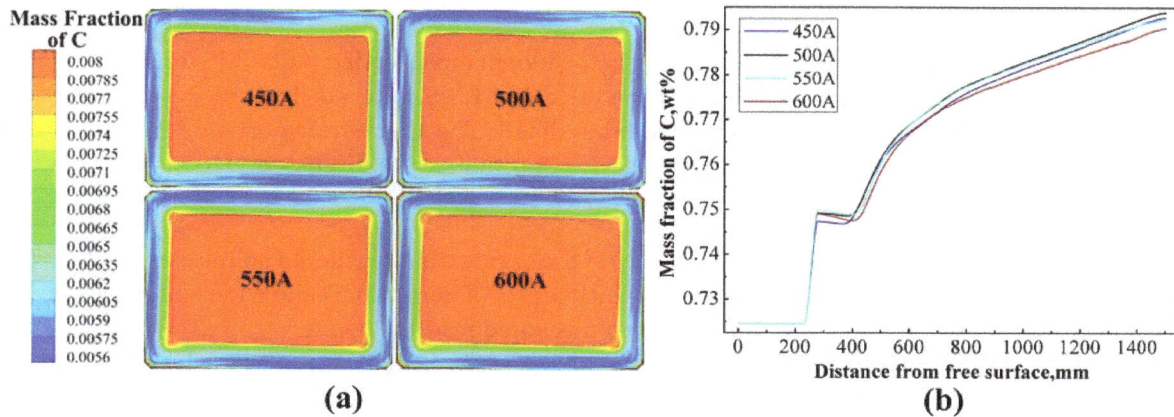

Figure 17. Distribution of element C at computational outlet (**a**) and at centerline (**b**).

Figure 18 displays the variation of surface temperature and shell thickness along with casting direction at different current intensities. With the increase of current intensity (from 450 A to 550 A), the surface temperature increases as well, when the current intensity is over 550 A, the surface temperature does not result in any obvious change. The surface temperatures at the mold outlet ($z = 700$ mm) for each case are 1325.7 K for 450 A, 1343.3 K for 500 A, 1359.4 K for 550 A, and 1355 K for 600 A, respectively. Also, 1391.5 K, 1400.9 K, 1405.8 K, and 1403.1 K at the outlet of the secondary cooling zone I ($z = 1.23$ m), which then increase to 1444.2 K, 1450.7 K, 1447.2 K and 1445 K respectively at the computational outlet due to the change of heat transfer conditions, respectively. With the increase of current intensity, the electromagnetic force rises, while the growth velocity of the solidifying shell in the EMS effect zone deceases. The solidified shell in the thickness direction at the mold outlet at four different current intensities is about 15.3 mm for 450 A, 15.2 mm for 500 A, 14.8 mm for 550 A, and 14.2 mm for 600 A, respectively, and increases to 27.55 mm, 27.2 mm, 27.57 mm, and 26.86 mm at the outlet of computational zone ($z = 1500$ mm), respectively.

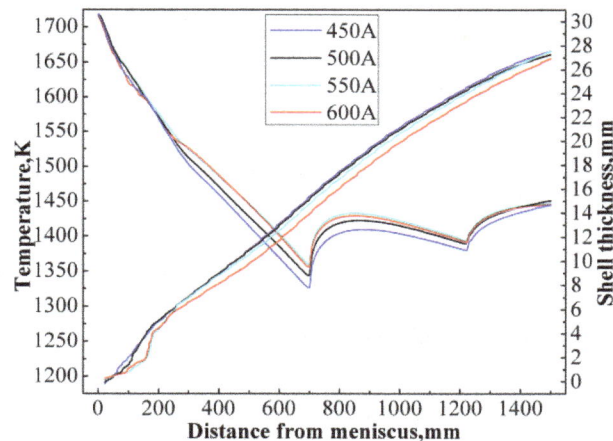

Figure 18. Variation of surface temperature and shell thickness along with casting direction at different current intensities.

By comprehensively considering the mold metallurgical behavior at different current intensities, the M-EMS with a current intensity of 600 A was found to be more suitable for big bloom castings than the other cases.

5. Conclusions

The multi-physical metallurgical behavior in the bloom mold loaded with EMS was investigated by a three-dimensional numerical model. The effects of EMS induced flow on the heat transfer, solidification, and solute transport were studied in detail. The solute transport model was verified and metallurgical differences under different EMS conditions were discussed. The conclusions are summarized as follows:

(1)   The basically consistent variation tendency of the segregation profiles of solute element C in the region of the initial solidified shell with a thickness of 30 mm at both the wide and narrow sides can be observed between the simulated and measured results.

(2)   Compared with the case without EMS, the bloom mold loaded with EMS is beneficial to the elimination of steel superheat, reduces the breadth of the mushy zone, and aggravates the level fluctuation from 4.5 mm to 6.2 mm. The distribution of temperature, solute, and solidified shell is more uniform in the EMS effective zone, the highest degree of negative segregation at the mold corner decreases from 0.78 to 0.74, but increases from 0.84 to 0.738 at the narrow and wide sides. The mass fraction of solute element C at the computational outlet increases from 0.7743% to 0.7904%. The EMS mold is not beneficial to the improvement of centerline segregation for big bloom casting.

(3)   With the increase of EMS current intensity (from 450 A to 600 A), the stirring effect and tangential velocity at the solidification front around the center of the wide and narrow sides increases, the level fluctuation is aggravated from 5.3 mm to 6.2 mm, the surface temperature in the EMS effective zone, the uniformity degree of temperature, and the solute distribution in the molten steel all increase as well, while the growth velocity of the solidifying shell thickness in the EMS effective zone decreases. The mass fraction of solute element C at the center of the computational outlets (z = 1.5 m) decreases from 0.7925% to 0.7904%. The M-EMS with a current intensity of 600 A is more suitable for big bloom castings.

(4)   The model has great application potential for a qualitative study of multi-physical phenomena in the bloom mold coupled with EMS, especially for the solute transport and solidification process coupled with turbulent flow. However, the present model would apply only to part of the caster, particularly the turbulent flow zone. To enhance the inner quality of the final products, the heat transfer and solute transport behavior below the computational domain need further investigation, especially for efficient ways to alleviate central segregation.

Acknowledgments: The authors would like to express their gratitude for the financial support by the National Natural Science Fund of China (51604200).

Author Contributions: Hongwei Ni and Hua Zhang conceived and designed the study. Qing Fang performed the experiments and simulations. Bao Wang contributed to the result analysis and paper preparation. Qing Fang and Fei Ye wrote the manuscript.

Conflicts of Interest: The authors declare no conflicts of interest.

References

1.   Wu, H.J.; Wei, N.; Bao, Y.P.; Wang, G.X.; Xiao, C.P.; Liu, J.J. Effect of M-EMS on the solidification structure of a steel billet. *Int. J. Miner. Metall. Mater.* **2011**, *18*, 159–164. [CrossRef]

2.   Sun, T.; Yue, F.; Wu, H.J.; Guo, C.; Li, Y.; Ma, Z.C. Solidification structure of continuous casting large round billets under mold electromagnetic stirring. *J. Iron Steel Res. Int.* **2016**, *23*, 329–337. [CrossRef]

3.   Jiang, D.B.; Zhu, M.Y. Solidification Structure and Macrosegregation of Billet Continuous Casting Process with Dual Electromagnetic Stirrings in Mold and Final Stage of Solidification: A Numerical Study. *Metall. Mater. Trans. B.* **2016**, *47*, 3446–3458. [CrossRef]

4.   Geng, X.; Li, X.; Liu, F.B.; Jiang, Z.H. Optimisation of electromagnetic field and flow field in round billet continuous casting mould with electromagnetic stirring. *Ironmak. Steelmak.* **2015**, *42*, 675–682. [CrossRef]

5. Yu, H.Q.; Zhu, M.Y. 3-D Numerical simulation of flow field and temperature field in a round billet continuous casting mold with electromagnetic stirring. *Acta Metal. Sin.* **2008**, *44*, 1465.

6. Liu, H.P.; Xu, M.G.; Qiu, S.T.; Zhang, H. Numerical simulation of fluid flow in a round bloom mold with in-mold rotary electromagnetic stirring. *Metall. Mater. Trans. B* **2012**, *43*, 1657–1675. [CrossRef]

7. Singh, R.; Thomas, B.G.; Vanka, S.P. Large eddy simulations of double-ruler electromagnetic field effect on transient flow during continuous casting. *Metall. Mater. Trans. B* **2014**, *45*, 1098–1115. [CrossRef]

8. Singh, R.; Thomas, B.G.; Vanka, S.P. Effects of a magnetic field on turbulent flow in the mold region of a steel caster. *Metall. Mater. Trans. B* **2013**, *44*, 1201–1221. [CrossRef]

9. Timmel, K.; Eckert, S.; Gerbeth, G. Experimental investigation of the flow in a continuous-casting mold under the influence of a transverse, direct current magnetic field. *Metall. Mater. Trans. B* **2011**, *42*, 68–80. [CrossRef]

10. Yang, Z.G.; Wang, B.; Zhang, X.F.; Wang, Y.T.; Dong, H.B. Effect of electromagnetic stirring on molten steel flow and solidification in bloom mold. *J. Iron Steel Res. Int.* **2014**, *21*, 1095–1103. [CrossRef]

11. Ren, B.Z.; Chen, D.F.; Wang, H.D.; Long, M.J.; Han, Z.W. Numerical analysis of coupled turbulent flow and macroscopic solidification in a round bloom continuous casting mold with electromagnetic stirring. *Steel Res. Int.* **2015**, *86*, 1104–1115. [CrossRef]

12. Tian, X.Y.; Zou, F.; Li, B.W.; He, J.C. Numerical analysis of coupled fluid flow, heat transfer and macroscopic solidification in the thin slab funnel shape mold with a new type EMBr. *Metall. Mater. Trans. B* **2010**, *41*, 112–120. [CrossRef]

13. Ren, B.Z.; Chen, D.F.; Wang, H.D.; Long, M.J.; Han, Z.W. Numerical simulation of fluid flow and solidification in bloom continuous casting mould with electromagnetic stirring. *Ironmak. Steelmak.* **2015**, *42*, 401–408. [CrossRef]

14. Sun, H.B.; Zhang, J.Q. Effect of feeding modes of molten steel on the mould metallurgical behavior for round bloom casting. *ISIJ Int.* **2011**, *51*, 1657–1663. [CrossRef]

15. Aboutalebi, M.R.; Hasan, M.; Guthrie, R.I.L. Coupled turbulent flow, heat, and solute transport in continuous casting processes. *Metal. Mater. Trans. B.* **1995**, *26*, 731–744. [CrossRef]

16. Yang, H.L.; Zhao, L.G.; Zhang, X.Z.; Deng, K.W.; Li, W.C.; Gan, Y. Mathematical simulation on coupled flow, heat, and solute transport in slab continuous casting process. *Metal. Mater. Trans. B* **1998**, *29*, 1345–1356. [CrossRef]

17. Lei, H.; Zhang, H.W.; He, J.C. Flow, solidification, and solute transport in a continuous casting mold with electromagnetic brake. *Chem. Eng. Technol.* **2009**, *32*, 991–1002. [CrossRef]

18. Kang, K.G.; Ryou, H.S.; Hur, N.K. Coupled turbulent flow, heat, and solute transport in continuous casting processes with an electromagnetic brake. *Numer. Heat Transf. Part A* **2005**, *48*, 461–481. [CrossRef]

19. Sun, H.B.; Zhang, J.Q. Macrosegregation improvement by swirling flow nozzle for bloom continuous castings. *Metal. Mater. Trans. B.* **2014**, *45*, 936–946. [CrossRef]

20. Asai, S.; Nishio, N.; Muchi, I. Theoretical analysis and model experiments on electromagnetically driven flow in continuous casting. *ISIJ Int.* **1982**, *22*, 126–133. [CrossRef]

21. Jones, W.P.; Launder, B.E. The calculation of low-Reynolds-number phenomena with a two-equation model of turbulence. *Int. J. Heat Mass Trans.* **1973**, *16*, 1119–1130. [CrossRef]

22. Lam, C.K.G.; Bremhorst, K. A modified form of the k-ε model for predicting wall turbulence. *ASME Trans. J. Fluids Eng.* **1981**, *103*, 456–460. [CrossRef]

23. Poirier, D.R. Permeability of flow of interdentritic liquid in columnar-dendritic alloys. *Metal. Mater. Trans. B* **1987**, *18*, 245–255. [CrossRef]

24. Li, W.S.; Shen, B.Z.; Shen, H.F.; Liu, B.C. Modelling of macrosegregation in steel ingots: Benchmark validation and industrial application. *IOP Conf. Ser. Mater. Sci. Eng.* **2012**, *33*, 1–8. [CrossRef]

25. Hrenya, C.M.; Bolio, E.J.; Chakrabarti, D.; Sinclair, J.L. Comparison of low Reynolds number k-ε turbulence models in predicting fully developed pipe flow. *Chem. Eng. Sci.* **1995**, *50*, 1923–1941. [CrossRef]

26. Spitzer, K.H.; Dubke, M.; Schwerdtfeger, K. Rotational electromagnetic stirring in continuous casting of round strands. *Metall. Mater. Trans. B* **1986**, *17*, 119–131. [CrossRef]

27. Deng, A.Y.; Xu, L.; Wang, E.G.; He, J.C. Numerical analysis of fluctuation behavior of steel/slag interface in continuous casting mold with static magnetic field. *J. Iron Steel Res. Int.* **2014**, *21*, 809–816. [CrossRef]

28. Zhao, Z.F.; Ni, H.W.; Zhang, H.; Chen, G.Y.; Yi, W.; Hong, J. Technology and process optimisation of bloom casting of ultrahigh speed rail steel. *Ironmak. Steelmak.* **2014**, *41*, 539–546. [CrossRef]

29. Lai, K.Y.M.; Salcudean, M.; Tanaka, S.; Guthrie, R.I.L. Mathematical modeling of flows in large tundish systems in steelmaking. *Metall. Mater. Trans. B* **1986**, *17*, 449–459. [CrossRef]

30. Alizadeh, M.; Jahromi, A.J.; Abouali, O. A new semi-analytical model for prediction of the strand surface temperature in the continuous casting of steel in the mold region. *ISIJ Int.* **2008**, *48*, 161–169. [CrossRef]

31. Yu, H.Q.; Zhu, M.Y. Influence of electromagnetic stirring on transport phenomena in round billet continuous casting mould and macrostructure of high carbon steel billet. *Ironmak. Steelmak.* **2012**, *39*, 574–584. [CrossRef]

32. Zeng, J.; Chen, W.; Wang, Q.; Wang, G. Improving inner quality in continuous casting rectangular billets: Comparison between mechanical soft reduction and final electromagnetic stirring. *Trans. Indian Inst. Metals* **2016**, *69*, 1–10. [CrossRef]

8

Electropolishing Behaviour of 73 Brass in a 70 vol % H₃PO₄ Solution by Using a Rotating Cylinder Electrode (RCE)

Ching An Huang [1,2,3,*], **Jhih You Chen** [1] **and Ming Tsung Sun** [1]

[1] Department of Mechanical Engineering, Chang Gung University, Taoyuan 33302, Taiwan; d0322001@stmail.cgu.edu.tw (J.Y.C.); mtsun@mail.cgu.edu.tw (M.T.S.)

[2] Department of Mechanical Engineering, Ming Chi University of Technology, New Taipei 24301, Taiwan

[3] Bone and Joint Research Center, Chang Gung Memorial Hospital, Taoyuan 33305, Taiwan

* Correspondence: gfehu@mail.cgu.edu.tw

Academic Editor: Hugo Lopez

Abstract: The electropolishing behaviour of 73 brass was studied by means of a rotating cylinder electrode (RCE) in a 70 vol % H_3PO_4 solution at 27 °C. Owing to the formation of a blue Cu^{2+}-rich layer on the brass-RCE, an obvious transition peak was detected from kinetic- to diffusion-controlled dissolution in the anodic polarisation curve. Electropolishing was conducted at the potentials located at the transition peak, the start, the middle, and the end positions in the limiting-current plateau corresponding to the anodic polarisation curve of the brass-RCE. A well-polished surface can be obtained after potentiostatic electropolishing at the middle position in the limiting-current plateau. During potentiostatic etching in the limiting-current plateau, a blue Cu^{2+}-rich layer was formed on the brass-RCE, reducing its anodic dissolution rate and obtaining a levelled and brightened brass-RCE. Moreover, a rod climbing phenomenon of the blue Cu^{2+}-rich layer was observed on the rotating brass-RCE. This enhances the coverage of the Cu^{2+}-rich layer on the brass-RCE and improves its electropolishing effect obviously.

Keywords: electropolishing; brass-RCE; Cu^{2+}-rich layer; rod climbing

---

1. Introduction

Due to having suitable mechanical properties and corrosion resistance, 73 brass has been widely used in many applications, such as in condensation tubes, radiators, coins, musical instruments, ornaments, etc. [1,2]. It is a type of Cu alloy with a Zn content of approximately 30 wt %, and it consists of a single α phase. In general, a brass component can be mechanically or electrochemically polished to obtain a shiny and lustrous appearance. It is well known that a brass component with a complicated profile or a fine structure can be easily polished by using an electrochemical method in which the component is anodically polarised in the limiting current plateau in an appropriate electrolyte. During electropolishing (EP), two processes, levelling and brightening, occur on the metallic surface simultaneously or independently [3,4]. The purpose of EP is, in general, to achieve a levelled and brightened surface through anodic dissolution in a suitable electrolyte. Many researchers [5–11] have noted that Cu and its alloys can be well electropolished in aqueous H_3PO_4 solutions with H_3PO_4 concentration from 50 to 100 vol %.

Although the EP behaviour of Cu and its alloys has been investigated for over eighty years [12,13], its EP mechanism in aqueous H_3PO_4 solutions is still not well recognised. There are two EP mechanisms of Cu and Cu alloys. One is the salt-film mechanism [4,6,14] and the other is the acceptor mechanism [15–17]. According to the former mechanism, a salt film with saturated metal ions is developed on the anode surface during the EP process. Based on the latter mechanism, the mass transport is controlled by the transportation of complex ions from electrode into the anode surface. In the acceptor mechanism, the mass transport is limited by the acceptor species, such as water molecules or anions [18,19], adjacent to the anode in which the concentration of acceptor species is exhausted through reaction with the metallic cations.

In our previous work [20], we noted that a blue Cu^{2+}-rich layer on the brass is essential to obtain a levelled and brightened surface during potentiostatic etching in the limiting-current plateau. This finding was evidenced from our study with a rotating disc electrode (RDE), on which the Cu^{2+}-rich layer impeded its anodic dissolution and improved its EP effect obviously. In this study, the EP behaviour of 73 brass was studied with a rotating cylinder electrode (RCE) on which a linear velocity was constant at a rotating speed. It could be useful information when the EP behaviour of brass-RCE is recognized at different rotating speeds. The effect of the Cu^{2+}-rich layer formed on the 73 brass-RCE will be further clarified.

2. Experimental Procedure

A commercial 73 brass bar was used in this study. The 73 brass bar was shaped into a rotating cylinder electrode (RCE), which has a diameter of 10 mm and a length of 6 mm with an exposed area of 1.88 cm^2, to study its EP behaviour. Analytical-grade concentrated H_3PO_4 (86.1%) and de-ionised H_2O were mixed to obtain an aqueous 70 vol % H_3PO_4 solution for electrochemical and EP tests. The electrochemical and EP tests were performed at 27 ± 1 °C in an electrochemical three-electrode cell with a potentiostat/galvanostat (Potentiostat/Galvanostat Model 263A, EG & G Instruments, Oak Ridge, TN, USA). An RCE cell kit (Model 616, EG & G, Oak Ridge, TN, USA) containing 150 mL of the 70 vol % H_3PO_4 solution was used for each test. A platinised Ti mesh and an Ag/AgCl electrode in the saturated KCl solution were used as the counter and reference electrodes, respectively. The brass-RCE surface was mechanically ground with 600-grit emery paper, dried with air blaster, and then prepared for the electrochemical test or EP.

The anodic polarisation curve of the brass-RCE was measured by scanning a potential range from −250 mV (vs. open-circuit potential) to 3.0 V (vs. Ag/AgCl) with a scan rate of 5 mV/s. At the beginning of the anodic polarisation test or EP, the brass-RCE was immersed in the 70 vol % H_3PO_4 solution at the open-circuit potential for 300 s to reach a dynamically stable condition between the brass-RCE and electrolyte. EP of the brass-RCE was performed potentiostatically at the limiting-current plateau with a constant charge of 50 coulombs. Before EP, the surface of brass ECE was mechanically ground with 600-grit emery paper. After EP, the surface morphologies of brass-RCEs were examined with optical microscopy (OM, BH2-UMA, Olympus, Melville, NY, USA) and scanning electron microscopy (SEM, ZEISS DSM 982 Gemini, LEO Oberkochen, Germany).

3. Results and Discussion

3.1. Anodic Polarisation Behaviour

Figure 1 shows the anodic polarisation curves of the brass-RCE at different rotational speeds in the 70% H_3PO_4 solution. As shown in Figure 1, limiting-current plateaus in the anodic polarisation curves of the brass-RCE with a potential range of approximately 1 V can be clearly seen at different rotational speeds. The potential range of the limiting current plateau gradually shifts to relatively noble potential with an increase in the rotational speed of the brass-RCE. Based on the results of the anodic polarisation test, the current-density values of transition peaks and limiting-current plateaus at different rotating speeds are listed in Table 1. The limiting-current density of the brass-RCE increases with increasing rotational speed. These results are in agreement with the anodic dissolution behaviour of the RCE under a diffusion-controlled mechanism in which the mass transfer rate increases with increasing rotating speed [21]. As shown in Figure 1, a transition peak can be observed in each anodic polarisation curve, decreasing its anodic current density from kinetic- to diffusion-controlled dissolution. This suggests that there could be a viscous layer or a salt film formed on the brass-RCE surface, impeding its anodic dissolution rate in the transition from active- to diffusion-controlled dissolution. By visual observation, a blue layer, which could be regarded as a Cu^{2+}-rich layer in a transparent solution, developed on the brass-RCE surface at different rotational speeds when the polarised potential was at or above the transition peak.

Based on our previous study of the EP behaviour of 73 brass with a rotating disc electrode (RDE) [20], we confirmed that a transition peak can be found at a rotational speed lower than 1000 rpm. In this study, however, an obvious transition peak was detected with brass-RCE at 1500 rpm. This result implies that the Cu^{2+}-rich layer could exist on the brass-RCE surface at a rotational speed of 1500 rpm, but not on the brass-RDE. The effect of this Cu^{2+}-rich layer on the EP behaviour of the brass-RCE will be discussed in the following section.

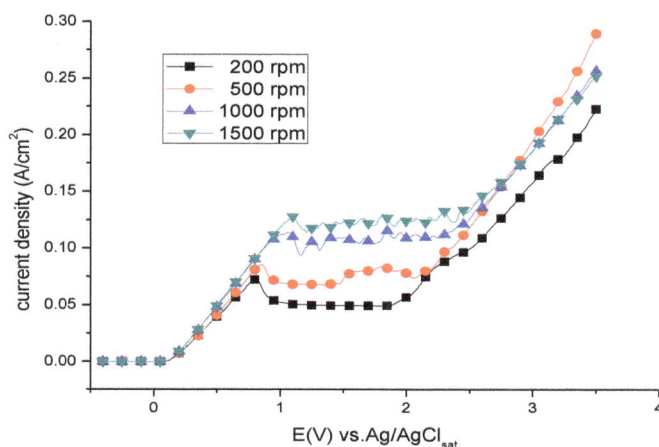

Figure 1. Anodic polarization curves of the brass-rotating cylinder electrode (RCE) at different rotational speeds in the 70 vol % H_3PO_4 solution with a scan rate of 5 mV/s.

Table 1. Values of limiting current densities (I_l) and transition current peaks (I_p) at different rotating speeds.

Rotating Speed (rpm)	I_l (A/cm^2)	I_p (A/cm^2)
200	0.0494	0.0727
500	0.0738	0.0855
1000	0.1049	0.1142
1500	0.1208	0.1278

3.2. Potentiostatic Polishing on the Limiting-Current Plateau

To study the EP behaviour, the brass-RCE was potentiostatically etched at four potentials located at the transition-peak and the start, middle, and end potentials corresponding to its limiting-current plateau. The polarised potentials were determined according to the anodic polarisation curves with a scan rate of 5 mV·s^{-1} and are shown in Figure 1. The EP potentials corresponding to the above-mentioned four specified potentials at different rotational speeds are listed in Table 2.

Table 2. Potentials corresponding to transition peak and the start, middle, and end of the limiting-current plateau at different rotating speeds.

Rotating Speed (rpm)	V_p (V)	V_s (V)	V_m (V)	V_e (V)
200	0.82	1.26	1.54	1.84
500	0.85	0.97	1.51	2.04
1000	1.07	1.21	1.72	2.22
1500	1.1	1.22	1.72	2.22

V_p: potential of transition-peak, V_s: start potential of limiting-current plateau, V_m: middle potential of limiting-current plateau, V_e: end potential of limiting-current plateau. All potentials shown in above table are compared to $Ag/AgCl_{sat}$ (0.197 V vs. SHE).

The surface morphologies in low and high magnifications for the left-side and right-side figures, respectively, of the brass-RCEs potentiostatically etched at their transition peaks at rotational speeds of 200, 500, 1000, and 1500 rpm are shown in Figure 2a–d, respectively. As shown in Figure 2a–d, a visually shiny surface of the brass-RCE could be achieved after anodic etching at the transition-peak potential. However, parts of the brass-RCE were not well levelled at a rotational speed lower than 500 rpm, developing a few shallow pits on the brass-RCE surface. Some deep pits can be found on the surface of brass-RCE after potentiostatic etching at 1500 rpm. Based on the above-mentioned results, the brass-RCE surface cannot be satisfactorily electropolished when the brass-RCE is potentiostatically etched at the transition-peak potential at rotational speeds lower than 500 and higher than 1500 rpm.

In our previous study [20], the grain boundaries were severely etched with a brass-RDE after potentiostatic etching at their transition peaks. Conversely, grain boundaries were faintly etched when the brass-RCE polarised at transition potentials at different rotational speeds. This obviously dissimilar effect can be attributed to the shape difference between the RDE and the RCE. According to our previous study, we found that the blue Cu^{2+}-rich layer is essential to obtain a levelled and brightened surface. This observation suggests that coverage of the Cu^{2+}-rich layer on the brass-RDE and brass-RCE could be different. A fully covered Cu^{2+}-rich layer could be expected on the surface of the brass-RCE because the EP effect of brass-RCE is obviously better than that of brass-RDE.

The surface morphologies of the brass-RCEs potentiostatically etched at the start of the limiting-current plateau with different rotational speeds are shown in Figure 3a–d. As for those polarised at transition potentials, all brass-RCEs were visually bright after potentiostatic etching at the start of the limiting-current plateau. A few small etched pits appeared on the surface of brass-RCE at a rotational speed of 1500 rpm. Unlike the EP effect on the brass-RDE at 1500 rpm, the grain boundaries were not etched with brass-RCE when EP was at the start potential of its limiting-current plateau. This observation indicates that EP at the start potential of its limiting-current plateau is available with RCE-brass at a rotational speed from 200 to 1500 rpm.

As shown in Figure 4a–d, a brightened and levelled surface could be obtained after potentiostatic etching in the middle of its limiting-current plateau at rotational speeds ranging from 200 to 1500 rpm. From experimental results shown in Figure 4a–d, a well electropolished surface could be achieved by means of brass-RCE when the EP is performed from the start to the middle potentials in its limiting-current plateau. According to the potentiostatic test, a well-polished surface can be achieved after electropolishing at potentials located at the transition peak and the start and middle in the limiting-current plateau. However, a few small pits the size of a few micrometers were observed on the brass-RCE surface. This is possibly the result of the enrichment of the Zn element beneath the oxide/hydroxide films on the brass surface [19].

Figure 5a–d shows the surface morphologies of the brass-RCEs potentiostatically etched at the end of their limiting-current plateau at different rotational speeds. A flow-streak feature and some pits were observed when the brass-RCE was potentiostatically etched at the end of the limiting-current plateau at rotational speeds of 200 and 500 rpm. However, these flow streaks were not observed on the brass-RCE surface potentiostatically etched at a rotational speed of 1000 rpm or higher, except for a few shallow pits. Because an oxygen-evolution reaction could occur through potentiostatic etching at the end of the limiting-current plateau, it can be expected that flow streaks and pits were formed due to the formation and evolution of oxygen bubbles during potentiostatic etching. The formation mechanism of flow streaks on brass-RDE was proposed in our previous study [20] in which oxygen bubbles move intermittently along the flow stream on the brass surface, leading to etching of a bubble-flow-streak feature. Because flow streaks were not found, this means that oxygen bubbles escape easily from the surface of brass-RCE at a rotational speed higher than 1000 rpm.

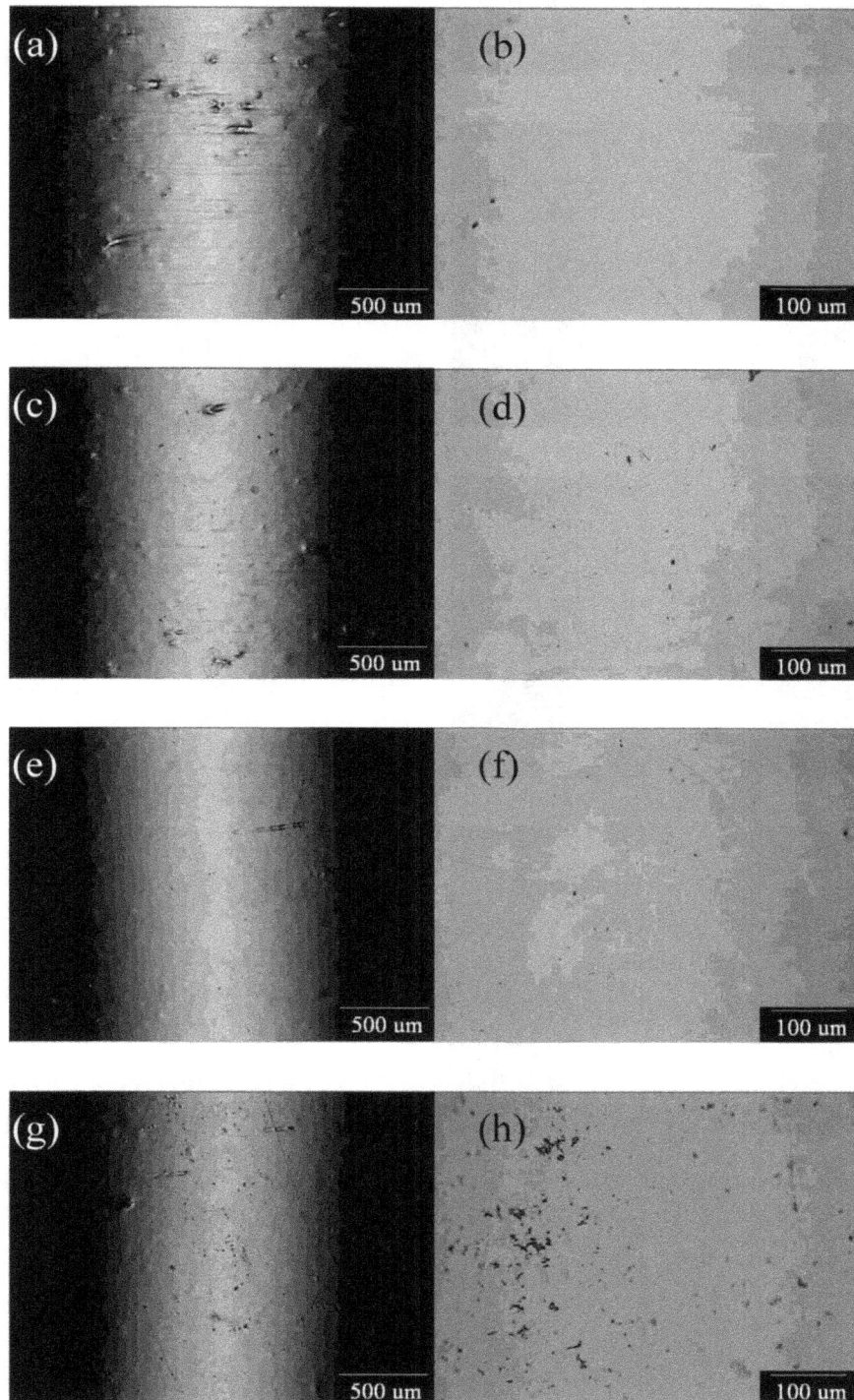

Figure 2. Surface morphologies of brass-RCEs potentiostatically etched at their transition peaks at rotational speeds of (**a,b**) 200; (**c,d**) 500; (**e,f**) 1000; and (**g,h**) 1500 rpm.

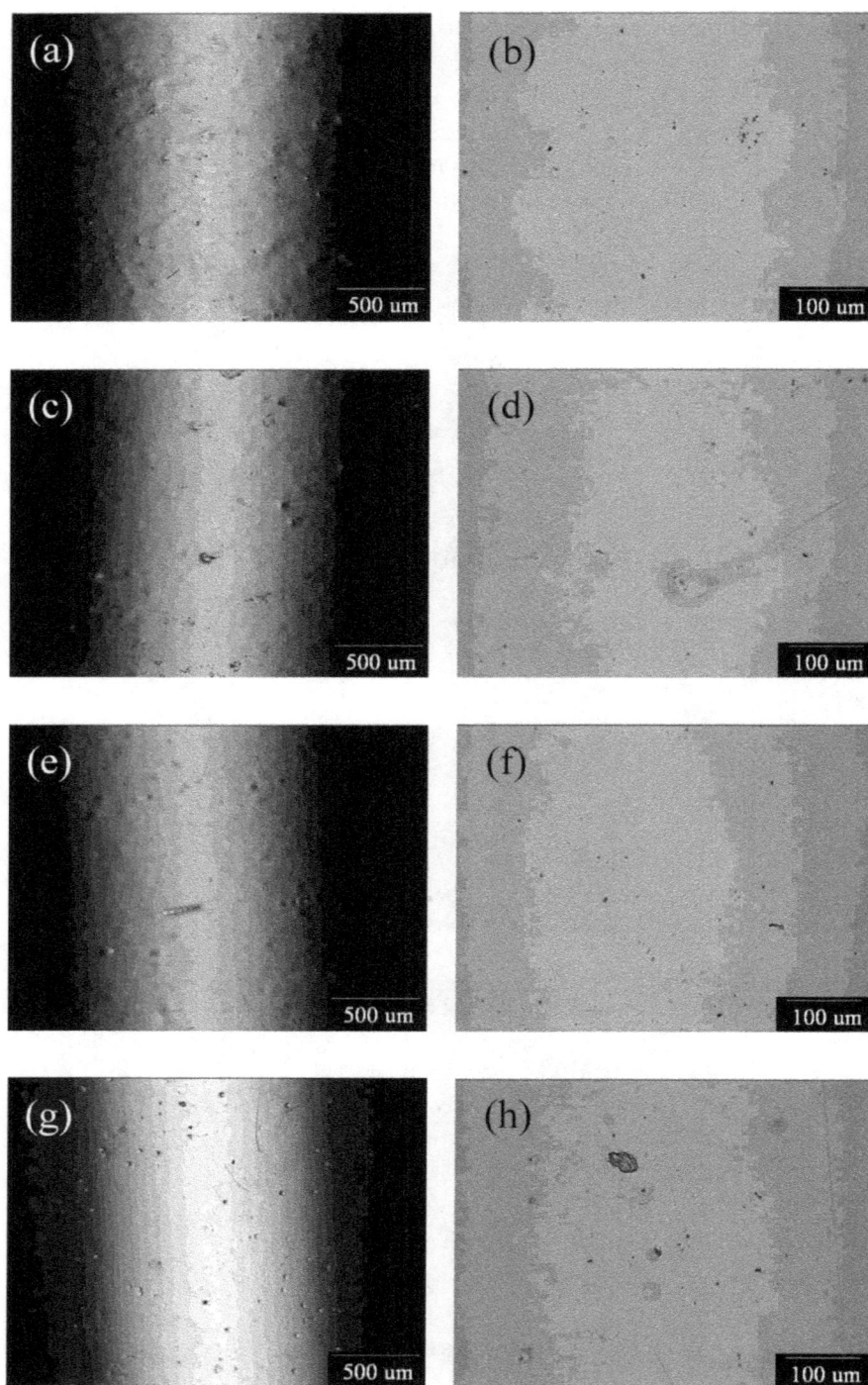

Figure 3. Surface morphologies of brass-RCEs potentiostatically etched at the start potentials of their limiting-current plateaus at rotational speeds of (**a,b**) 200; (**c,d**) 500; (**e,f**) 1000; and (**g,h**) 1500 rpm.

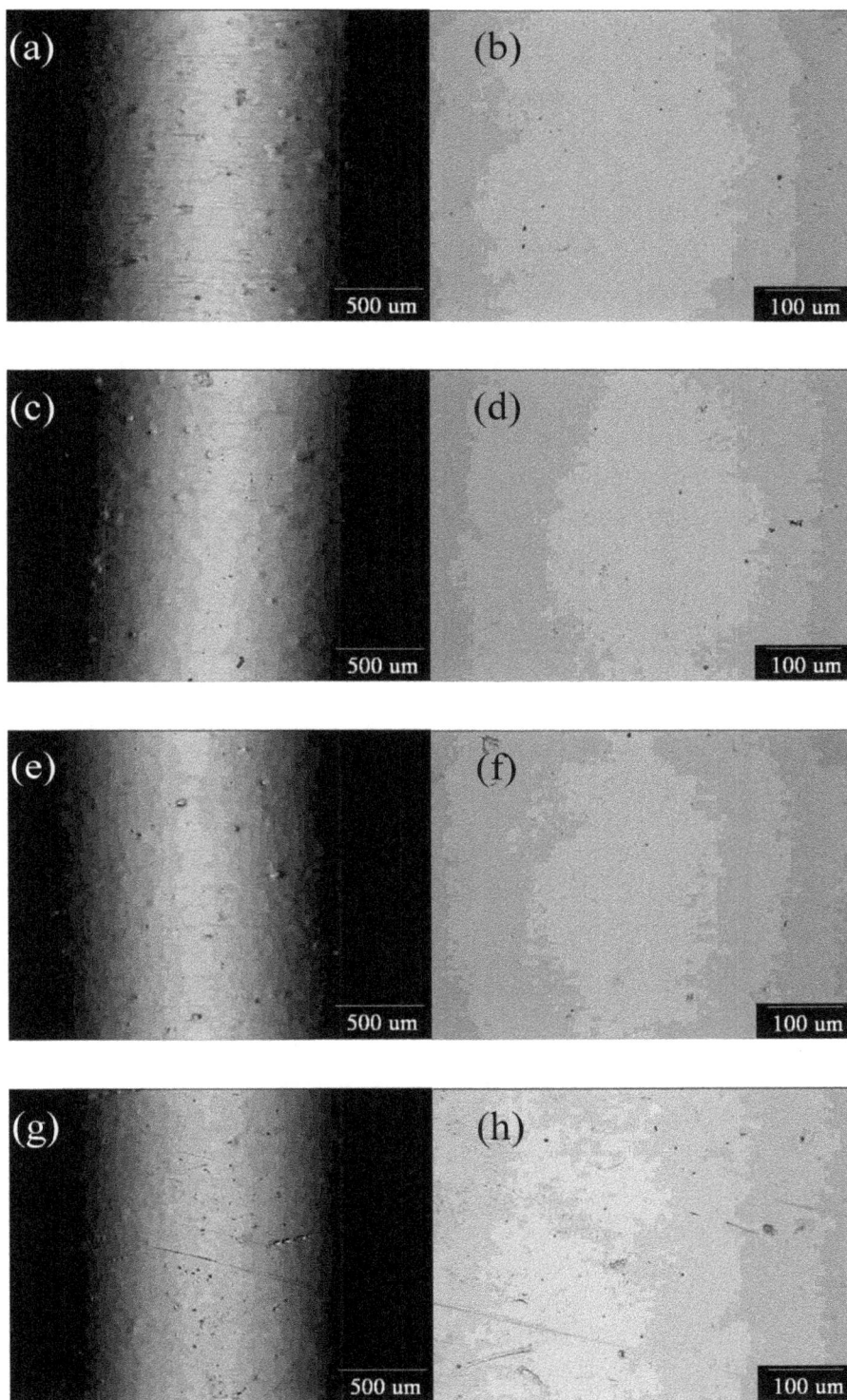

Figure 4. Surface morphologies of the brass-RCEs potentiostatically etched in the middle of their limiting-current plateaus at rotational speeds of (**a,b**) 200; (**c,d**) 500; (**e,f**) 1000; and (**g,h**) 1500 rpm.

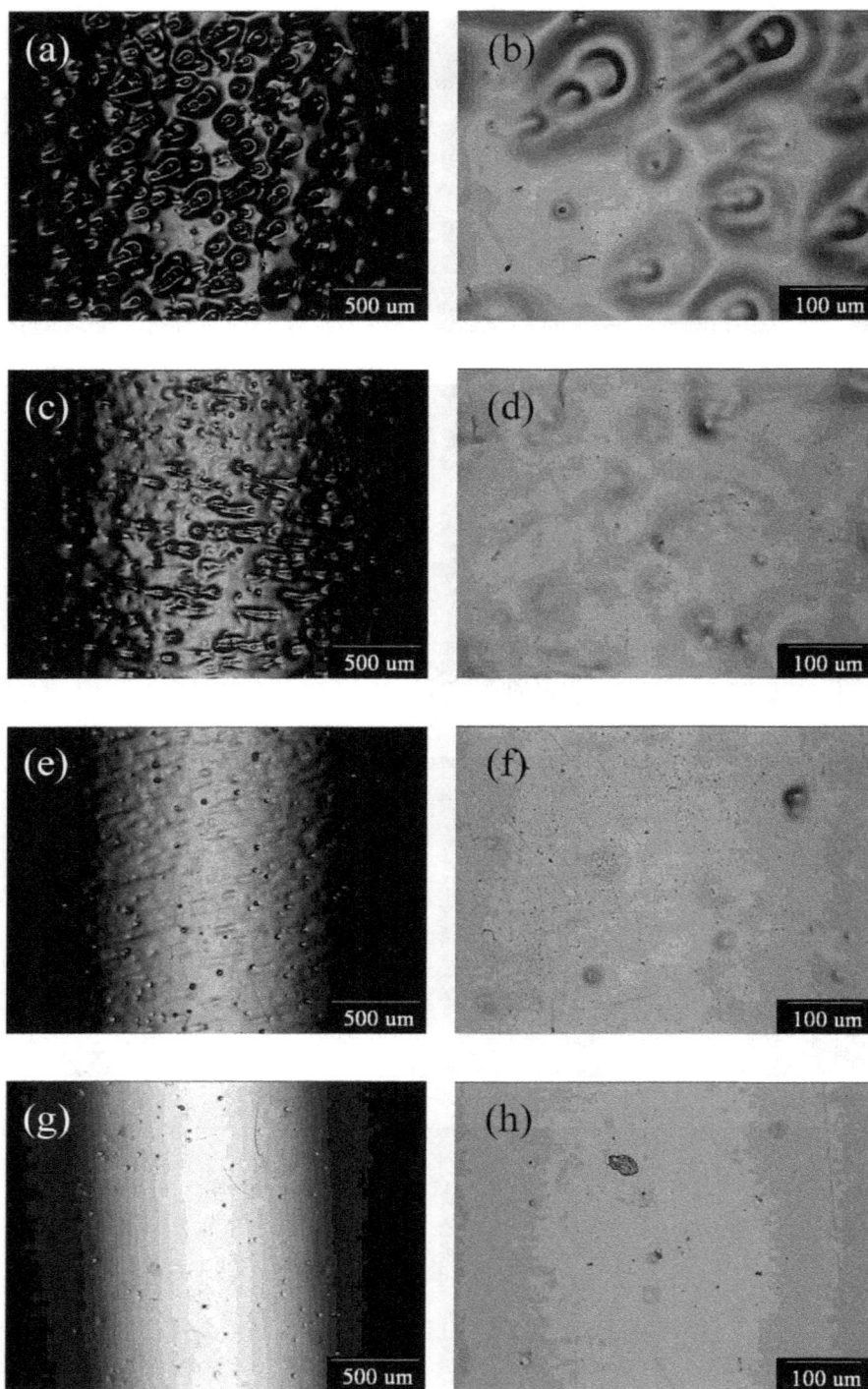

Figure 5. Surface morphologies of brass-RCEs potentiostatically etched at the end of their limiting-current plateaus at rotational speeds of (**a,b**) 200; (**c,d**) 500; (**e,f**) 1000; and (**g,h**) 1500 rpm.

3.3. Effect of the Blue Cu^{2+}-Rich Layer

Figure 6 shows the variation of the anodic current densities of brass-RCEs at rotational speeds of 0 and 100 rpm during potentiostatic etching in the middle of a limiting-current plateau at which the brass-RCE can be well electropolished in a rotational speed range from 200 to 1500 rpm. Interestingly, as shown in Figure 6, the anodic current density of the non-rotational brass-RCE is obviously higher than that of the brass-RCE at 100 rpm. It is well known that in EP a metal electrode is polarised in the limiting-current plateau in which the anodic dissolution is under a mass-transfer controlled process [22–24]. The limiting-current density of the RCE increases with increasing rotational speed or an increasing mass-transfer rate from the reacted chemical species in the solution to the RCE surface. This can be evidenced from the experimental results shown in Figure 1. This result implies that the EP mechanisms for the RCEs at 0 and 100 rpm must be different. Because the Cu^{2+}-rich layer formed on the brass-RCE impedes the anodic dissolution rate, a lack of the layer on the upper side of the non-rotational brass-RCE can be expected. This result leads to having a relatively high anodic current density during EP.

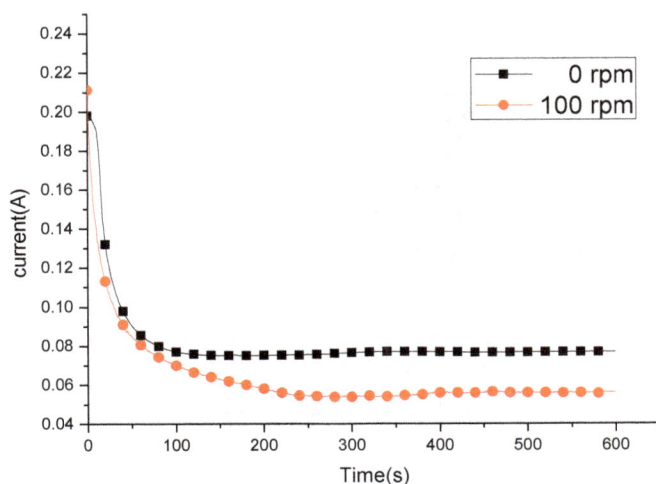

Figure 6. Variation in the anodic current densities of brass-RCEs potentiostatically etched in the middle of their limiting-current plateaus at rotational speeds of 0 and 100 rpm.

To clarify the effect of the blue Cu^{2+}-rich layer on the brass-RCE during EP, the brass-RCE was potentiostatically etched without rotation in the middle of the limiting-current plateau. Because the specific gravity of the Cu^{2+}-rich layer is obviously higher than that of the EP solution, 70 vol % H_3PO_4, the Cu^{2+}-rich layer formed on the upper site of the RCE will fall directly downward to the lower site of the brass-RCE during potentiostatic etching. That is, there will be a lack of the Cu^{2+}-rich layer on the upper side of stagnate brass-RCE during EP. Figure 7 shows the surface morphology of the brass-RCE electropolished without rotation. From the micrograph shown in Figure 7, a visually bright surface of brass-RCE could be achieved after EP. Moreover, a distinctly different effect of EP on the upper and lower sites of the brass-RCE was found. On the upper site, the grain boundaries of the brass-RCE were obviously etched, showing its grain structure, indicating that the upper site was not well electropolished. Alternatively, a brightened and levelled Cu deposited RCE was found on the lower site. Because the Cu^{2+}-rich layer fell from an upper site down to a lower site, the lower site of the non-rotational brass-RCE was well covered by abundant Cu^{2+} ions during EP. This behaviour indicates that the Cu^{2+}-rich layer on the brass-RCE is helpful for EP.

A Cu^{2+}-rich layer can be clearly seen on the brass-RCE with a rotational speed of 100 rpm or higher during EP. Due to the effect of gravity, the blue Cu^{+2}-rich layer cannot maintain its stick state on the surface of the brass-RCE without rotation. Based on the experimental results, the rotating brass-RCE must be fully covered with a blue Cu^{2+}-rich layer during EP. A rod-climbing phenomenon of the Cu^{+2}-rich layer can be observed during EP. This phenomenon can be illustrated from Figure 8a–f. During potentiostatic etching, a well-covered blue Cu^{2+}-rich layer was initially developed on the brass-RCE (see Figure 8b). Due to the effect of gravity, the outer part of the Cu^{+2}-rich layer fell down and concentrated to be a semi-sphere on the lower site of the brass-RCE by further potentiostatic etching (see Figure 8c,d). Interestingly, the semi-sphere went upwards along the surface of the brass-RCE, the rod climbing phenomenon, during potentiostatic etching (see Figure 8d). This rod-climbing phenomenon is evidenced in Figure 9a,b. As shown in Figure 8e, the semi-sphere climbed to approximately half the height of the brass-RCE, and it was unstable, scattering into a thin mushroom shape in the EP solution. When the semi-sphere diminished, a thin Cu^{+2}-rich layer was still found on the rotating brass-RCE (see Figure 8a,f), and the repeated formation sequences of semi-sphere, rod climbing, and scattering were observed. Because the Cu^{+2}-rich layer was found on the brass-RCE at 1500 rpm, an obvious transition peak was detected from its anodic polarisation curve (see Figure 1). Alternatively, the transition peak was not seen from the brass-RDE at 1500 rpm [20]. Thus, the EP effect on the brass-RCE potentiostatically etched at the transition peak is much better than that on the brass-RDE.

Figure 7. Surface morphology of the brass-RCE potentially etched in the middle of the limiting-current plateau.

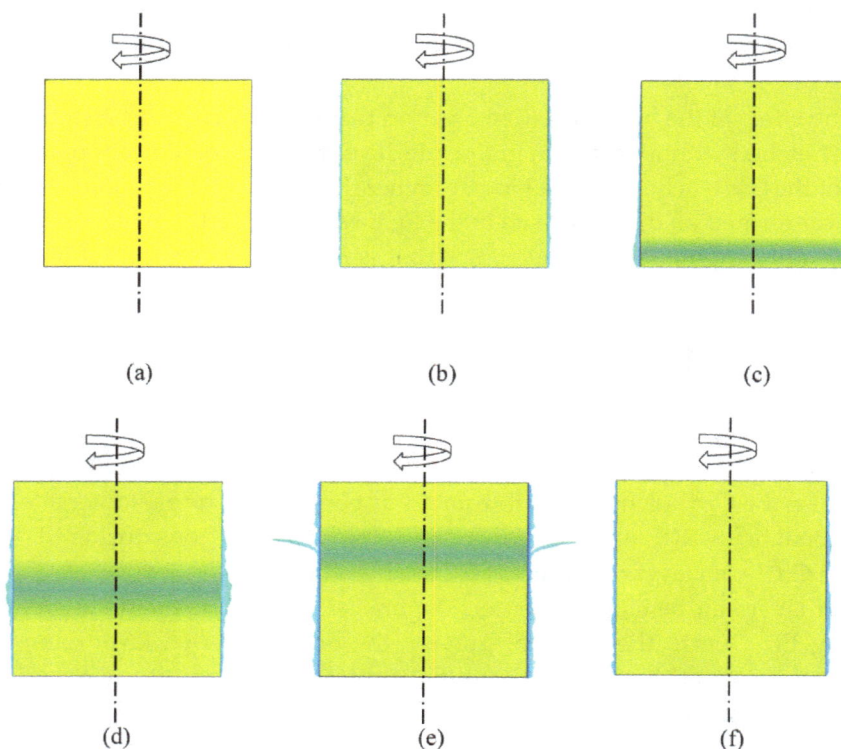

Figure 8. Schematic illustration of the rod-climbing phenomenon of the Cu^{+2}-rich layer (**a**) initial electropolishing; (**b**) full coverage; (**c,d**) rod-climbing; (**e**) formation a thin mushroom-shaped stream; (**f**) full coverage.

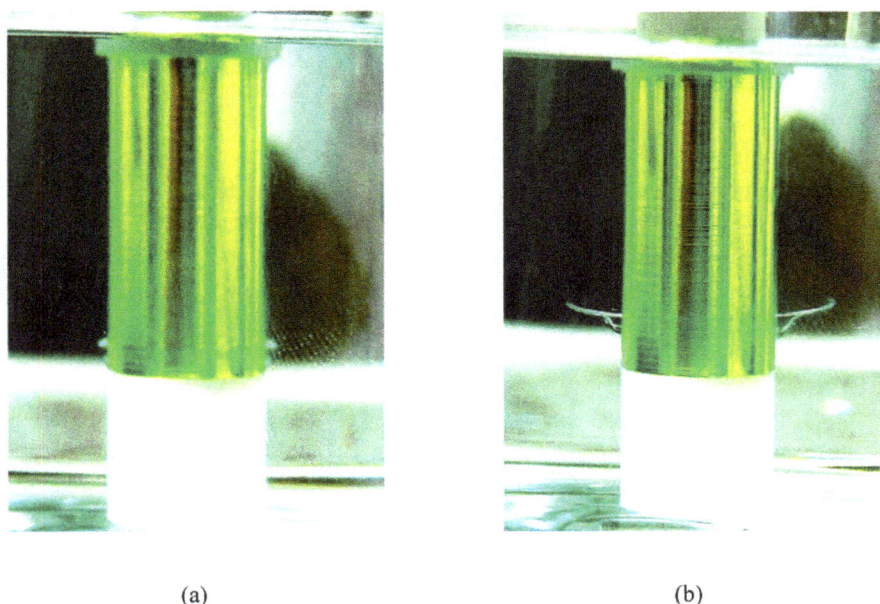

(a)   (b)

Figure 9. (**a**) Formation of the semi-sphere of the Cu^{2+}-rich layer in the bottom site of the brass-RCE; and (**b**) thin mushroom-shaped layer developed on the brass-RCE during electropolishing (EP).

The rod-climbing behaviour was studied using the flow-field simulation method [25]. According to the simulation, the Cu^{2+}-rich layer was rolled up by the upward flow. The rolled up semi-sphere went up with the upward flow until it met the downward flow, in which the outer part of Cu^{2+}-rich layer fell downward due to gravity. Next, the semi-sphere was flushed into a thin mushroom shape and spread into the 70 vol % H_3PO_4 solution. Detailed results of the simulation will be discussed and published in a related journal. Because the Cu^{2+}-rich layer can be developed on the rotating brass-RCE, a well electropolished surface can be easily achieved from the brass-RCE.

4. Conclusions

Potentiostatic etching of the brass-RCE was performed at potentials corresponding to the transition peak and in the start, the middle, and the end sites in its limiting-current plateau. Due to a stick layer of Cu^{2+} ions on the brass-RCE, the electropolishing effect was easier to achieve on the brass-RCE than it was on the brass-rotating disc electrode. During electropolishing with a rotating brass-RCE, rod climbing of the Cu^{2+}-rich layer was seen. An upward stream was developed along the brass-RCE, and a Cu^{2+}-rich semi-sphere climbed along the brass-RCE surface, resulting in good coverage of the Cu^{2+}-rich layer and improvement of its electropolishing effect.

Author Contributions: C.A. Huang wrote the manuscript, gave a discussion about experimental results, and responded reviewers' comments. J.Y. Chen had conducted experiments and observed the EP behaviour. M.T. Sun simulated the flow field to explain the rod climbing behaviour during EP of the rotating cylinder.

Conflicts of Interest: The authors declare no conflict of interest.

References

1. Zhang, J.; Tang, N.; Shang, Y.-J.; Xu, F. Effect of alloying elements on corrosion resistance of brass and function mechanisms. *Corros. Prot.* **2012**, *33*, 605–609.
2. Kwon, G.D.; Kim, Y.W.; Moyen, E.; Keum, D.H.; Lee, Y.H.; Baik, S.; Pribat, D. Controlled electropolishing of copper foils at elevated temperature. *Appl. Surf. Sci.* **2014**, *307*, 731–735. [CrossRef]
3. Landolt, D. Fundamental aspects of electropolishing. *Electrochim. Acta* **1987**, *32*, 1–11. [CrossRef]
4. Matlosz, M.; Landolt, D. Shape changes in electrochemical polishing: The effect of temperature on the anodic leveling of Fe-24Cr. *J. Electrochem. Soc.* **1989**, *136*, 919–929. [CrossRef]

5.   Gabe, D.R. Electropolishing of copper and copper-based alloys in ortho-phosphoric acid. *Corros. Sci.* **1972**, *12*, 113–120. [CrossRef]

6.   Iskander, S.S.; Mansour, I.A.S.; Sedahmed, G.H. Electropolishing of brass alloys in phosphoric acid. *Surf. Technol.* **1980**, *10*, 357–361. [CrossRef]

7.   Awad, A.M.; Ghany, N.A.A.; Dahy, T.M. Removal of tarnishing and roughness of copper surface by electropolishing treatment. *Appl. Surf. Sci.* **2010**, *256*, 4370–4375. [CrossRef]

8.   Vidal, R.; West, A.C. Copper electropolishing in concentrated phosphoric acid I. Experimental findings. *J. Electrochem. Soc.* **1995**, *142*, 2682–2689. [CrossRef]

9.   Varenko, E.S.; Loshkarev, Y.M.; Tarasova, L.P. Surface roughness of brass during electropolishing in orthophosphoric acid solutions. *Sov. Electrochem.* **1991**, *27*, 105–107.

10.   Li, D.; Li, N.; Xia, G.; Zheng, Z.; Wang, J.; Xiao, N.; Zhai, W.; Wu, G. An in-situ study of copper electropolishing in phosphoric acid solution. *Int. J. Electrochem. Sci.* **2013**, *8*, 1041–1046.

11.   Han, G.H.; Gunes, F.; Bae, J.J.; Kim, E.S.; Chae, S.J.; Shin, H.-J.; Choi, J.-Y.; Pribat, D.; Lee, Y.H. Influence of copper morphology in forming nucleation seeds for graphene growth. *Nano Lett.* **2011**, *11*, 4144–4148. [CrossRef] [PubMed]

12.   Jacquet, P.A. On the anodic behavior of copper in aqueous solutions of orthophosphoric acid. *J. Electrochem. Soc.* **1936**, *69*, 629–655. [CrossRef]

13.   Jacquet, P.A. Electrolytic method for obtaining bright copper surfaces. *Nature* **1935**, *135*, 1076. [CrossRef]

14.   Logie, S.; Chick, J.; Campbell, M.; Bilbao, S. The influence of bore profile on slurred transients in brass instruments. In Proceedings of the Forum Acusticum 2011, Aalborg, Denmark, 26 June–1 July 2011.

15.   Du, B.; Suni, I.L. Mechanistic studies of cu electropolishing in phosphoric acid electrolytes. *J. Electrochem. Soc.* **2004**, *151*, 375–378. [CrossRef]

16.   Huo, J.; Solanki, R.; McAndrew, J. Study of anodic layers and their effects on electropolishing of bulk and electroplated films of copper. *J. Appl. Electrochem.* **2004**, *34*, 305–314. [CrossRef]

17.   Glarum, S.H.; Marshall, J.H. The anodic dissolution of copper into phosphoric acid. *J. Electrochem. Soc.* **1985**, *132*, 2872–2878. [CrossRef]

18.   West, A.C.; Deligianni, H.; Andricacos, P.C. Electrochemical planarization of interconnect metallization. *IBM J. Res. Dev.* **2005**, *49*, 37–48. [CrossRef]

19.   Gentile, M.; Koroleva, E.V.; Skeldom, P.; Thompson, G.E.; Bailey, P.; Noakes, T.C.Q. Influence of pre-treatment on the surface composition of Al-Zn alloys. *Corros. Sci.* **2010**, *52*, 688–694. [CrossRef]

20.   Huang, C.A.; Chang, J.H.; Zhao, W.J.; Huang, S.Y. Examination of the electropolishing behaviour of 73 brass in a 70% H_3PO_4 solution using a rotating disc electrode. *Mater. Chem. Phys.* **2014**, *146*, 230–239. [CrossRef]

21.   Pérez, T.; Nava, J.L. Simulation of turbulent flow of a rotating cylinder electrode. Influence of using plates and concentric cylinder as counter electrodes. *Int. J. Electrochem. Soc.* **2013**, *8*, 4690–4699.

22.   Datta, M.; Landolt, D. On the role of mass transport in high rate dissolution of iron and nickel in ECM electrolytes—I. Chloride solutions. *Electrochim. Acta* **1980**, *25*, 1255–1262. [CrossRef]

23.   Datta, M.; Landolt, D. On the role of mass transport in high rate dissolution of iron and nickel in ECM electrolytes—II. Chlorate and nitrate solutions. *Electrochim. Acta* **1980**, *25*, 1263–1271. [CrossRef]

24.   Datta, M.; Vega, L.F.; Romankiw, L.T.; Duby, P. Mass transport effects during electropolishing of iron in phosphoric-sulfuric acid. *Electrochim. Acta* **1992**, *37*, 2469–2475. [CrossRef]

25.   Sun, M.T.; Huang, C.A.; Huang, S.Y. The rod-climbing phenomenon of the viscous film on the surface of brass cylinder during electropolishing in the aqueous phosphoric acids. In Proceedings of the 211th ECS Meeting, Chicago, IL, USA, 6–10 May 2007; p. 552.

9

A Novel Method for Fracture Toughness Evaluation of Tool Steels with Post-Tempering Cryogenic Treatment

Ramona Sola [1],*, Roberto Giovanardi [1], Giovanni Parigi [2] and Paolo Veronesi [1]

[1] Department of Engineering "E. Ferrari", University of Modena and Reggio Emilia, Modena 41125, Italy; roberto.giovanardi@unimore.it (R.G.); paolo.veronesi@unimore.it (P.V.)

[2] Stav, Barberino del Mugello, Florence 50031, Italy; parigi@stav.biz

* Correspondence: ramona.sola@unimore.it

Academic Editor: Filippo Berto

Abstract: Cryogenic treatments are usually carried out immediately after quenching, but their use can be extended to post tempering in order to improve their fracture toughness. This research paper focuses on the influence of post-tempering cryogenic treatment on the microstructure and mechanical properties of tempered AISI M2, AISI D2, and X105CrCoMo18 steels. The aforementioned steels have been analysed after tempering and tempering + cryogenic treatment with scanning electron microscopy, X-ray diffraction for residual stress measurements, and micro- and nano-indentation to determine Young's modulus and plasticity factor measurement. Besides the improvement of toughness, a further aim of the present work is the investigation of the pertinence of a novel technique for characterizing the fracture toughness via scratch experiments on cryogenically-treated steels. Results show that the application of post-tempering cryogenic treatment on AISI M2, AISI D2, and X105CrCoMo18 steels induce precipitation of fine and homogeneously dispersed sub-micrometric carbides which do not alter hardness and Young's modulus values, but reduce residual stresses and increase fracture toughness. Finally, scratch test proved to be an alternative simple technique to determine the fracture toughness of cryogenically treated steels.

Keywords: fracture toughness; scratch test; residual stress; tool steel; cryogenic treatment

1. Introduction

Cryogenic treatment is widely used to enhance the mechanical and physical properties of tool steels, hot work steels, and high carbon steels. According to literature [1–4], the greatest improvement in properties is obtained by carrying out the deep cryogenic treatment between quenching and tempering. However, in the case of tool steels, an improvement can be obtained even by performing cryogenic treatment at the end of the usual heat treatment cycle (i.e., treating the finished tools). This last solution is more flexible than the previous one, and can extend the use of the treatment to many practical applications [5,6].

Patil et al. [7] demonstrated that the application of cryogenic treatment to cutting tools improves wear resistance, hardness, dimensional stability, cutting tool durability, and tool life, and it reduces tool consumption, leading to a general reduction in production cost. Perez et al. [8] reported the importance of cryogenic treatments to increase toughness, thermal fatigue resistance, and wear resistance of hot work steel (AISI H13 as example) in order to maximize their lifetime. These benefits are achieved by deep cryogenic treatment because it decreases retained austenite content and it promotes the precipitation of fine carbides uniformly dispersed in martensite matrix, as reported by Sola et al. [9] and Gavriliuk et al. [10]. Retained austenite is a soft and unstable phase that reduces steel hardness and can be converted into martensite in working conditions and under stress, forming brittle (not tempered) martensite, with an increase of volume of 4%, inducing local stresses. Cryogenic treatment—by transforming retained austenite to martensite—improves dimensional stability. In addition to the transformation of retained austenite to martensite, secondary and fine carbides are formed in the structure, increasing mechanical properties, toughness, and wear resistance. According to Perez [8], quenching and cryogenic treatment generate a high internal stress state due to thermal stresses and the transformation of martensite into austenite. Furthermore, thermal stresses increase the number of structural defects and the carbon-supersaturated martensite becomes unstable. Carbon atoms move towards the new structural defects created, martensite is decomposed, and carbide precipitation takes place during the warming up phase to room temperature, producing a reduction of residual stress and resulting in a homogeneously dispersed network of tiny carbides.

The evolution of carbides precipitation in chromium-containing steels, molybdenum-containing steels, and chromium–molybdenum-containing steels was discussed by Perez et al. [8,11], Gavriliuk et al. [10], and Villa et al. [12,13], studying the low-temperature martensitic transformation in tool steels and high-carbon steels. With internal friction analysis, Mossbauer spectroscopy, and synchrotron X-ray diffraction, these authors demonstrated that the carbon atoms are immobile at temperatures below $-100\ °C$, and the possibility of their diffusion exponentially decreases with decreasing temperature. Instead, during the heating up to room temperature from cryogenic temperature, an ageing of carbon-supersaturated martensite (starting from $-50\ °C$) leads to martensite decomposition (for example, in a spinodal-like decomposition of a supersaturated solution) into carbon-rich areas which could induce precipitation of nanometric carbides.

Cryogenic treatment barely changes the tensile mechanical properties and hardness of tool steel and hot work steel [8,9,13–15]. However, it is worth noting that cryogenic treatment notably improved the fracture toughness of such steels because a fine, homogeneously dispersed carbide precipitation and a tougher martensite matrix are formed (with lower carbon content).

In this framework, toughness measurement is an important tool to assess the effectiveness of the cryogenic treatment on such steels, but standard methods require careful sample preparation and dedicated equipment, while a simpler technique could be easily adopted as a quality control tool, as an alternative to ASTM E399 e BS 5447 standard method [16]. The most popular alternative method is the Vickers indentation fracture test, where the fracture toughness, Kc, is determined throughout a Vickers probe and according to Equation (1):

$$Kc = \alpha \left[\frac{E}{H}\right]^{\frac{1}{2}} \left[\frac{P}{c_0^{1/2}}\right] \tag{1}$$

where P is the indentation load, E is the Young's modulus, H is the hardness, c_o is the average length of radial cracks generated during the indentation, and α is a dimensionless constant. Several authors proposed refinements to Equation (1) [17–19] derived from a combination of empirical tests and dimensional analysis. All these expressions account for the residual stress, the plastic dissipation inside the material, and the nature of cracks. Moreover, during indentation fracture testing, it is fundamental to take considerable care to measure the average length of the cracks that begin from the four corners of the probe. Despite advances in microscopic analysis, considerable uncertainties could occur because of the possibility of spalling around indentation impression and/or the skill or subjectivity of the observer. Akono et al. in [20] proposed an alternative novel technique to measure the fracture toughness by scratch testing. The authors derived the fracture toughness expression from linear elastic fracture mechanics, and in [21] applied these techniques to ceramics, metals, polymers, and in [22], to micro-particulate composites. Akono et al. in [21] proposed a detailed description of an analytical model (with theoretical hypothesis and pertinence), materials surface preparation (the same for nano-indentation or micro-indentation), and the equipment and testing procedure, and showed that the Kc values measured via scratch test were in agreement with literature values, with a relative error of 2%–8% for ceramics and 3%–7% for metals.

In general terms, the scratch test consists of pulling a probe across the surface of the material under a controllable applied normal stress, and it is relevant nowadays to several fields of science and engineering [23], ranging from strength characterization of ceramics [24] to adhesion of thin films and coatings [25–27] and wear and damage resistance of metals, especially adhesion resistance of nitrided and nitrocarburized steels [28,29] and polymers [30]. Akono et al. [23] demonstrated that the failure mode (fracture or plastic yielding) is influenced by the materials properties as well as geometry of the scratching tool. According to the authors, it is possible to link the forces acting on scratch tip and the tool geometry to the plane strain fracture toughness Kc, according to the following equation:

$$Kc = \frac{F_t}{\sqrt{2wd(w + 2d)}} \tag{2}$$

where F_t is the horizontal (tangential) force necessary for the movement of the indenter, w is the blade (indenter tip) width, and d is the measured penetration depth.

Hence, the aim of the current study is to investigate the effect of post-tempering cryogenic treatment on the microstructure and mechanical properties of three different steels—the tool steel AISI M2, the hot work steel AISI D2, and the high chromium knife steel X105CrCoMo18 steel—as well as to investigate the application of a novel technique for characterizing the fracture toughness via scratch test experiments, simpler than standard method.

2. Materials and Methods

Standard bars of AISI D2, AISI M2, and X105CrCoMo18 tool steels were cut to obtain samples of the required size (40 mm diameter). Chemical compositions are given in Table 1. The samples were treated as summarized in Table 2. The cryogenic treatment investigated was carried out in liquid nitrogen (LN2) after tempering, using the following critical parameters: cryogenic temperature −193 °C, cooling rate 40 °C/h, soaking time 24 h, heating rate to room temperature 40 °C/h.

Fracture toughness tests were carried out on a CSM Instrument Revetest Micro scratch tester (Neuchatel, Switzerland). Before the test, the surface samples were prepared per the procedure described by Akono et al. in [21]. The specimens were tested with a 200 μm Rockwell C diamond indenter at a scratching speed of 6 mm/min with vertical force equal to 30 N, and the scratch length was 6 mm. The scratch tester measures the penetration depth. Compared to a standard fracture toughness test (ASTM E399: standard Test Method for Linear-Elastic Plane-Strain Fracture Toughness K_{Ic} of Metallic Materials), the scratch test is a non-destructive test and it can be replicated on different zones of the same sample, and it is easier to apply and more flexible because it is not necessary to manufacture a standard sample—only a proper preparation of the surface of the specimen is needed (polishing).

Table 1. Nominal chemical composition of steels used in the investigation (wt %).

Material	C	Si	Mn	Cr	Mo	V	W	Co	Fe
AISI D2	1.50	0.30	0.30	11.50	0.70	1.00	-	-	bal.
AISI M2	0.9	0.3	0.25	4.10	5.00	1.80	6.40	-	bal.
X105CrCoMo18	1.09	0.40	0.40	17.30	1.10	0.10	-	1.50	bal.

The microstructure of the treated samples was studied using a NOVA NanoSEM450, FEI Company—Bruker corporation (Hillsboro, OR, USA), scanning electron microscope (SEM) and the residual stresses were measured using Z-ray $\sin^2\psi$ method (ENIXE-TTX Residual Stress Diffractometer) with a Co tube radiation, 24.5 kV as tension, and 5.5 mA as current, 7 acquisition in ψ on 156 degree 2θ angle. To obtain samples suitable for the microstructural analysis, the specimens were properly polished and etched with Murakami's reagent (10 g $K_3Fe(CN)_6$, 10 g KOH, 100 mL water). Vickers HV1 microhardness tests were performed with a Vickers 432-SVD, Wolpert Wilson Instruments, INSTRON Company (Norwood, MA, USA), microhardness tester applying 9.8 N as normal force and a dwell time equal to 10 s.

Table 2. List of treatment conditions considered.

Material	Sample Code	Treatment
AISI D2	D2-C	Vacuum quenching at 1080 °C, vacuum tempering at 480 °C, cryogenic treatment at −80 °C for 2 h in liquid nitrogen (LN2), tempering at 480 °C, cryogenic treatment at −193 °C in LN2 for 24 h
	D2	Vacuum quenching at 1080 °C, vacuum tempering at 480 °C, cryogenic treatment at −80 °C for 2 h in LN2, tempering at 480 °C
AISI M2	M2-C	Vacuum quenching at 1080 °C, three vacuum tempering at 550 °C for 2 h, cryogenic treatment at −193 °C in LN2 for 24 h
	M2	Vacuum quenching at 1080 °C, three vacuum tempering at 550 °C for 2 h
X105CrCoMo18	X105-C	Vacuum quenching at 1030 °C, vacuum tempering at 500 °C for 2 h, cryogenic treatment at −193 °C in LN2 for 24 h
	X105	Vacuum quenching at 1030 °C, vacuum tempering at 500 °C for 2 h

Nanoindentation tests were carried out in load control mode on a calibrated Ultra Nanoindenter (UNHT) by CSM Instrument (Neuchatel, Switzerland) equipped with a Berkovich diamond tip at a constant loading rate of 200 $\mu N \cdot s^{-1}$, up to a maximum load of 30,000 mN, and the resolution of displacement 1 nm. The 50 s total indentation time was divided into three segments, consisting of 20 s loading and unloading and 10 s holding time. The tests were performed by creating three 10×10 grids of indents spaced 100 μm for a total of 300 indents for samples. More details of the nanoindentation test are discussed by Bocchini et al. in [31]. During the test, the nanoindenter records the penetration depth h and the load w. The slope dw/dh of the unloading curve at the beginning of unloading can be used to measure E as described by Fougere et al. [32], Chen et al. [33], and Balijepalli et al. [34,35]. Chen et al. in [33] showed a typical nanoindentation test load–indentation depth curve, and the author explained how it is possible to calculate the plasticity factor η_p, defined as the ratio of plastic deformation work to total deformation work. A low value of η_p means a high resistance to plastic deformation.

3. Results and Discussion

Figures 1–3 show the microstructure of non-cryogenically treated and post-tempering cryogenically treated samples.

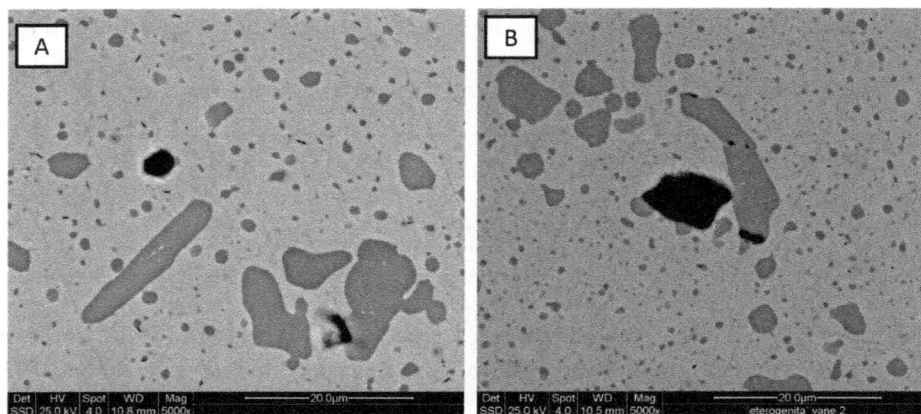

Figure 1. SEM micrographs of (**A**) D2 and (**B**) D2-C samples.

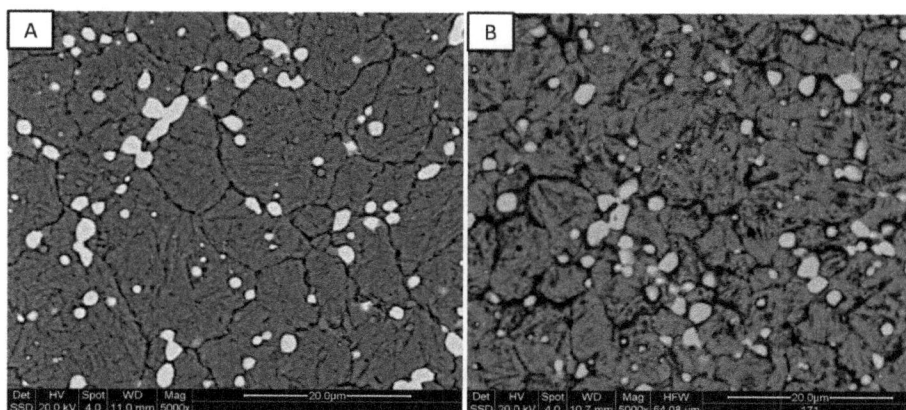

Figure 2. SEM micrographs of (**A**) M2 and (**B**) M2-C samples.

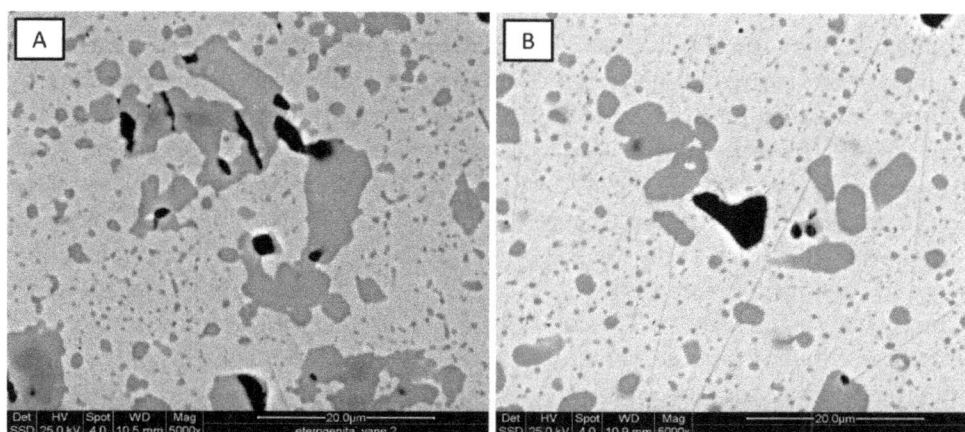

Figure 3. SEM micrographs of (**A**) X105 and (**B**) X105-C samples.

The AISI D2 samples (Figure 1) contain large primary carbides and smaller spherical carbides distributed homogeneously in a ferrite matrix parallel to the working direction. AISI M2 (Figure 2) cryogenically and non-cryogenically treated samples exhibit a martensitic matrix in which spheroidal carbides are distributed. X105CrCoMo18 samples (Figure 3) show a microstructure similar to AISI D2 samples. Image processing was performed using the public domain software ImageJ, and it was possible to estimate the average particle size and their volume fraction, as reported in Table 3.

Table 3. Particle size and volume fraction (%) estimated using image analysis software.

Sample	Particle Size (μm)	Volume Fraction (%)
D2-C	0.444 ± 0.1	19.7 ± 1
D2	0.555 ± 0.1	14.5 ± 1
M2-C	0.592 ± 0.1	9.1 ± 1
M2	0.617 ± 0.1	7.0 ± 1
X105-C	0.394 ± 0.1	15.9 ± 1
X105	0.472 ± 0.1	11.1 ± 1

By analysing ten representative images for each sample, it was found that in all the cryogenically treated steels, the volume fraction of submicrometric carbides was higher compared to untreated samples. Moreover, the carbides were finer and more homogeneously distributed in the cryogenically treated samples. Some authors [4,11] attribute the effect to the activation of the tempering transformation of the martensite because of its oversaturation attained at $-196\,^{\circ}C$. Because of this, the carbide precipitation occurs during the subsequent heating to room temperature from cryogenic temperature, with higher activation energy, thus leading to higher nucleation rate and in turn to finer dimensions and a more homogeneous distribution.

A possible advantage resulting from the precipitation of fine carbides as result of cryogenic treatment is the improvement of the fracture toughness of the steel; an increase of Kc was obtained in the cryogenically treated samples, as shown in Table 4, where the fracture toughness values estimated via scratch test are reported. Optical micrographs of scratches are visible in Table 5, and a scratch at high magnification is reported in Figure 4. The Kc values reported in Table 4 agree with the fracture toughness values measured with the standard method, as reported by Molinari et al. in a paper [5] where the effect of deep cryogenic treatment carried out after tempering on the mechanical properties of AISI M2 and AISI H3 was studied. In all of the tool steels investigated, the cryogenic treatment increased the fracture toughness value because the reduction in microcracking tendency resulted from reduced internal stress when the fine carbide precipitation occurs [14]. This is visible in Tables 3 and 4, where the increment of carbides content and decrease of residual stresses are reported. The reduction in temperature reduced density lattice defects (dislocations) and thermodynamic instability of the martensite, which drives carbon and alloying elements to nearby defects. These clusters act as nuclei for the formation of fine carbides when stress is subsequently relieved. The precipitation of carbides that also occurs during heating from cryogenic treatment temperature is responsible for the residual stress relaxation [36]. The present investigation in tool steel favours this hypothesis for two reasons: (1) the distribution of the carbides in the cryogenically treated samples was more homogeneous than in the non-cryogenically treated samples, and (2) the carbide volume fraction in the cryogenically treated samples was higher than in the non-cryogenically treated ones. The precipitation of more hard carbides in the cryogenically treated samples can reduce the carbon—and also supersaturation—in the matrix, improving its toughness. The combination of higher carbides content and the reduction of residual stresses enhanced the steel fracture toughness. A higher carbides content decreased the microcracking tendency. Moreover, it is well known [15] that a reduction in carbide size reduces the probability of carbide fracture, and can therefore increase the fracture toughness under specific contact conditions. The microscopic analysis confirms all of these observations. Indeed, in Figure 4, a SEM micrograph of the scratch of the AISI D2-C sample is reported. It is clearly visible that the large carbides inside the scratch are cracked, and other cracks are around the smaller carbides. Along the propagation front, cracks breaks or surrounds the carbides, and for this the crack propagation slows down, with benefits on fracture toughness.

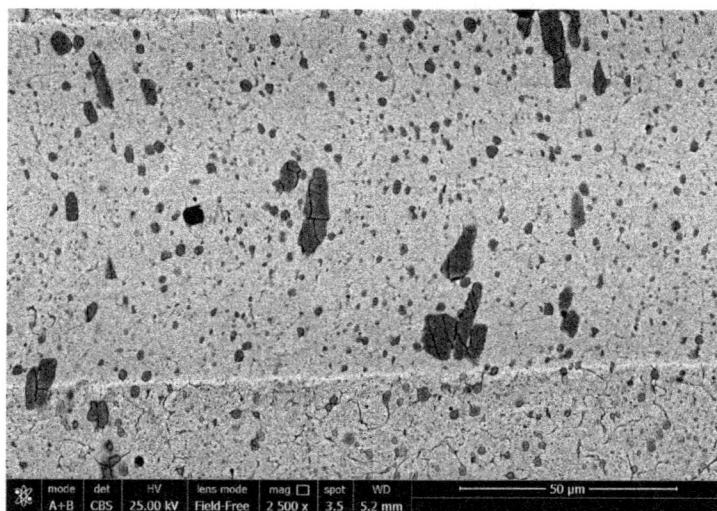

Figure 4. SEM micrograph of the scratch on the AISI D2-C sample.

Microhardness (HV1) values are reported in Table 4. The results show that the cryogenic treatment—when carried out after the usual heat treatment—increased fracture toughness without affecting the hardness of the steel. Cryogenically treated samples were a little less hard than non-cryogenically treated ones, and the differences in hardness values were not significant; in other words, the increase in toughness was attained without reducing hardness, but this is ascribed to fine and homogeneously dispersed carbides that also decreased the residual stresses.

The obtained elastic modulus (E) and plasticity factor (η_p %) are listed in Table 4. The studied treatments did not essentially modify the Elastic Modulus value of all the steels investigated, but post-tempering cryogenic treatment incremented the plasticity factor (η_p %) value with respect to non-cryogenically treated materials, indicating that the martensite transformation and carbon precipitation enhanced the plastic deformation work and the toughness.

Table 4. Results of fracture toughness tests, residual stresses analysis, and Vickers hardness test.

Sample	Kc (MPa·m$^{1/2}$)	Residual Stresses (MPa)	HV1	E (GPa)	η_p %
D2-C	47.06 ± 1	105 ± 35	807 ± 6	205 ± 6	57.7
D2	36.24 ± 1	159 ± 36	814 ± 23	208 ± 5	50.1
M2-C	47.91 ± 2	−66 ± 31	899 ± 22	211 ± 1	61.3
M2	36.74 ± 2	110 ± 39	902 ± 16	209 ± 3	49.4
X105-C	45.91 ± 3	281 ± 28	708 ± 11	212 ± 2	55.5
X105	29.96 ± 1	324 ± 36	699 ± 9	211 ± 4	33.4

Table 5. Optical micrographs of scratches. 100× magnifications.

Sample	Scratch 100×
D2-C	
D2	
M2-C	
M2	
X105-C	
X105	

4. Conclusions

A novel method for fracture toughness measurements via scratch test was applied, and the estimated values were reasonably in agreement with the literature values measured with standard methods. This novel method for fracture toughness measurements is a non-destructive test that is easy to apply, flexible, and does not need a standard sample. Post-tempering cryogenic treatment barely changed the hardness of AISI M2, AISI D2, and X105CrCoMo18 steels, but it significantly influenced fracture toughness and residual stresses. The precipitation of finer carbides homogeneously dispersed in the martensite matrix due to cryogenic treatment did not alter the Young's Modulus, as measured by nanoindentation test in load control mode on, it decreased martensite residual stress, and improved plasticity factor and toughness.

Author Contributions: Ramona Sola and Giovanni Parigi designed and performed the experiments, Roberto Giovanardi and Paolo Veronese contributed the design of experiments and the results analysis.

Conflicts of Interest: The authors declare no conflict of interest.

References

1. Podgornik, B.; Paulin, I.; Zajec, B.; Jacobson, S.; Leskovsek, V. Deep Cryogenic treatment of tool steels. *J. Mater. Process. Technol.* **2016**, *229*, 398–406. [CrossRef]

2. Baldissera, P.; Delprete, D. Deep Cryogenic treatment: A bibliografic review. *Open Mech. Eng. J.* **2008**, *2*, 1–11. [CrossRef]

3. Leskovsek, V.; Podgornik, B. Vacuum heat treatment, deep cryogenic treatment and simultaneous pulse plasma nitriding and tempering of P/M S390MC steel. *J. Mater. Sci. Eng. A* **2012**, *531*, 119–129. [CrossRef]

4. Das, D.; Sarkar, R.; Dutta, A.K.; Ray, K.K. Influence of sub-zero treatments on fracture toughness of AISI D2 steel. *Mater. Sci. Eng. A* **2010**, *528*, 589–603. [CrossRef]

5. Molinari, A.; Pellizzari, M.; Gialanella, S.; Straffelini, G.; Stiasny, K.H. Effect of deep cryogenic treatment on the properties of tool steel. *J. Mater. Process. Technol.* **2001**, *118*, 350–355. [CrossRef]

6. Pellizzari, M.; Molinari, A.; Giardini, L.; Maldarelli, L. Deep Cryogenic treatment of AISI M2 high speed steel. *Int. J. Microstruct. Mater.* **2008**, *3*, 383–390. [CrossRef]

7. Patil, N.; Kakkeri, S.; Sangamesh. Effect of cryogenic treated and untreated tool on its tool life—Review. *Int. J. Sci. Res.* **2012**, *3*, 141–145.

8. Perez, M.; Belzunce, F.J. The effect of deep cryogenic treatments on the mechanical proprieties of an AISI H13 steel. *Mater. Sci. Eng. A* **2015**, *624*, 32–40. [CrossRef]

9. Sola, R.; Poli, G.; Giovanardi, R.; Veronesi, P.; Parigi, G. Effect of deep cryogenic treatment on the properties of AISI M2 steel. In Proceedings of the European Conference on Heat Treatment 2015 and 22nd Heat Treatment and Surface Engineering from Tradition to Innovation Congress, Venice, Italy, 20–22 May 2015.

10. Gavrilijuk, V.; Theisen, W.; Sirosh, V.V.; Polshin, E.V.; Kortmann, A.; Mogilny, G.S.; Petrov, Y.N.; Tarusin, Y.V. Low-temperature martensitic transformation in tool steels in relation to their deep cryogenic treatment. *Acta Mater.* **2013**, *61*, 1705–1715. [CrossRef]

11. Perez, M.; Rodriguez, C.; Belzunce, F.J. The use of cryogenic treatment to increase fracture toughness of a hot work tool steel used to make forging dies. *Procedia Mater. Sci.* **2014**, *3*, 604–609. [CrossRef]

12. Villa, M.; Pantleon, K.; Somers, M.A.J. Evolution of compressive strains in retained austenite during sub-zero Celsius martensite formation and tempering. *Acta Mater.* **2014**, *65*, 383–392. [CrossRef]

13. Villa, M.; Grumsen, F.B.; Pantleon, K.; Somers, M.J. Martensitic transformation and stress partitioning in a high-carbon steel. *Scr. Mater.* **2012**, *67*, 621–624. [CrossRef]

14. Huang, J.Y.; Zhu, Y.T.; Liao, X.Z.; Beyerlein, I.J.; Bourke, M.A.; Mitchell, T.E. Microstructure of cryogenic treated M2 tool steel. *Mater. Sci. Eng. A* **2003**, *339*, 241–244. [CrossRef]

15. Singh, M.; Singh, H. Influence of deep-cryogenic treatment on the wear behavior and mechanical properties of mild steel. *Int. J. Res. Eng. Technol.* **2014**, *4*, 169–173.

16. ASTM International. *ASTM E399-12e3: Standard Test Method for Linear-Elastic Plane-Strain Fracture Toughness K_{Ic} of Metallic Materials*; ASTM International: West Conshohocken, PA, USA, 2012.

17. Quinn, G.D.; Bradt, R.C. On Vickers fracture Toughness Test. *J. Am. Ceram. Soc.* **2007**, *90*, 673–680. [CrossRef]

18. Harding, D.S.; Oliver, W.C.; Pharr, G.M. Cracking during nanoindentation and its use in the measurement of fracture toughness. In Proceedings of the Fall meeting of the Materials Research Society (MRS), Boston, MA, USA, 28 November–9 December 1994; pp. 663–668.

19. Widjaja, S.; Yip, T.H.; Limarga, A.M. Measurement of creep-induced localized residual stress in soda-lime glass using nano-indentation technique. *Mater. Sci. Eng. A* **2001**, *318*, 211–215. [CrossRef]

20. Akono, A.T.; Ulm, F.J. Scratch test model for the determination of fracture toughness. *Eng. Fract. Mech.* **2011**, *78*, 334–342. [CrossRef]

21. Akono, A.T.; Randall, N.X.; Ulm, F.J. Experimental determination of the fracture toughness via microscratch tests: Application to polymers, ceramics, and metals. *J. Mater. Res.* **2012**, *27*, 485–493. [CrossRef]

22. Bouché, G.A.; Akono, A.T. Micromechanics-based estimates on the macroscopic fracture toughness of micro-particulate composites. *Eng. Fract. Mech.* **2015**, *148*, 243–257. [CrossRef]

23. Akono, A.T.; Alm, F.J. An improved technique for characterizing the fracture toughness via scratch test experiments. *Wear* **2014**, *313*, 117–124. [CrossRef]

24. ASTM International. *ASTM Standard C1624: Standard Test Method for Adhesion Strenght and Mechanical Failure Modes of Ceramics Coatings by Quantitative Single Point Scratch Testing*; ASTM International: West Conshohocken, PA, USA, 2015.

25. Bull, S.J.; Berasetegui, E.G. An overview of the potential of quantitative coating adhesion measurement by scratch testing. *Tribol. Int.* **2006**, *39*, 91–114. [CrossRef]

26. Nohava, J.; Bonferroni, B.; Bolelli, G.; Lusvarghi, L. Interesting aspects of indentation and scratch methods for characterization of thermally-sprayed coatings. *Surf. Coat. Technol.* **2010**, *205*, 1127–1131. [CrossRef]

27. Taurino, R.; Barbieri, L.; Bondioli, F. Surface properties of new green building material after TiO$_2$–SiO$_2$ coatings deposition. *Ceram. Int.* **2016**, *42*, 4866–4874. [CrossRef]

28. Sola, R.; Poli, G.; Veronesi, P.; Giovanardi, R. Effects of surface morphology on the wear and corrosion resistance of post-treated Nitrided and nitrocarburized 42CrMo4 steel. *Metall. Mater. Trans. A* **2014**, *45*, 2827–2833. [CrossRef]

29. Sola, R. Post-treatment surface morphology effect on the wear and corrosion resistance of nitrided and nitrocarburized 41CrAlMo7 steel. *Metall. Ital.* **2010**, *102*, 21–31.

30. Taurino, R.; Fabbri, E.; Pospiech, D.; Synytska, A.; Messori, M. Preparation of scratch resistant superhydrophobic hybrid coatings by sol-gel process. *Prog. Org. Coat.* **2014**, *77*, 1635–1641. [CrossRef]

31. Bocchini, G.F.; Poli, G.; Sola, R.; Veronesi, P. Comparison—By nanoindentation—Among PM steels obtained from diffusion-bonded powders (nominally equivalent). In Proceedings of the World Powder Metallurgy Congress and Exhibition, Florence, Italy, 10–14 October 2010; Volume 3.

32. Fougere, G.E.; Riester, L.; Ferber, M.; Weetman, J.R.; Siegal, R.W. Young's modulus of nanocrystalline Fe measured by nanoindetation. *Mater. Sci. Eng. A* **1995**, *204*, 1–6. [CrossRef]

33. Chen, H.T.; Yan, M.F.; Fu, S.S. Martensite transformation induced by plasma nitrocarburizing on AISI304 austenitic stainless steel. *Vacuum* **2014**, *105*, 33–38. [CrossRef]

34. Balijepalli, S.K.; Colantoni, I.; Donnini, R.; Kaciulis, S.; Lucci, M.; Montanari, R.; Ucciardello, N.; Varone, A. Modulo elastico della fase S in un acciaio 316 L kolsterizzato. *Metall. Ital.* **2013**, *1*, 42–47.

35. Balijepalli, S.K.; Colantoni, I.; Donnini, R.; Kaciulis, S.; Montanari, R.; Varone, A. Young's modulus profile in kolsterized AISI 316 L steel. *Mater. Sci. Forum* **2013**, *762*, 183–188. [CrossRef]

36. Senthilkumar, D. Influence of shallow and deep cryogenic treatment on residual state of stress of 4140 steel. *J. Mater. Process. Technol.* **2011**, *211*, 396–401. [CrossRef]

10

Study of Adsorption of Hydrogen on Al, Cu, Mg, Ti Surfaces in Al Alloy Melt via First Principles Calculation

Yu Liu [1,2,*], Yuanchun Huang [1,2,3,*], Zhengbing Xiao [1,2,3] and Xianwei Reng [1,2]

[1]   Research Institute of Light Alloy, Central South University, Changsha 410083, China; xiaozb@csu.edu.cn (Z.X.); renxianweichina@126.com (X.R.)

[2]   Nouferrous Metal Oriented Advanced Structural Materials and Manufacturing Cooperative Innovation Center, Central South University, Changsha 410083, China

[3]   College of Mechanical and Electrical Engineering, Central South University, Changsha 410083, China

*   Correspondence: csuliuyu@csu.edu.cn (Y.L.); science@csu.edu.cn (Y.H.)

Academic Editor: Hugo F. Lopez

Abstract: Adsorption of hydrogen on Al(111), Cu(111), Mg(0001), and Ti(0001) surfaces have been investigated by means of first principles calculation. The calculation of surface energy indicates that Mg(0001) is the most stable surface, while Ti(0001) is the most unstable surface among all the four calculated surfaces. The obtained adsorption energy shows that the interaction between Al and H atoms should be energetically unfavorable, and the adsorption of hydrogen on Mg(0001) surface was found to be energetically preferred. Besides, the stability of hydrogen adsorption on studied surfaces increased in the order of Al(111), Ti(0001), Cu(111), Mg(0001). Calculation results also reveal that hydrogen adsorption on fcc and hcp sites are energetically stable compared with top and bridge sites for Ti(0001), Cu(111), and Mg(0001), while hydrogen adsorbing at the top site of Al(111) is the most unstable state compared with other sites. The calculated results agreed well with results from experiments and values in other calculations.

Keywords: surface energy; hydrogen adsorption; stability; first principles calculation

---

1. Introduction

Aluminum alloy has been widely used in the aeronautics industry, space industry, nuclear industry, and military industry for its low density, high strength, and anti-corrosive ability [1–3]. Numerous efforts have been devoted to improving the quality of aluminum alloy products, so as to meet the increasing demands for high performance from composition and processing technology.

It is widely recognized that the first step to ensure the properties of an aluminum component is a high-quality ingot, since it plays a key role in determining the microstructure evolution in subsequent processing steps. In order to prepare such an ingot, the purification process of liquid aluminum alloys prior to casting has a crucial importance. The purpose of purification is to remove impurities from molten metals, which may be in the form of gases, inclusions, or dissolved metals, which may lead to casting defects.

Porosity—a frequent casting defect—has proven to be a challenge for its formation and evolution. At present, it is believed that one of the main reasons for porosity formation in aluminum alloys is attributable to hydrogen evolution. In fact, hydrogen is supposed to be the only gas dissolved in aluminum melt [4,5]. When the hydrogen content reaches a critical value in the liquid metal, molecular hydrogen pores form and may grow depending on the local hydrogen concentration levels and the diffusion rate [6]. This kind of gas porosity is normally observed only as small distributed pores, which are also termed microporosity. The formed microporosities have detrimental effects on both the service and processing properties of metals. Hence, the removal of the hydrogen from the aluminum melt is crucial for the production of high-quality castings.

Several approaches have been developed for degassing in the last several decades; for example, re-melting degassing [7], vacuum degassing [8,9], ultrasonic degassing [10,11], and spray degassing [12,13], as well as rotary impeller degassing with nitrogen, argon, or a mixture of the inert gases and chlorine as a purge gas [14–16]. Nevertheless, the research on technology and equipment of high-efficiency degassing depends on reliable physical property data and parameters of degassing mechanisms, which are still incompletely understood.

In terms of hydrogen removal, it is necessary to have a good understanding of the interaction between hydrogen and the components of the aluminum melt (such as alloying elements and inclusions) for an efficient hydrogen removal. Anyalebechi analyzed the effects of alloying elements on the solubility of hydrogen in liquid aluminum via Wagner's interaction parameter [17]. However, investigation of the interaction between hydrogen and the component of the aluminum melt is limited to experimental equipment and technology. Recently, first-principles calculations based on density functional theory have been widely used in interfacial research for their high reliability and accuracy.

To our knowledge, few systematic and theoretical studies regarding the interaction between hydrogen and alloying elements in molten aluminum have been reported. Therefore, the aim of the present work is to study the adsorption of hydrogen on the alloying element surfaces of Cu, Mg, and Ti from first-principles calculations, and the matrix element is also calculated for comparison.

2. Computational Methods

The first-principles calculation was carried out by means of the Cambridge Sequential Total Energy Package (CASTEP, Accelrys, San Diego, CA, USA), and the calculation was conducted in a plane-wave basis using the projector-augmented wave (PAW) method [18]. The interaction between ions and electrons was described using the ultrasoft pseudopotentials introduced by Vanderbilt and provided by Kresse and Hafner [19]. The exchange correlation functional was described by general gradient approximation (GGA) of Perdew-Burke-Ernzerhof (PBE), convergence tolerance of total energy per atom is 2×10^{-6} eV, and the Brodygen-Fletcher-Gplldfarb-Shanno (BFGS) [19,20] method was used for optimization.

The present studies of surface energy and H adsorption are focused on low-index surfaces as typical examples owing to their close-packed nature and lower energy state; i.e., (0001) of Mg and Ti, (111) of Al and Cu. Accordingly, a $2 \times 2 \times 1$ unit cell was selected for (111) of Cu, (0001) of Mg and Ti, and 1×1 unit cell for Al(111).

After the test, a seven-layer slab of Al, Mg, and Ti, with a vacuum thickness between any two successive metal slabs of 2.7, 2.1, and 1.75 nm, respectively, and a six-layer slab of Cu with a vacuum thickness of 2.7 nm were used for the calculation of surface energy and adsorption energy.

The influence of different k-point sampling and plane-wave cutoff energy were performed in a series of test calculations, and this led to the calculations being performed with $17 \times 17 \times 1$ and $21 \times 21 \times 1$ k-point sampling and a cutoff energy of 450 for Al(111) and Cu(111), and 480 eV for Mg(0001) and Ti(0001), respectively. The above parameters were determined to be high enough to ensure accuracy.

For the adsorption of H atoms, both of FCC and HCP structures have four adsorption positions. Accordingly, Figure 1 shows the schematic illustrations of various adsorption positions of H; i.e., the top (A), fcc (B), bridge (C), and hcp (D) sites of FCC (111) or HCP (0001) surface. For H adsorption on each slab, the atoms in the top two layers and the adsorbate were allowed to relax, while the atoms in other layers were fixed at the bulk truncated positions.

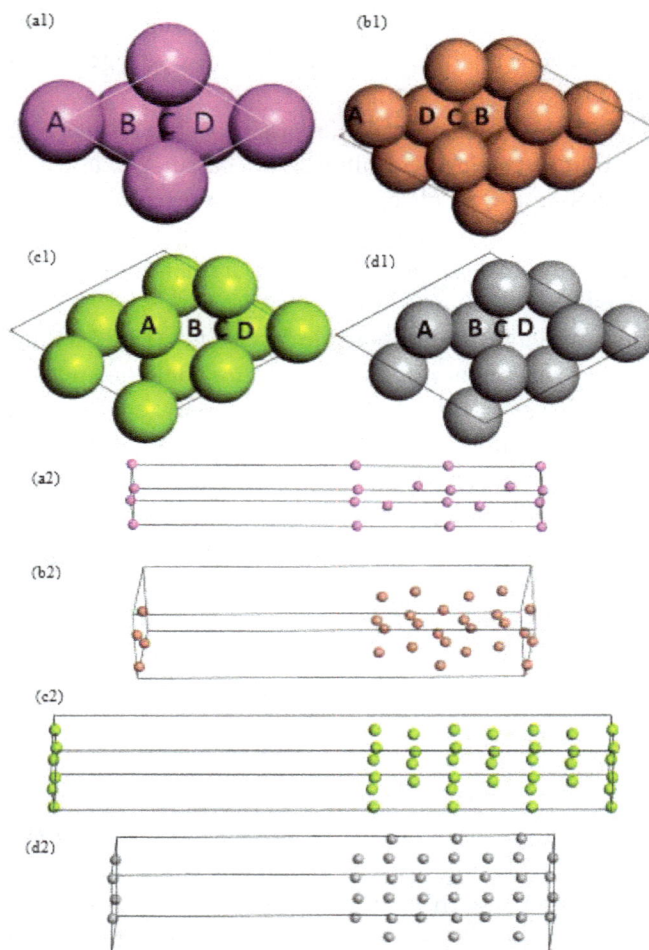

Figure 1. Schematic illustration of adsorption sites of H on (111) surface of (**a1**) Al and (**b1**) Cu; (0001) surface of (**c1**) Mg and (**d1**) Ti; together with the side structure of model surface of (**a2**) Al(111), (**b2**) Cu(111), (**c2**) Mg(0001), and (**d2**) Ti(0001). A—Top site, B—Fcc site, C—Bridge site, D—Hcp site.

3. Results and Discussion

3.1. Surface Energy and Work Function of Al, Cu, Mg, and Ti

Before surface calculation, geometry optimization was conducted on all the bulk structures, and the obtained lattice constants are listed in Table 1. It could be seen that the present lattice constants (Al: $a = 4.021$ Å; Cu: $a = 3.641$ Å; Mg: $a = 3.192$ Å, $c = 5.206$ Å; Ti: $a = 2.896$ Å, $c = 4.626$ Å) are consistent with corresponding experimental values in the literature and values in other calculations. Hence, the computation parameters and model setting above can provide reliable description for the calculation system, and the obtained bulks are used in the following interface modeling and computations of surface energy and adsorption energy.

Table 1. Lattice constants of Al, Cu, Mg, and Ti.

Elements	Lattice Constants (Å)		
	This Work	**Experiments**	**Other Calculation**
Al	$a = 4.021$	$a = 4.05$ [20]	$a = 3.982$ [21]
Cu	$a = 3.641$	$a = 3.61$ [22]	$a = 3.638$ [23]
Mg	$a = 3.192, c = 5.206$	$a = 3.21, c = 5.213$ [24]	$a = 3.19, c = 5.17$ [25]
Ti	$a = 2.896, c = 4.626$	$a = 2.904, c = 4.680$ [26]	$a = 2.864, c = 4.537$ [21]

Surface modeling of FCC and HCP structures were performed based on the obtained bulk structures, and the calculation of surface energy and work function of Al(111), Cu(111), Mg(0001), and Ti(0001) surfaces were performed by means of the following formulas [27]:

$$\gamma = 1/2A \times (E_{\text{slab}} - E_{\text{bulk}}) \tag{1}$$

$$\Phi = V_{\text{vac}} - E_f \tag{2}$$

where γ is surface energy, A is the surface area, E_{slab} is the total energy of the slab, E_{bulk} is the total energy of the corresponding bulk, and the factor $1/2$ in Equation (1) is owing to two equivalent surfaces in the slab, Φ is the work function, V_{vac} is the vacuum level in the vacuum region, and E_f stands for the Fermi energy of the slab.

The surface energy of Al(111), Cu(111), Mg(0001), and Ti(0001) were calculated via Equation (1), the computation results are listed in Table 2, and values from other sources are also presented for comparison. One can obviously deduce from Table 2 that the surface energy of Ti(0001) is the highest, which was calculated to be 2.034 J/m^2; the surface energy of Mg(0001) was lowest, which was calculated to be 0.716 J/m^2; and the surface energy of Al(111) and Cu(111) was located between the above two values. All of the calculation results agreed quite well with values from other sources, which again validates the computation details.

Surface energy is one of the basic quantities in surface physics, and is defined as the surface excess free energy per unit area of a particular crystal facet; that is, the energy required to cut a crystal into two parts or two surfaces. It is important in surface faceting, roughening, and crystal growth phenomena, and may be used to estimate surface segregation in binary alloys. On the other hand, the surface energy can be a measure of the stability of the surface—a lower surface energy indicates a more stable surface; that is to say, Mg(0001) is the most stable surface, and Al(111) follows, while Ti(0001) is the most unstable surface among all the four calculated surfaces.

Table 2. Surface energy of Al(111), Cu(111), Mg(0001), and Ti(0001).

Surface	Surface Energy (J/m^2)		
	This Work	Experiments	Other Calculation
Al(111)	0.864	1.14 [28]	0.988 [21]
Cu(111)	1.793	1.825 [28]	1.94 [29]
Mg(0001)	0.716	0.785 [28]	0.641 [30]
Ti(0001)	2.034	1.98 [31]	2.235 [21]

The work function of Al(111), Cu(111), Mg(0001), and Ti(0001) were calculated to be 4.26, 4.98, 3.80, and 4.58 eV, respectively, according to Equation (2), as listed in Table 3. The derived work functions are consistent with values in literature, and the maximum deviation of work function is 0.13 eV for Ti(0001).

Table 3. Work function of Al(111), Cu(111), Mg(0001), and Ti(0001).

Surfaces	Work Function (eV)		
	This Work	Experiments	Other Calculation
Al(111)	4.26	4.24 [32]	4.22 [21]
Cu(111)	4.98	4.94 [33]	5.1 [29]
Mg(0001)	3.80	3.86 [34]	3.76 [35]
Ti(0001)	4.58	4.45 [36]	4.67 [21]

The work function is a property of the surface of the material which is defined as the minimum thermodynamic energy needed to remove an electron from a solid to a point in the vacuum immediately outside the solid surface. In other words, it reflects the difficulty of losing an electron for a given surface, and the ability to lose electrons decreases with increasing work function. The work function of Cu(111) is the highest, which shows that it is the most difficult to lose electrons. Accordingly, the ability of the above surfaces to lose electrons is ranked as Mg(0001), Al(111), Ti(0001), Cu(111), in decreasing order.

3.2. Electronic Structures of Slabs

It is of great importance to investigate the electronics structures of Al(111), Cu(111), Mg(0001), and Ti(0001) surfaces. Figure 2 shows the comparison of total density of states (DOS) of Al, Cu, Mg, and Ti atoms in the bulk and surface. It can be seen that for all atoms—Al, Cu, Mg, and Ti—the bandwidths of DOS of the surface atoms are smaller than those of the bulk atom. Another feature that can be deduced from Figure 2 is that the DOSs of analyzed atoms in the surface have higher values around the Fermi level than those atoms in bulk. It is easy to derive the conclusion that some changes of electronic structures have occurred during surface formation for Al(111), Cu(111), Mg(0001), and Ti(0001).

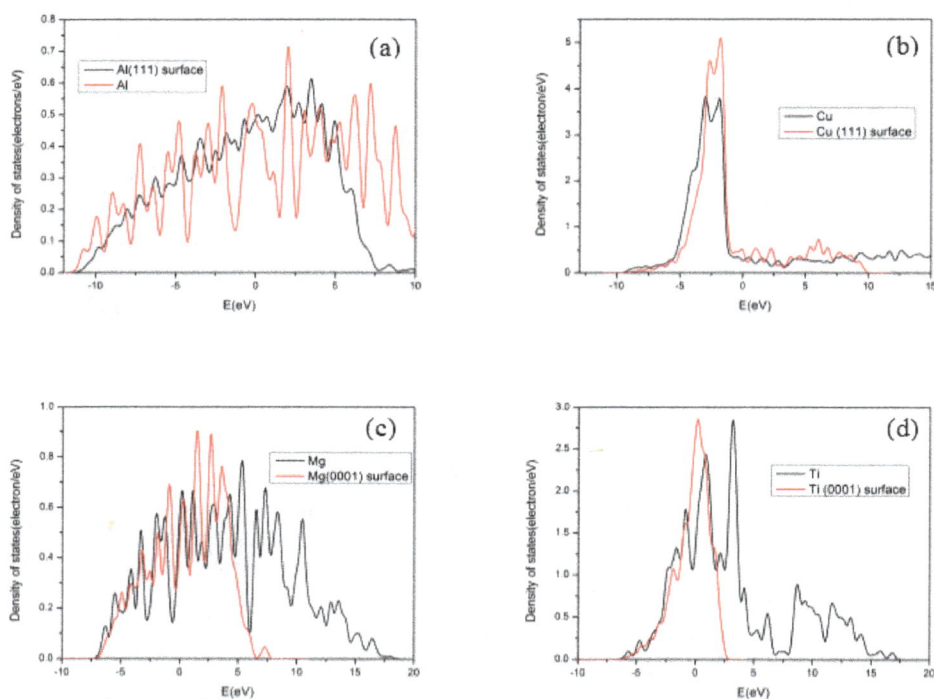

Figure 2. Density of states (DOS) of (**a**) Al; (**b**) Cu; (**c**) Mg; and (**d**) Ti in the bulk and surface.

3.3. H Adsorption on Al(111), Cu(111), Mg(0001), and Ti(0001) Surfaces

The adsorption energy (E_{ads}) of H on Al(111), Cu(111), Mg(0001), and Ti(0001) surfaces are derived according to the following formula:

$$E_{ads} = E_{\text{H-slab}} - E_{slab} - E_{H_2}/2 \tag{3}$$

where $E_{\text{H-slab}}$ and E_{slab} are total energies of the slab with and without H adsorption, respectively, and E_{H2} is the total energy of a H_2 molecule.

After a series of calculations, the adsorption energy of H is obtained for various adsorption sites of Al(111), Cu(111), Mg(0001), and Ti(0001) surfaces as Figure 1 shows, and the results are summarized in Table 4. Though experimentally-derived values of the adsorption energy of hydrogen atoms on element surfaces are still missing, we list values from references for comparison in the present paper to verify the reliability of this study.

It should be noted that during the calculation of H adsorption on the bridge site of Al(111) and Ti(0001), the adsorbates will move to fcc site and hcp site, respectively, after relaxation calculation, which reveals that hydrogen adsorption at bridge sites of Al(111) and Ti(0001) is energetically unstable.

Table 4. Adsorption energy (E_{ads}) of hydrogen on various adsorption sites of Al(111), Cu(111), Mg(0001), and Ti(0001) surfaces.

Surface	Sites	E_{ads} (eV)	
		This Work	Other Calculation
Al(111)	Top	0.235	0.226 [a]
	Fcc	0.081	0.069 [a]
	Hcp	0.132	0.122 [a]
Cu(111)	Top	−1.667	−1.83 [b]
	Fcc	−2.385	−2.37 [b]
	Hcp	−2.353	−2.36 [b]
	Bridge	−2.178	−2.22 [b]
Mg(0001)	Top	−1.21	−1.04 [c]
	Fcc	−2.546	−2.56 [c]
	Hcp	−2.471	−2.52 [c]
	Bridge	−1.90	−1.89 [c]
Ti(0001)	Top	0.610	0.400 [a]
	Fcc	−1.486	−1.342 [a]
	Hcp	−1.502	−1.396 [a]

[a] Reference [21]; [b] Reference [37]; [c] Reference [25].

Several highlights could be deduced from Table 4. Firstly, our calculation results agree quite well with values from other sources. Secondly, one sees clearly from Table 4 that for Al(111) surface, the E_{ads} values of hydrogen adsorbed on all sites are positive, suggesting that the interaction between Al and H atoms should be mainly repulsive and energetically unfavorable. Thirdly, it seems more likely that the adsorption of hydrogen on the Mg(0001) surface was found to be energetically preferred, due to the E_{ads} of −2.546, −2.471, and −1.90 eV for hydrogen adsorbed on fcc, hcp, and bridge sites of Mg(0001), respectively. As a whole, the stability of hydrogen adsorption on the studied surfaces increased in the order of Al(111), Ti(0001), Cu(111), Mg(0001). The stability of hydrogen on fcc and hcp sites is energetically stable compared with top and bridge sites for Ti(0001), Cu(111), and Mg(0001). Hydrogen adsorbing at the top site of Al(111) is the most unstable state compared with other site of Al(111).

On the other hand, we also calculate the heat of formation of hydrogen adsorption (ΔH_e) to compare the stability of hydrogen on Al(111), Cu(111), Mg(0001), and Ti(0001) surfaces. Accordingly, ΔH_f is calculated by means of the following formula:

$$\Delta H_f = \frac{1}{N}(E_{\text{H-slab}} - E_{\text{slab}} - E_{H_2}/2) \tag{4}$$

where N is the total number of atoms in the system, $E_{\text{H-slab}}$, E_{slab}, and E_{H2} have the same meaning as mentioned above. The calculated ΔH_f are presented in Table 5.

Table 5. Heat of formation (ΔH_f) of hydrogen on various adsorption sites of Al(111), Cu(111), Mg(0001), and Ti(0001) surfaces.

Sites		Top	Fcc	Hcp	Bridge
ΔH_f (kJ/mol)	Al_8H	5.04	1.74	2.83	
	$Cu_{24}H$	−12.86	−18.40	−18.15	−16.80
	$Mg_{29}H$	−7.78	−16.37	−15.89	−12.22
	$Ti_{28}H$	4.06	−9.88	−9.99	

It can be seen that the ΔH_f of hydrogen adsorbed on Al(111) are all positive, while in other systems, only hydrogen adsorbed on the top site of Ti(0001) is positive. ΔH_f of hydrogen adsorbed on Cu(111) are much closer to each other. In terms of $Mg_{29}H$, the lowest and highest ΔH_f occur in the Fcc site and top sites, respectively, and the ΔH_f of hydrogen adsorbed on Fcc and Hcp site of Ti(0001) are almost equal. A lower heat of formation suggests a stronger interaction between H and the surface; that is to say, the interaction between H and the studied surfaces decreased in the order Al(111), Ti(0001), Cu(111), Mg(0001). The conclusion deduced from ΔH_f agrees well with what is indicated by E_{ads}.

Accordingly, it is more likely for hydrogen atoms to adsorb on Cu and Mg surfaces than Al and Ti. On the other hand, we can conclude that relatively more energy is required to remove a hydrogen atom from Mg and Cu surfaces, and less energy to remove a hydrogen atom from Al surfaces. Besides, for the studied elements Al, Cu, Mg, and Ti, the hydrogen solubility of molten aluminum suffers most from Cu and Mg, in short range atomic group, and Ti follows. Based on the obtained calculation results, theoretical guidance can be provided for hydrogen removal from aluminum melt.

4. Summary and Conclusions

The adsorption of hydrogen on Al(111), Cu(111), Mg(0001), and Ti(0001) surfaces has been investigated by means of first principles calculation, and several remarks could be drawn from the computation results. Firstly, Mg(0001) is the most stable surface, while Ti(0001) is the most unstable surface among all the four calculated surfaces, and the ability of the above surfaces to lose electrons is ranked as Mg(0001), Al(111), Ti(0001), and Cu(111) in decreasing order. Secondly, the interaction between Al and H atoms should be energetically unfavorable, and the adsorption of hydrogen on the Mg(0001) surface were found to be energetically preferred, due to the E_{ads} of -2.546, -2.471, and -1.90 eV for hydrogen adsorbed on fcc, hcp, and bridge sites of Mg(0001), respectively. Thirdly, the stability of hydrogen adsorption on the studied surfaces increased in the order Al(111), Ti(0001), Cu(111), Mg(0001). Fourthly, hydrogen adsorption on fcc and hcp sites are energetically stable compared with top and bridge sites for Ti(0001), Cu(111), and Mg(0001). Hydrogen adsorbing at the top site of Al(111) is the most unstable state compared with other sites. Finally, the calculated results agree well with results from experiments and values in other calculations.

Acknowledgments: The authors are grateful for the financial supported by the Fundamental Research Funds for the Central Universities of Central South University(2015zzts042), the National Basic Research Program of China (2015CB057305).

Author Contributions: Yu Liu and Yuanchun Huang conceived and designed the calculations; Yu Liu performed the calculations; Yu Liu and Yuanchun Huang analyzed the data; Zhengbing Xiao and Xianwei Ren contributed analysis tools; Yu Liu wrote the paper.

Conflicts of Interest: The authors declare no conflict of interest.

References

1.   Wanhill, R.J.H.; Bray, G.H. Chapter 2—Aerostructural Design and Its Application to Aluminum-Lithium Alloys. *Alum. Lithium Alloy.* **2014**, *14*, 27–58.

2.   Li, Y.-P. Application and Prospect of Aluminum Alloys Automobile industry. *Alum. Fabr.* **2007**, *173*, 23–24.

3.   Shi, Q.; Xiong, W. Application and Development of Aluminium Alloys in Automobile Industry. *New Technol. New Process* **2006**, *12*, 55–58.

4.   Poirier, D.R.; Yeum, K.; Maples, A.L. A thermodynamic prediction for microporosity formation in aluminum-rich Al-Cu alloys. *Metall. Trans. A* **1987**, *18*, 1979–1987. [CrossRef]

5.   Han, Q.; Viswanathan, S. Hydrogen evolution during directional solidification and its effect on porosity formation in aluminum alloys. *Metall. Mater. Trans. A* **2002**, *33*, 2067–2072. [CrossRef]

6.   Felberbaum, M.; Désy, E.L.; Weber, L.; Rappaz, M. Effective hydrogen diffusion coefficient for solidifying aluminium alloys. *Acta Mater.* **2011**, *59*, 2302–2308. [CrossRef]

7.   Zhao, L.; Pan, Y.; Liao, H.; Wang, Q. Degassing of aluminum alloys during re-melting. *Mater. Lett.* **2012**, *66*, 328–331. [CrossRef]

8.   Xu, H.; Meek, T.T.; Han, Q. Effects of ultrasonic field and vacuum on degassing of molten aluminum alloy. *Mater. Lett.* **2007**, *61*, 1246–1250. [CrossRef]

9.   Zhu, X.; Jiang, D.; Tan, S. Improvement in the strength of reticulated porous ceramics by vacuum degassing. *Mater. Lett.* **2001**, *51*, 363–367. [CrossRef]

10. Éskin, G.I. Prospects of ultrasonic (cavitational) treatment of the melt in the manufacture of aluminum alloy products. *Metallurgist* **1998**, *42*, 284–291. [CrossRef]

11. Eskin, G.I. Principles of Ultrasonic Treatment: Application for Light Alloys Melts. *Adv. Perform. Mater.* **1997**, *4*, 223–232. [CrossRef]

12. Wu, R.; Shu, D.; Sun, B.; Wang, J.; Li, F.; Chen, H.; Lu, Y. Theoretical analysis and experimental study of spray degassing method. *Mater. Sci. Eng. A* **2005**, *408*, 19–25. [CrossRef]

13. Wu, R.; Qu, Z.; Sun, B.; Shu, D. Effects of spray degassing parameters on hydrogen content and properties of commercial purity aluminum. *Mater. Sci. Eng. A* **2007**, *456*, 386–390. [CrossRef]

14. Warke, V.S.; Tryggvason, G.; Makhlouf, M.M. Mathematical modeling and computer simulation of molten metal cleansing by the rotating impeller degasser: Part I. Fluid flow. *J. Mater. Process. Technol.* **2005**, *168*, 112–118. [CrossRef]

15. Warke, V.S.; Shankar, S.; Makhlouf, M.M. Mathematical modeling and computer simulation of molten aluminum cleansing by the rotating impeller degasser: Part II. Removal of hydrogen gas and solid particles. *J. Mater. Process. Technol.* **2005**, *168*, 119–126. [CrossRef]

16. Wang, L.; Guo, E.; Huang, Y.; Lu, B. Rotary impeller refinement of 7075Al alloy. *Rare Met.* **2009**, *28*, 309–312. [CrossRef]

17. Anyalebechi, P.N. Analysis of the effects of alloying elements on hydrogen solubility in liquid aluminum alloys. *Scri. Metall. Mater.* **1995**, *33*, 1209–1216. [CrossRef]

18. Kresse, G.; Joubert, D. From Ultrasoft Pseudopotentials to the Projector Augmented-Wave Method. *Phys. Rev. B Condens. Matter* **1999**, *59*, 1758–1775. [CrossRef]

19. Kresse, G.; Hafner, J. Norm-conserving and ultrasoft pseudopotentials for first-row and transition elements. *J. Phys. Condens. Matter* **1994**, *6*, 8245. [CrossRef]

20. Zope, R.R.; Mishin, Y. Interatomic potentials for atomistic simulations of the Ti-Al system. *Phys. Rev. B* **2003**, *68*, 024102. [CrossRef]

21. Wang, J.W.; Gong, H.R. Adsorption and diffusion of hydrogen on Ti, Al, and TiAl surfaces. *Int. J. Hydrog. Energy* **2014**, *39*, 6068–6075. [CrossRef]

22. Straumanis, M.E.; Yu, L.S. Lattice parameters, densities, expansion coefficients and perfection of structure of Cu and of Cu—In α phase. *Acta Crystallogr.* **1969**, *25*, 676–682. [CrossRef]

23. Ganne, J.P.; Lebourgeois, R.; Paté, M.; Dubreuil, D; Pinier, L.; Pascard, H. The electromagnetic properties of Cu-substituted garnets with low sintering temperature. *J. Eur. Ceram. Soc.* **2007**, *27*, 2771–2777. [CrossRef]

24. Kittel, C. *Introduction to Solid State Physics*, 8th ed.; Addison-Wiley: New York, NY, USA, 2005.

25. Wu, G.-X.; Zhang, J.Y.; Wu, Y.-Q.; Li, Q.; Chou, K.; Bao, X.H. First-Principle Calculations of the Adsorption, Dissociation and Diffusion of Hydrogen on the Mg(0001) Surface. *Acta Phys. Chim. Sin.* **2008**, *24*, 55–60. [CrossRef]

26. Martin, A.S.; Manchester, F.D. The H-Ti (Hydrogen-Titanium) system. *Bull. Alloy Phase Diagr.* **1987**, *8*, 30–42. [CrossRef]

27. Halas, S.; Durakiewicz, T.; Joyce, J.J. Surface energy calculation—Metals with 1 and 2 delocalized electrons per atom. *Chem. Phys.* **2002**, *278*, 111–117. [CrossRef]

28. Tyson, W.R.; Miller, W.A. Surface free energies of solid metals: Estimation from liquid surface tension measurements. *Surf. Sci.* **1977**, *62*, 267–276. [CrossRef]

29. Polatoglou, H.M.; Methfessel, M.; Scheffler, M. Vacancy-formation energies at the (111) surface and in bulk Al, Cu, Ag, and Rh. *Phys. Rev. B* **1993**, *48*, 1877–1883. [CrossRef]

30. Wright, A.F.; Feibelman, P.J.; Atlas, S.R. First-principles calculation of the Mg(0001) surface relaxation. *Surf. Sci.* **1994**, *302*, 215–222. [CrossRef]

31. Murr, L.E. *Interfacial Phenomena in Metals and Alloys*; Addison-Wesley: Upper Saddle River, NJ, USA, 1974.

32. Michaelson, H.B. The work function of the elements and its periodicity. *J. Appl. Phys.* **1977**, *48*, 4729–4733. [CrossRef]

33. DeBoer, F.R.; Boom, R.; Mattens, W.C.M.; Miedema, A.R.; Niessen, A.K. *Cohesion in Metals*; HoIIand, N., Ed.; North Holland: Amsterdam, The Netherlands, 1988.

34.   Skriver, H.L.; Rosengaard, N.M. Surface energy and work function of elemental metals. *Phys. Rev. B* **1992**, *46*, 7157–7168. [CrossRef]

35.   Ji, D.-P.; Zhu, Q.; Wang, S.-Q. Detailed first-principles studies on surface energy and work function of hexagonal metals. *Surf. Sci.* **2016**, *651*, 137–146. [CrossRef]

36.   Krumbein, A.D.; Malamud, H. Measurement of the effect of chlorine treatment on the work function of titanium and zirconium. *J. Appl. Phys.* **1954**, *25*, 591–592.

37.   Strömquist, J.; Bengtsson, L.; Persson, M.; Hammer, B. The dynamics of H absorption in and adsorption on Cu(111). *Surf. Sci.* **1998**, *397*, 382–394. [CrossRef]

11

The Effect of Iron Content on Glass Forming Ability and Thermal Stability of Co–Fe–Ni–Ta–Nb–B–Si Bulk Metallic Glass

Aytekin Hitit * and Hakan Şahin

Department of Materials Science and Engineering, Afyon Kocatepe University, Afyonkarahisar 03200, Turkey; hakansahin@aku.edu.tr
* Correspondence: hitit@aku.edu.tr

Academic Editor: Akihiko Hirata

Abstract: In this study, change in glass forming ability (GFA) and thermal stability of Co–Fe-based bulk metallic glasses were investigated as a function of iron content. Cylindrical samples of alloys with diameters of up to 4 mm were synthesized by a suction casting method in an arc furnace. Structures and thermal properties of the as-cast samples were determined by X-ray diffraction (XRD) and differential scanning calorimetry (DSC), respectively. It was found that the critical casting thickness of the alloys reduced as iron content was increased and cobalt content was decreased. It was determined that GFA parameters, reduced glass transition temperature (T_g/T_l) and δ ($= T_x/(T_l - T_g)$), show a very good correlation with critical casting thickness values. It was also observed that changing iron content did not effect thermal properties of the alloys.

Keywords: metallic glasses; glass forming ability; thermal analysis; microhardness

---

1. Introduction

Co- and Fe-based bulk metallic glasses have received considerable attention because of their good soft magnetic properties [1–10], ultrahigh fracture strength values [1,3,6,7,9–16] and high corrosion resistance [17,18]. Besides, they have a potential to be used in biomedical applications [17,19,20]. In addition to these attractive properties, they also have very high thermal stability [1–16,21–23]. Additionally, it is known that Co- and Fe-based metallic glasses that contain high amounts of boron can be used as precursors to develop nanocomposites that have ultrahigh hardness values [8,24,25]. Because such ultrahigh hardness values of these composites result from the precipitation of borides which form upon devitrification, it is reasonable to expect that, if the boron content of Co- or Fe-based bulk metallic glass is increased, the hardness of the composite becomes higher due to increased volume fraction of borides.

A comparison of boron contents of Co- and Fe-based bulk metallic glasses shows that, in general, boron contents of Co-based bulk metallic glasses [1,14,15] are higher than those of Fe-based bulk metallic glasses [2,4,6,7,13]. In fact, the boron content of Co-based bulk metallic glasses can be as high as 37.5 atom % [14]. However, the highest boron content of a Fe-based bulk metallic glass is 25 atom % [6]. When compared in terms of cost, it is obvious that Fe-based bulk metallic glasses are more attractive than Co-based bulk metallic glasses. Therefore, replacing cobalt with iron in a Co-based bulk metallic glass containing high amount of boron without degrading its GFA will definitely make the resulting metallic glasses more cost-effective precursors to develop composites having ultrahigh hardness values. In addition, the cost of composites can be lowered further by using low-cost industrial raw materials such as ferro-boron, ferro-niobium, and ferro-tantalum.

In this study, we report the effect of iron content on GFA and the thermal properties of a Co-based bulk metallic glass, $Co_{41}Fe_{20}Ni_2Ta_{2.75}Nb_{2.75}B_{26.5}Si_5$, which has a critical casting thickness of 4 mm [26].

For this reason, cobalt is partially replaced with iron and $Co_{41-x}Fe_{20+x}Ni_2Ta_{2.75}Nb_{2.75}B_{26.5}Si_5$ alloys (where $x = 10$, 20, and 30) were synthesized. The critical casting thicknesses, the thermal properties, and the microhardnesses of the alloys were determined as a function of iron content.

2. Materials and Methods

Co–Fe-based alloy ingots with composition of $Co_{41-x}Fe_{20+x}Ni_2Ta_{2.75}Nb_{2.75}B_{26.5}Si_5$ (where $x = 10$, 20, 30) were prepared by arc melting the mixtures of pure Co (99.8 wt %), Fe (99.9 wt %), Ni (99.9 wt %), Ta (99.9 wt %), Nb (99.8 wt %), and Si (99.9 wt %) metals and pure crystalline B (98 wt %) in a Ti-gettered ultrahigh purity argon atmosphere. Master alloys were melted three times in order to ensure homogeneity. Cylindrical samples of the alloys with diameters up to 4 mm and a length of 40 mm were synthesized by suction casting method in a vacuum arc furnace. Structures of the samples were examined by X-ray diffraction (XRD, Shimadzu XRD-6000, Kyoto, Japan) with Cu Kα radiation. The glass transition temperatures (T_g), the crystallization temperatures (T_x), the solidus temperatures (T_m), and the liquidus temperatures (T_l) of the alloys were determined by differential scanning calorimetry (DSC, Netzsch STA 449 F3, Selb, Germany) at a heating rate of 0.33 K/s. Microhardness measurements were carried out with a Vickers microhardness tester (Shimadzu HMV 2 L, Kyoto, Japan) under a load of 2.94 N. For each alloy, microhardnesses of as-cast samples were measured. Twenty microhardness measurements were carried out for each sample, and the arithmetic mean of the measurements were considered as the microhardness of the alloy.

3. Results

XRD patterns of samples are given in Figure 1. The base alloy, $Co_{41}Fe_{20}Ni_2Ta_{2.75}Nb_{2.75}B_{26.5}Si_5$, has a critical casting thickness of 4 mm. For the casting thickness of 5 mm, $(Co,Fe)_{21}Ta_2B_6$ and $(Co,Fe)_2B$ phases form. The $Co_{31}Fe_{30}Ni_2Ta_{2.75}Nb_{2.75}B_{26.5}Si_5$ alloy has a critical casting thickness of 3 mm. For the casting thickness of 4 mm, the precipitation of $(Co,Fe)_2B$ phase was observed for this alloy. The critical casting thicknesses of alloys $Co_{21}Fe_{40}Ni_2Ta_{2.75}Nb_{2.75}B_{26.5}Si_5$ and $Co_{11}Fe_{50}Ni_2Ta_{2.75}Nb_{2.75}B_{26.5}Si_5$ were found to be 2 mm and 0.5 mm, respectively. For both of these alloys, the precipitation of the $(Co,Fe)_2B$, $(Co,Fe)_{16}Ta_6Si_7$, and Fe_3Si phases was observed in the samples having diameters larger than the critical casting thicknesses.

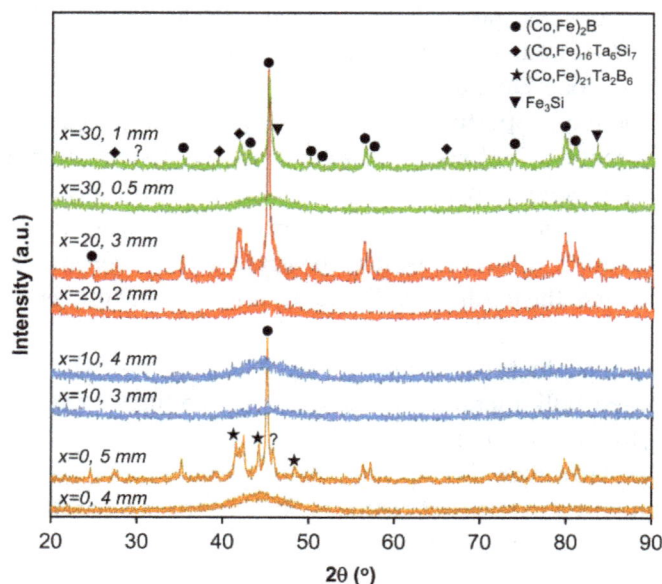

Figure 1. XRD patterns of the $Co_{41-x}Fe_{20+x}Ni_2Ta_{2.75}Nb_{2.75}B_{26.5}Si_5$ ($x = 10$, 20, and 30) alloys.

Thermal properties of the alloys were determined by DSC (Figure 2). During heating, all the DSC traces showed an endothermic event, which is the indication of the glass transition and followed by exothermic reactions, which are the signs of crystallization of the glassy structure. T_g and T_x of the base alloy, $Co_{41}Fe_{20}Ni_2Ta_{2.75}Nb_{2.75}B_{26.5}Si_5$, are 891 and 947 K, respectively [21]. The T_g of $Co_{31}Fe_{30}Ni_2Ta_{2.75}Nb_{2.75}B_{26.5}Si_5$ alloy is 890 K. Additionally, the T_x of $Co_{31}Fe_{30}Ni_2Ta_{2.75}Nb_{2.75}B_{26.5}Si_5$ alloy is 957 K, which is about 10 K higher than the T_x of $Co_{41}Fe_{20}Ni_2Ta_{2.75}Nb_{2.75}B_{26.5}Si_5$ alloy. The T_g of $Co_{21}Fe_{40}Ni_2Ta_{2.75}Nb_{2.75}B_{26.5}Si_5$ and $Co_{11}Fe_{50}Ni_2Ta_{2.75}Nb_{2.75}B_{26.5}Si_5$ alloys are found to be 895 and 894 K, respectively. Moreover, the T_x of $Co_{21}Fe_{40}Ni_2Ta_{2.75}Nb_{2.75}B_{26.5}Si_5$ and $Co_{11}Fe_{50}Ni_2Ta_{2.75}Nb_{2.75}B_{26.5}Si_5$ alloys are determined as 960 and 964 K, respectively. In addition, the T_l of the base alloy is 1443 K [26], and the T_l of 30, 40, and 50 atom % iron-containing alloys are 1458, 1488, and 1532 K, respectively. Thermal properties of the alloys are given in Table 1. In addition, the microhardnesses of the alloys were determined to be around 1200 H_v.

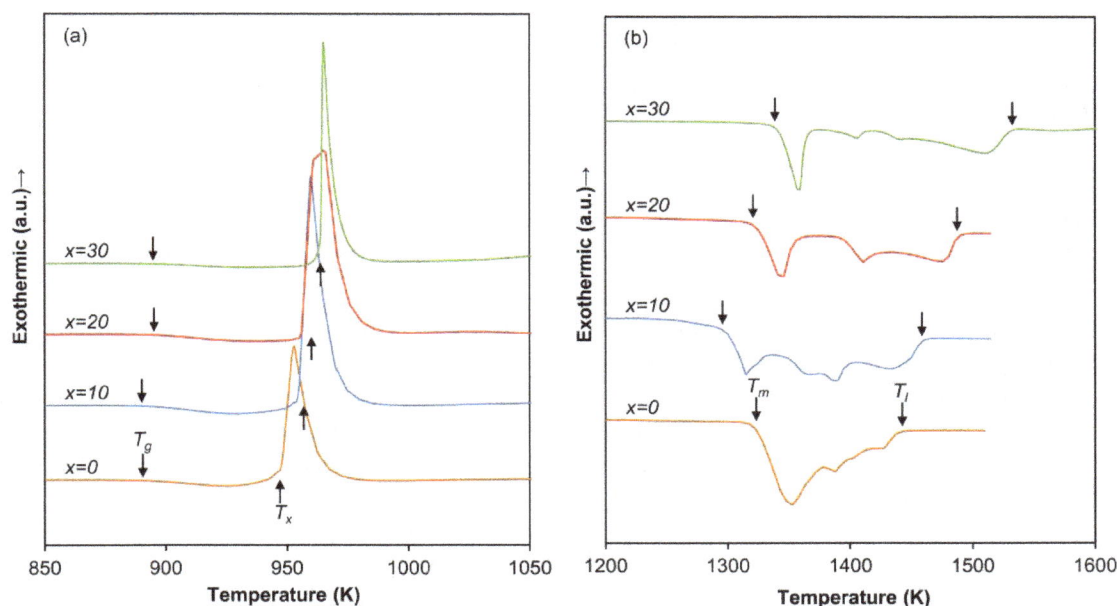

Figure 2. DSC curves of the $Co_{41-x}Fe_{20+x}Ni_2Ta_{2.75}Nb_{2.75}B_{26.5}Si_5$ ($x = 10, 20$ and 30) alloys. (**a**) low temperature; (**b**) high temperature.

Table 1. Thermal properties (T_g, T_x, T_l, T_m), parameters for GFA, critical casting thickness and microhardnesses of Co–Fe–Ni–Ta–Nb–B–Si alloys.

Alloy	T_g (K)	T_x (K)	T_m (K)	T_l (K)	T_g/T_l	$T_x/(T_l-T_g)$	$T_x/(T_g+T_l)$	D_{max} (mm)	H_v
$Co_{41}Fe_{20}Ni_2Ta_{2.75}Nb_{2.75}B_{26.5}Si_5$ [26]	891	947	1323	1443	0.617	1.716	0.406	4	1197
$Co_{31}Fe_{30}Ni_2Ta_{2.75}Nb_{2.75}B_{26.5}Si_5$	890	957	1295	1458	0.610	1.685	0.408	3	1242
$Co_{21}Fe_{40}Ni_2Ta_{2.75}Nb_{2.75}B_{26.5}Si_5$	895	960	1318	1488	0.601	1.619	0.403	2	1240
$Co_{11}Fe_{50}Ni_2Ta_{2.75}Nb_{2.75}B_{26.5}Si_5$	894	964	1343	1532	0.584	1.511	0.397	0.5	1238

4. Discussion

The critical casting thickness decreases as iron content is increased. It is quite obvious that the reduction in critical casting thickness results from the fact that the T_l of the alloys increases with iron content. For a constant T_g, if liquidus temperature increases, the minimum cooling rate necessary to obtain a completely amorphous structure also increases. As a result, the casting thickness must be decreased to achieve this cooling rate. Well-known GFA parameters, reduced glass transition temperature (T_g/T_l) [27] and δ ($= T_x/(T_l - T_g)$), show a very good correlation with the critical casting thickness values (Figure 3). However, the correlation between the GFA parameter $T_x/(T_g + T_l)$ and the critical casting thickness values is not as satisfactory as those of the others.

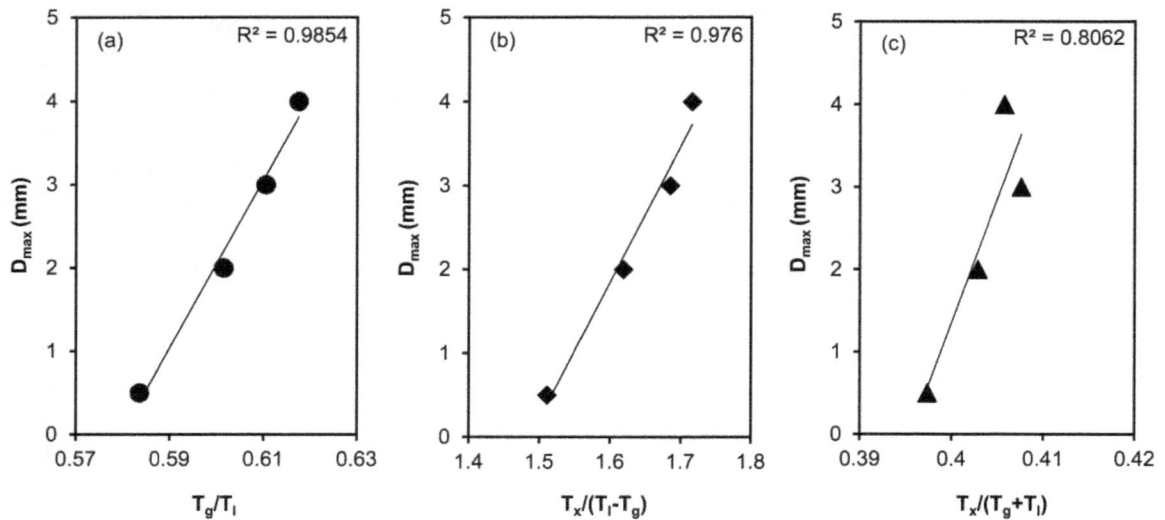

Figure 3. Relationship between the critical casting thickness (D_{max}) for the formation of a glassy phase and GFA parameters. (**a**) Reduced glass transition temperature (T_g/T_l); (**b**) $T_x/(T_l - T_g)$; (**c**) $T_x/(T_g + T_l)$.

$(Co,Fe)_2B$, $(Co,Fe)_{16}Ta_6Si_7$, and Fe_3Si phases form during cooling. It is shown that $(Co,Fe)_2B$ phase forms at temperatures higher than temperatures at which the $(Co,Fe)_{16}Ta_6Si_7$ phase forms [25]. In other words, $(Co,Fe)_2B$ is the first phase that precipitates during cooling. Since the melting temperature of the Fe_2B phase is higher than the Co_2B phase, increasing iron content increases the melting temperature of the $(Co,Fe)_2B$ phase. As a result, the liquidus temperatures of the alloys increase.

Glass transition temperatures of the alloys do not change with iron content. However, crystallization temperatures increase slightly. Similarly, microhardnesses of the alloys remain almost constant. Cohesive energies of Fe–Fe and Co–Co bonds are 413 and 424 kJ/mol, respectively [28]. It is quite reasonable to assume that cohesive energy of Fe–Co bond is close to these values. When the iron content of the alloys is increased, the number of Co–Co bonds decreases, but the number of Fe–Co bonds increases. Since cohesive energies of these bond are almost the same, the total cohesive energy of the amorphous structure remains constant. For this reason, the thermal properties and the microhardnesses of the alloys do not change.

The $Co_{21}Fe_{40}Ni_2Ta_{2.75}Nb_{2.75}B_{26.5}Si_5$ alloy has a reasonably high iron content and critical casting thickness. Therefore, it is believed that it can used as a precursor to develop composites with ultrahigh hardness values at a low cost.

5. Conclusions

The following conclusions can be reached from this study:

1. The critical casting thicknesses of the alloys decrease with iron content due to the increase in liquidus temperatures.
2. The critical casting thicknesses of the alloys show a very good correlation with reduced glass transition temperature, T_g/T_l and $T_x/(T_l - T_g)$.
3. Thermal properties and microhardnesses of the alloys do not change with iron content because of the fact that cohesive energies of the Co–Co- and Fe–Fe bonds are almost the same.

Acknowledgments: This study was supported by grant No. 10.MUH.10 from the Research Council of Afyon Kocatepe University.

Author Contributions: Aytekin Hitit and Hakan Şahin synthesized the alloys and cast the samples. Hakan Şahin performed XRD and DSC experiments and microhardness measurements. Aytekin Hitit and Hakan Şahin analyzed the results. Aytekin Hitit wrote and edited the paper.

Conflicts of Interest: The authors declare no conflict of interest.

References

1. Inoue, A.; Shen, B.L.; Koshiba, H.; Kato, H.; Yavari, A.R. Ultra-high strength above 5000 MPa and soft magnetic properties of Co–Fe–Ta–B bulk glassy alloys. *Acta Mater.* **2004**, *52*, 1631–1637. [CrossRef]

2. Liu, D.Y.; Sun, W.S.; Zhang, H.F.; Hu, Z.Q. Preparation, thermal stability and magnetic properties of Fe–Co–Ni–Zr–Mo–B bulk metallic glass. *Intermetallic* **2004**, *12*, 1149–1152. [CrossRef]

3. Song, D.S.; Kim, J.H.; Fleury, E.; Kim, W.T.; Kim, D.H. Synthesis of ferromagnetic Fe-based bulk glassy alloys in the Fe–Nb–B–Y system. *J. Alloy. Compd.* **2005**, *389*, 159–164. [CrossRef]

4. Lee, S.; Kato, H.; Kubota, T.; Yubuta, K.; Makino, A.; Inoue, A. Excellent Thermal Stability and Bulk Glass Forming Ability of Fe–B–Nb–Y Soft Magnetic Metallic Glass. *Trans. Jpn. Inst. Met.* **2008**, *49*, 506–512. [CrossRef]

5. Tiberto, P.; Piccin, R.; Lupu, N.; Chiriac, H.; Baricco, M. Magnetic properties of Fe–Co-based bulk metallic glasses. *J. Alloy. Compd.* **2009**, *483*, 608–612. [CrossRef]

6. Jia, F.; Zhang, W.; Zhang, X.; Xie, G.; Kimura, H.; Makino, A.; Inoue, A. Effect of Co concentration on thermal stability and magnetic properties of (Fe,Co)–Nb–Gd–B glassy alloys. *J. Alloy. Compd.* **2010**, *504S*, 129–131. [CrossRef]

7. Dong, Y.; Wang, A.; Man, Q.; Shen, B. $(Co_{1-x}Fe_x)_{68}B_{21.9}Si_{5.1}Nb_5$ bulk glassy alloys with high glass-forming ability, excellent soft-magnetic properties and super high fracture strength. *Intermetallics* **2012**, *23*, 63–67. [CrossRef]

8. Han, J.J.; Wang, C.; Kou, S.; Liu, X. Thermal stability, crystallization behavior, Vickers hardness and magnetic properties of Fe−Co−Ni−Cr−Mo−C−B−Y bulk metallic glasses. *Trans. Nonferr. Met. Soc. China* **2013**, *23*, 148–155. [CrossRef]

9. Li, J.W.; He, A.N.; Shen, B.L. Effect of Tb addition on the thermal stability, glass-forming ability and magnetic properties of Fe–B–Si–Nb bulk metallic glass. *J. Alloy. Compd.* **2014**, *586*, 46–49. [CrossRef]

10. Li, J.W.; Estevez, D.; Jiang, K.M.; Yang, W.M.; Man, Q.K.; Chang, C.T.; Wang, X.M. Electronic-structure origin of the glass-forming ability and magnetic properties in Fe–RE–B–Nb bulk metallic glasses. *J. Alloy. Compd.* **2014**, *617*, 332–336. [CrossRef]

11. Inoue, A.; Shen, B.L.; Chang, C.T. Super-high strength of over 4000 MPa for Fe-based bulk glassy alloys in $[(Fe_{1-x}Co_x)_{0.75}B_{0.2}Si_{0.05}]_{96}Nb_4$ system. *Acta Mater.* **2004**, *52*, 4093–4099. [CrossRef]

12. Inoue, A.; Shen, B.L.; Chang, C.T. Fe- and Co-based bulk glassy alloys with ultrahigh strength of over 4000 MPa. *Intermetallics* **2006**, *14*, 936–944. [CrossRef]

13. Chang, Z.Y.; Huang, X.M.; Chen, L.Y.; Ge, M.Y.; Jiang, Q.K.; Nie, X.P.; Jiang, J.Z. Catching Fe-based bulk metallic glass with combination of high glass forming ability, ultrahigh strength and good plasticity in Fe–Co–Nb–B system. *Mater. Sci. Eng. A* **2009**, *517*, 246–248. [CrossRef]

14. Dun, C.; Liu, H.; Shen, B. Enhancement of plasticity in Co–Nb–B ternary bulk metallic glasses with ultrahigh strength. *J. Non Cryst. Solids* **2012**, *358*, 3060–3064. [CrossRef]

15. Wang, J.; Wang, L.; Guan, S.; Zhu, S.; Li, R.; Zhang, T. Effects of boron content on the glass-forming ability and mechanical properties of Co–B–Ta glassy alloys. *J. Alloy. Compd.* **2014**, *617*, 7–11. [CrossRef]

16. Yazici, Z.Ö.; Hitit, A.; Yalcin, Y.; Ozgul, M. Effects of Minor Cu and Si Additions on Glass Forming Ability and Mechanical Properties of Co-Fe-Ta-B Bulk Metallic Glass. *Met. Mater. Int.* **2016**, *22*, 50–57. [CrossRef]

17. Li, S.; Wei, Q.; Li, Q.; Jiang, B.; Chen, Y.; Sun, Y. Development of Fe-based bulk metallic glasses as potential biomaterials. *Mater. Sci. Eng. C* **2015**, *52*, 235–241. [CrossRef] [PubMed]

18. Souza, C.A.C.; Ribeiro, D.V.; Kiminami, C.S. Corrosion resistance of Fe-Cr-based amorphous alloys: An overview. *J. Non Cryst. Solids* **2016**, *442*, 56–66. [CrossRef]

19. Zhou, Z.; Wei, Q.; Li, Q.; Jiang, B.; Chen, Y.; Sun, Y. Development of Co-based bulk metallic glasses as potential biomaterials. *Mater. Sci. Eng. C* **2016**, *69*, 46–51. [CrossRef] [PubMed]

20. Li, H.F.; Zheng, Y.F. Recent advances in bulk metallic glasses for biomedical applications. *Acta Biomater.* **2016**, *36*, 1–20. [CrossRef] [PubMed]

21. Men, H.; Pang, S.J.; Zhang, T. Thermal stability and microhardness of new Co-based bulk metallic glasses. *Mater. Sci. Eng. A* **2007**, *449–451*, 538–540. [CrossRef]

22. Hitit, A.; Talaş, Ş.; Kara, R. Effects of silicon and chromium additions on glass forming ability and microhardness of Co-based bulk metallic glasses. *Indian J. Eng. Mater. Sci.* **2014**, *21*, 111–115.

23. Li, J.; Law, J.Y.; Ma, H.; He, A.; Man, Q.; Men, H.; Huo, J.; Chang, C.; Wang, X.; Li, R.W. Magnetocaloric effect in Fe–Tm–B–Nb metallic glasses near room temperature. *J. Non Cryst. Solids* **2015**, *425*, 114–117. [CrossRef]

24. Fornell, J.; González, S.; Rossinyol, E.; Suriñach, S.; Baro, M.D.; Louzguine-Luzgin, D.V.; Perepezko, J.H.; Sort, J.; Inoue, A. Enhanced mechanical properties due to structural changes induced by devitrification in Fe–Co–B–Si–Nb bulk metallic glass. *Acta Mater.* **2010**, *58*, 6256–6266. [CrossRef]

25. Hitit, A.; Geçgin, M.; Öztürk, P. Effect of Annealing on Microstructure and Microhardness of Co–Fe–Ni–Ta–B–Si Bulk Metallic Glass. *J. Mater. Sci. Technol.* **2015**, *31*, 148–152. [CrossRef]

26. Çolak, F. Synthesis and Characterization of Nickel, Silicon and Niobium Containing Cobalt-Iron Based Bulk Metallic Glasses. Ph.D. Thesis, Afyon Kocatepe University, Afyonkarahisar, Turkey, 2011.

27. Turnbull, D. Under What Conditions can a Glass be Formed? *Contemp. Phys.* **1969**, *10*, 473–478. [CrossRef]

28. Kittel, C. Energy Bands. In *Introduction to Solid State Physics*, 8th ed.; Stuart, C., Patricia, M.F., Martin, B., Eds.; John Wiley and Sons, Inc.: Hoboken, NJ, USA, 2005; Volume 7, pp. 161–182.

12

Weibull Statistical Reliability Analysis of Mechanical and Magnetic Properties of FeCuNb$_x$SiB Amorphous Fibers

Ruixuan Li *, Haiying Hao, Yangyong Zhao and Yong Zhang

State Key Laboratory for Advanced Metals and Materials, University of Science and Technology Beijing, No. 30, Xueyuan Road, Beijing 100083, China; haiying-0823@163.com (H.H.); zhaoyangyongwd@126.com (Y.Z.); drzhangy@ustb.edu.cn (Y.Z.)
* Correspondence: liruixuan1208@163.com

Academic Editor: K.C. Chan

Abstract: Glass-coated Fe$_{76.5-x}$Cu$_1$Nb$_x$Si$_{13.5}$B$_9$ (x = 0, 1, 2, 3, 3.5) fibers were successfully fabricated by a modified Taylor method. The fibers showed circular morphology and smooth surface with different diameters. The mechanical properties of the fibers were evaluated and the Weibull statistical analysis has been introduced to characterize the strength reliability of the fibers, with the modulus m of the amorphous fibers reaching above 20. The magnetic properties were also studied. Lower coercivity was found for the fibers with amorphous, nanocrystalline, and microcrystalline structures rather than that for the coarse crystalline ones. The glass-coated FeCuNbSiB amorphous fibers have excellent comprehensive performance compared with the other kind of fibers.

Keywords: amorphous fibers; modified Taylor technique; mechanical properties; Weibull statistical analysis

1. Introduction

Fe-based soft magnetic materials have developed recent years. Because of the low cost and their excellent soft magnetic properties—e.g., high saturation magnetization and low coercivity—they have received considerable attention [1]. The glass-coated fibers with a diameter ranging from a few microns to a few hundred microns can be prepared by a Taylor-type technique—i.e., a glass-coated melt spinning method. The preparation process allows the control of microstructures and geometrical characteristics by adjusting the technological parameters, such as the feed-in rate of the glass-tube and the winding speed [2,3]. Previous investigations have focused on the magnetic properties. Most typically, the composition of FeCuNbSiB with trademark "Finemet" is widely studied [4–6] while the studies of the mechanical properties are less reported, relatively. Most of the works focus on the mechanical performance of the amorphous bare fibers. However, the coated-glass offers an advantage of high corrosion resistant, high temperature resistance, and high insulation in many practical applications [7,8]. This paper mainly studies the mechanical and magnetic properties of the glass-coated Fe-based fibers and the effect of structures to those properties.

2. Experiment

The ingots of Fe-based system with the compositions of Fe$_{76.5-x}$Cu$_1$Nb$_x$Si$_{13.5}$B$_9$ (x = 0, 1, 2, 3, 3.5) (at %) were prepared by arc-melting appropriate amounts of Fe (99.9%), Cu (99.9%), Nb (99.999%), Si (99.95%), and industrial Fe–B alloy (which consists of 78.543 wt % Fe, 20.41 wt % B) under a Ti-gettered argon atmosphere.

The fibers were produced by a modified-Taylor technique as shown in Figure 1 [9,10]. Firstly, about 1 g of master alloy was placed in a closed-end Boron-added Pyrex glass tube with 10 mm in diameter and 1 mm in thickness. Then the tube and alloy are heated by an induction coil to around 1300 °C. Lastly, the softened glass tube is drawn into a fine capillary by a pre-prepared glass bar with cutting-edge and rolled onto a rotating cylinder using a motor coupled to the winding system. A water jet was directed at the fibers in order to enhance the quenching. As the diameter cannot be controlled exactly, the glass-coated Fe-based fibers which have diameters ranging from 10–30 μm have been obtained by the equipment (Figure 2) developed in our group with the experimental parameters controlled into the same. One qualified fiber of each of the five compositions is selected and then they are cut into several pieces for different experiments. Although the measuring error and different influential factors may lead to different diameters even in one fiber, the diameter of each fiber is relatively small and the fibers are all completely cooled. Consequently, the effect of the diameter and the resulting cooling rate can be ignored.

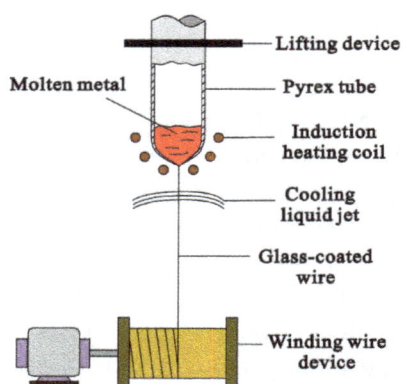

Figure 1. Schematic diagram of the glass-coated melt-spinning process [8,9].

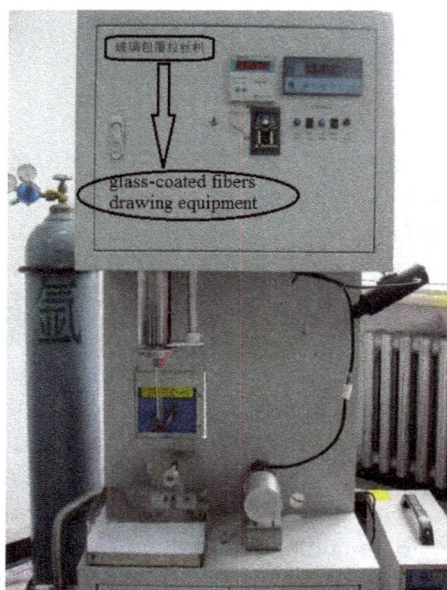

Figure 2. The picture of the glass-coated fibers drawing equipment.

The structure of the five different composition fibers was studied by an X-ray diffraction (XRD, TTRIII, Tokyo, Japan) using Cu K_α radiation. The surface and fracture morphology of the fibers were examined by scanning electron microscopy (SEM, Zeiss Supra 55, Oberkochen, Germany). The tension tests were performed by using an Instron-5900 machine at room temperature. The gauge length was set as 20 mm and the specimen was elongated at a rate of 1 mm/min. Figure 3 shows the geometry of the tensile sample. The magnetic characterization was carried out using a vibration sample magnetometer (VSM, Quantum Design, San Diego, CA, USA).

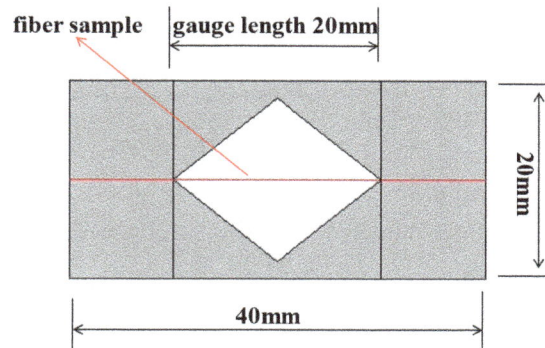

Figure 3. Schematic diagram of the tensile sample.

3. Results

3.1. Surface Morphology and Structure of the Fe-Based Fibers

Figure 4 shows the scanning electron micrograph of typical glass-coated fibers ($x = 2$) where its composite characteristic can be clearly seen. The fiber shows circular morphology and smooth surface and the glass layer is only 0.5 μm, implying that the modified Taylor technique is very effective in producing the high quality fibers [11,12].

Figure 5 shows the XRD analysis for the different Nb-content fibers. The broadened XRD patterns of $x = 0$, 2 and 3.5 samples show amorphous nature, while there are BCC α-Fe crystalline peaks in the $x = 1$ and 3 samples. The degree of crystallization of the $x = 1$ sample is higher than that of the $x = 3$ one. Ignoring the effect of diameters, experimental investigations show that the different structures can be related to the composition. It has also been proven that the appropriate experimental parameters, such as the high drawing speed and reasonable cooling water position, tend to form fully amorphous structures [13].

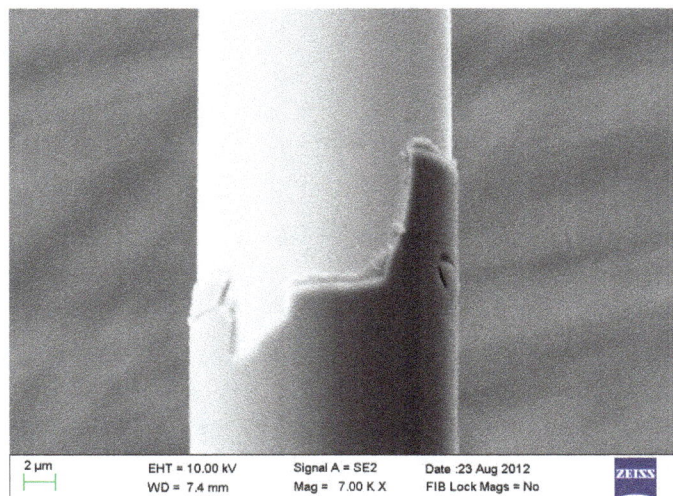

Figure 4. Scanning electron microscopy (SEM) micrographs of the glass–coated fibers ($x = 2$).

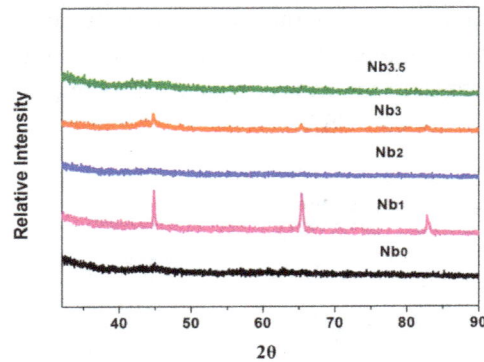

Figure 5. X-ray diffraction patterns of $Fe_{76.5-x}Cu_1Nb_xSi_{13.5}B_9$ (x = 0, 1, 2, 3, 3.5) fibers.

Adding Cu and Nb elements to FeSiB alloy tends to form nanocrystalline structures. Copper can promote the nucleation rate of grains, and Niobium can hinder their growth and make grain distribution homogeneous [14]. It is not certain whether there is nanocrystalline precipitation in the amorphous fibers from the XRD pattern. The longitudinal section morphology of fibers was observed in the SEM. In order to contrast the crystal and amorphous fibers, two group of fibers (x = 0, x = 3) were chosen. Ten fibers of each group were selected and prepared with cool mosaic method. A few hours later, they were burnished using sand paper and mechanical polished. It can be seen that carbon was sprayed on the sample surface. Figure 6 shows the longitudinal section morphology of the fibers (a, x = 3, b, x = 0). It is clear that there are circular crystalline grains generating in the x = 3 fibers and the size of the grain is less than one micron. For the fibers of x = 0, many fine nanocrystalline grains shaped as thin strips can be seen in the amorphous matrix. It is easy to form microcrystalline or nanocrystalline for FeCuNbSiB mircowires produced by the glass-coated melt spinning method.

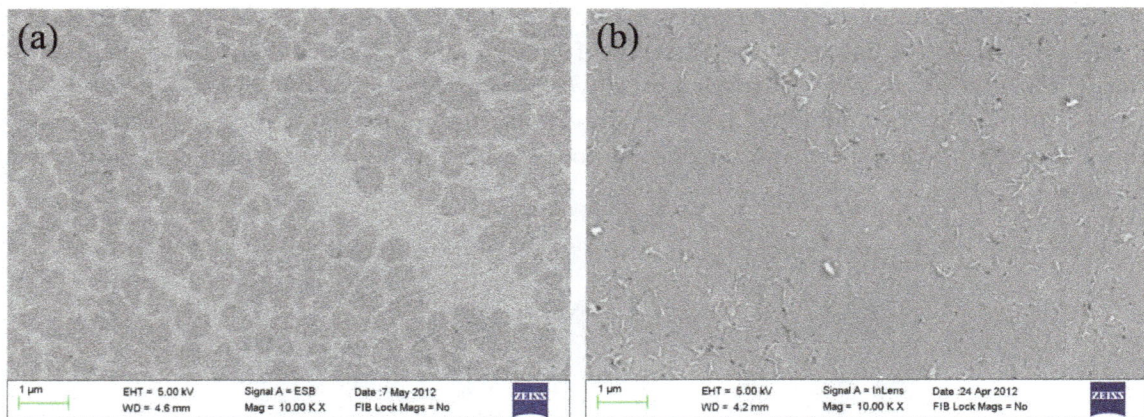

Figure 6. Longitudinal section morphology of fibers (**a**) x = 3 (**b**) x = 0.

3.2. Tensile Tests Analysis

The mechanical properties of the glass-coated fibers were measured in tension tests. Ten samples with relatively the same diameter in each composition were made. The results were obtained by computer automatically. Then five curves in each group with relatively small fluctuation were selected. Figure 7 shows the typical tensile stress-strain curves of $Fe_{76.5-x}Cu_1Nb_xSi_{13.5}B_9$ (x = 0, 1, 2, 3, 3.5) (at %) fibers. There is no plastic deformation stage following the elastic deformation, which identifies typical brittle fracture.

Figure 8 shows the strength distribution of the glass-coated fibers with different content of Nb. The amorphous fibers (x = 0, 2, 3.5) show higher tensile fracture strength between 2.5 GPa to 3.0 GPa than that of the crystalline ones ($x = 1, 3$). The fibers of $x = 1$ show the worst mechanical performance with the tensile fracture reaching about 1.5 GPa and elongation below 2%, as shown in Figure 6b. The tensile strength of the glass-coated amorphous fibers in this test are close to that of general Fe-based bare fibers [15].

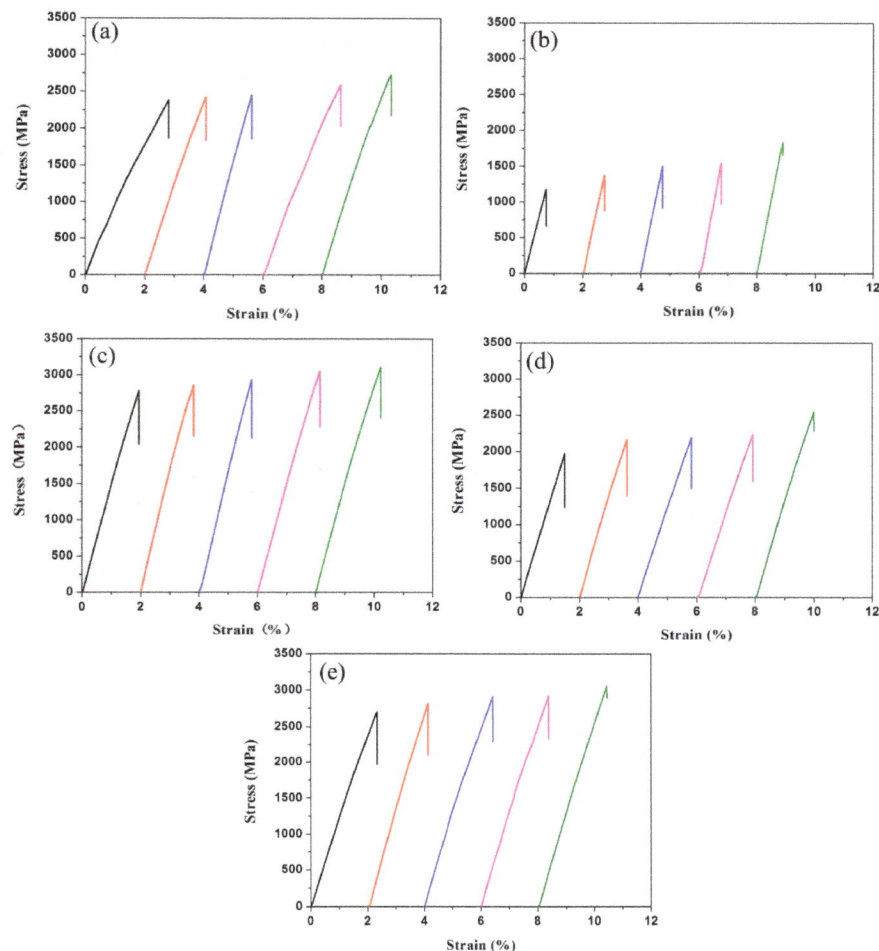

Figure 7. Tensile stress-strain curves of $Fe_{76.5-x}Cu_1Nb_xSi_{13.5}B_9$ fibers: **(a)** $x = 0$, $d = 25$ μm; **(b)** $x = 1$, $d = 15$ μm; **(c)** $x = 2$, $d = 15$ μm; **(d)** $x = 3$, $d = 25$ μm; **(e)** $x = 3.5$, $d = 15$ μm.

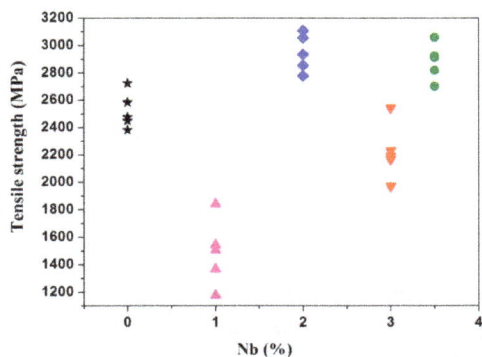

Figure 8. The strength distribution of the glass-coated fibers with different content of Nb.

3.3. Weibull Statistical Analysis

In the tensile tests, we assume that the cross section of the fibers is a perfect circle with uniform diameter. In fact, the fibers inevitably have geometric fluctuation and surface defects. In addition, because of the small sizes of the fibers, it is difficult to ensure that the process of the sample preparation, clamping, and loading operation is completely the same. A fluctuation in fracture strength is observed.

Weibull-statistical analysis has been introduced to characterize the strength reliability of the glass-coated fibers [16–18]. The Weibull equation is usually used to determine the Weibull modulus in the double logarithmic form

$$\ln\{\ln[1/(1-P)]\} = \ln V + m \ln \sigma - m \ln \sigma_0 \tag{1}$$

where m is the Weibull-modulus that represents the strength reliability, σ_0 is a scaling parameter, and V is the volume of the tested samples. P is the fracture probability at a given uniaxial stress, and can be calculated using the equation

$$P_i = (i - 0.5)/n \tag{2}$$

where n is the total number of the samples tested, and i is the sample ranking in ascending order of the failure stress [17].

Figure 9 shows the Weibull plots of the samples calculated by Equation (1). The Weibull plots for the raw data of these samples are represented very well by a linear least-square fit, as can be seen in Figure 9. The Weibull modulus is 20.54, 25.06, and 25.30 for the amorphous fibers ($x = 0, 2, 3.5$), and is 7.07 and 11.63 for the crystalline ones ($x = 1, 3$), respectively. The Weibull modulus m actually reflects the reliability of the test samples. A higher m value represents a narrow dispersion of the fracture strength, and thus a higher reliability. The results show that the strength reliability of the amorphous fibers is higher than that of the crystalline ones in the texts. Crystalline fibers which have grain-boundaries and surface defects lead to larger fluctuations in the fracture strength. The m value for the glass-coated amorphous Fe-based fibers in the present study is close to that of the Mg-based [16] and smaller than that of the Co-based [17] and Zr-based amorphous fibers [18].

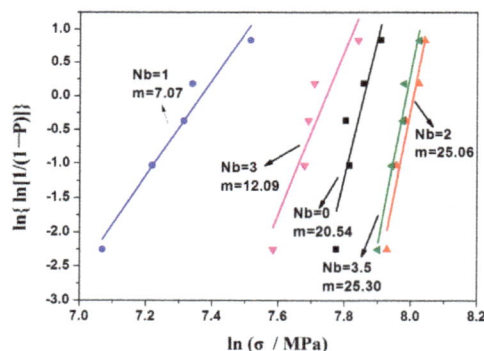

Figure 9. Weibull plots of the tensile fracture strength and corresponding fits to the data for the five types FeCuNbSiB fibers showing an inclined fracture.

3.4. SEM Analysis of Fractured Samples

Figure 10 illustrates the fracture morphology observed in the sample of $x = 2$. The fracture surfaces exhibit two morphologically distinct zones. One zone is smooth and featureless, while the other is a pronounced vein pattern, as can be seen in the Figure 10a. The tensile fracture angle θ between the tensile axis and the fracture plane can be readily measured on the surface of the sample, as marked in Figure 10b. It is found that the tensile fracture angle θ is equal to 54°, which obviously deviates from the angle of the maximum shear stress plane (45°). The glass layer is only 0.5 μm, which can be seen from Figure 3, only accounting for 5.88% of the total fiber. This phenomenon can easily explain the reason for the glass-coated Fe-based fibers with such a high tensile strength of about 3 GPa, which is close to the tensile strength of the general Fe-based bare fiber and exceeds that of the glass-coated fibers reported.

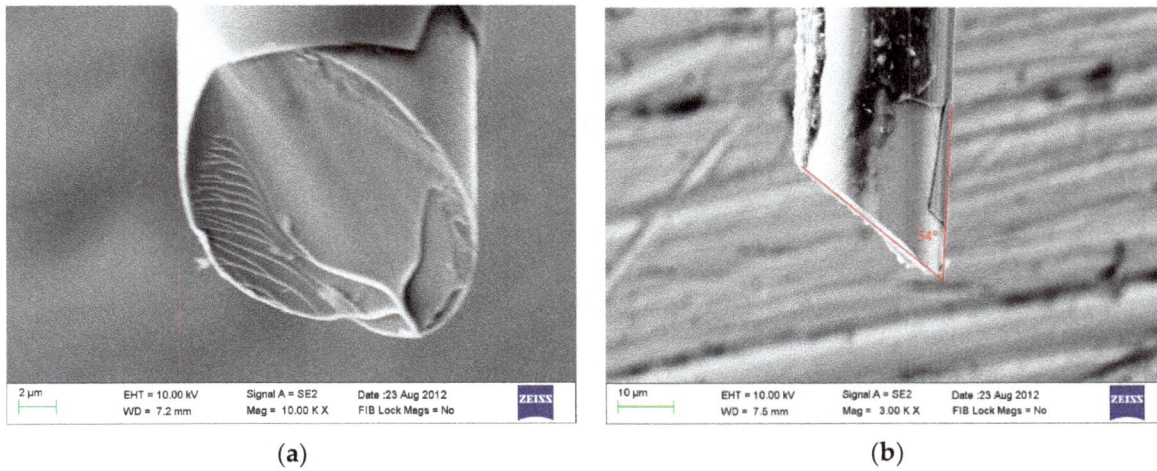

Figure 10. SEM micrograph of the fractured samples. (**a**) Fracture surface shows vein pattern; (**b**) fracture angle is about $54°$.

3.5. Magnetic Properties of the Samples

The room-temperature hysteresis loops for the five different compositions in the series of FeCuNb$_x$SiB fibers are presented in Figure 11. As observed, the fibers of $x = 0$, 3, and 3.5 show long narrow shapes, while the fibers of $x = 1$ and 2 show nearly rectangular shape and are still not saturated in the applied magnetic field of 500 Oe. It can be inferred that the magnetic behavior is greatly influenced by the composition and structure of the fibers.

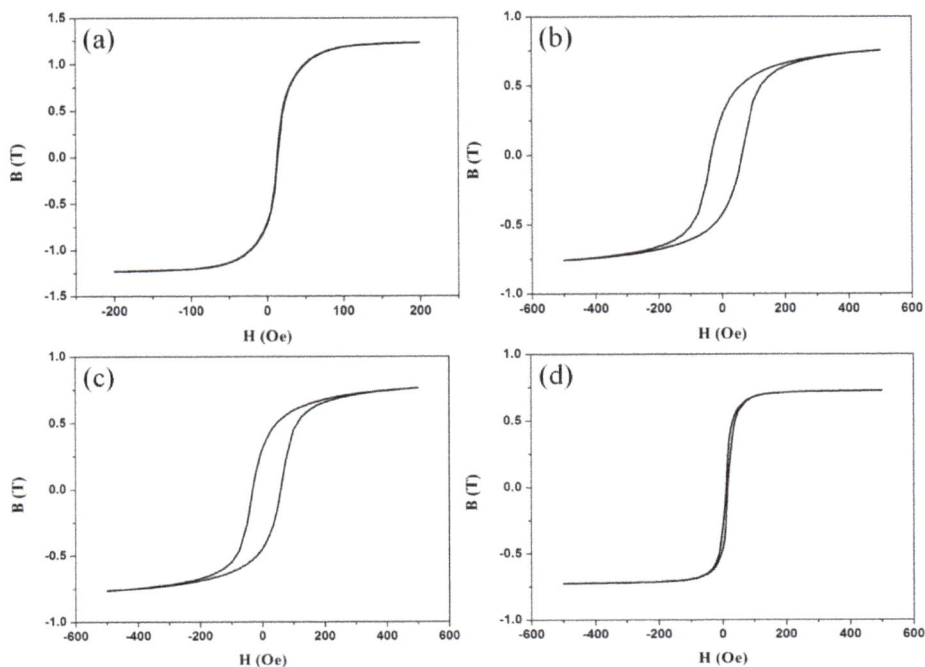

Figure 11. *Cont.*

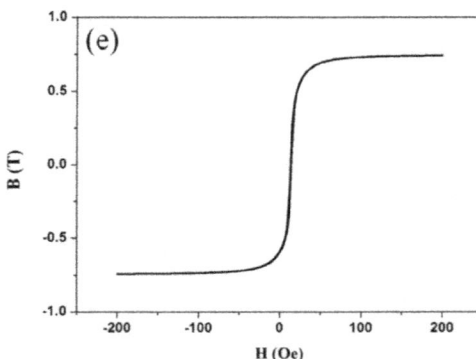

Figure 11. Hysteresis loops of different compositions of $Fe_{76.5-x}Cu_1Nb_xSi_{13.5}B_9$ fibers, (**a**) $x = 0$, $d = 25$ μm; (**b**) $x = 1$, $d = 30$ μm; (**c**) $x = 2$, $d = 30$ μm; (**d**) $x = 3$, $d = 20$ μm; (**e**) $x = 3.5$, $d = 15$ μm.

The impact of Nb on the saturation induction density Bs, coercivity Hc and remanence ratio Mr/Ms of the Fe-based wires is shown in Figure 12. The coercivity result corresponds to the XRD analysis except the fibers of $x = 2$. The amorphous fibers exhibit particularly low coercivity value. While in coarse crystalline samples, the value is almost dozens of times more than that of amorphous and nanocrystalline ones, where the major controlling factor is changed and magneto-crystalline anisotropy is reduced. The change rule of remanence ratio Mr/Ms is the same with that of the coercivity. However, the content of Nb has different influence on the saturation induction density. It is clear that the sample of $x = 0$ shows the highest value of 1.2 T. The remaining samples all show lower value between 0.7 T and 0.8 T.

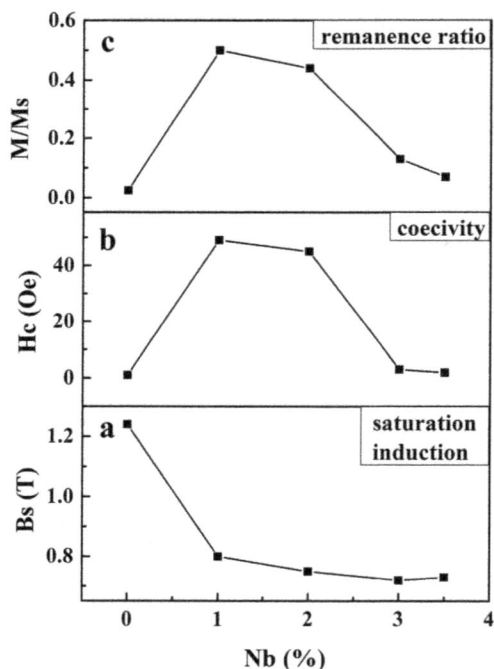

Figure 12. The content of Nb dependence of Bs (**a**), Hc (**b**), and Mr/Ms (**c**).

The effect of the glass layer on the magnetic properties was also studied. The glass was removed by hydrofluoric acid with the mass fraction of 20%. Figure 13 shows the hysteresis loops of the as-cast ($x = 3$) and glass-removed fibers with the diameters of 30 μm. The coercivity decreased from 3.26 Oe to 1.20 Oe. In the practical applications, we can consider whether to remove the glass or not according to different conditions.

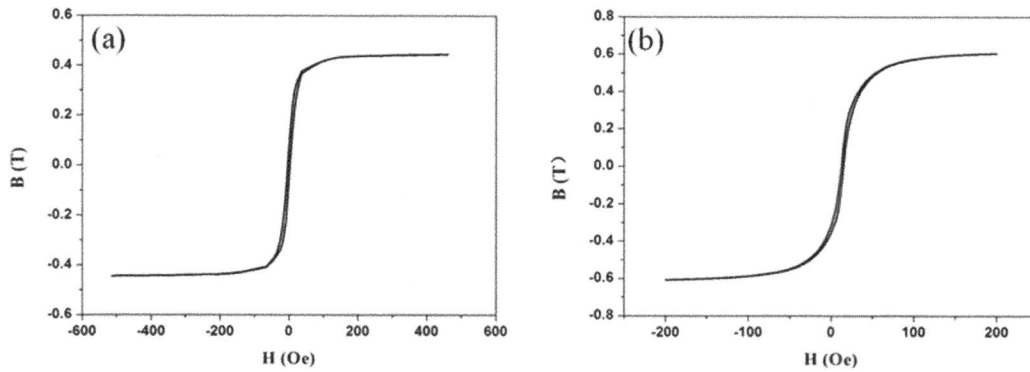

Figure 13. Hysteresis loops of (**a**) as-cast and (**b**) glass-removed $Fe_{73.5}Cu_1Nb_3Si_{13.5}B_9$ fibers.

4. Discussion

It has been verified above that the glass-coated FeCuNbSiB fibers made of modified-Taylor method show excellent properties such as high strength and low coercivity, and the high-strength effect is highlighted especially in amorphous ones. The amorphous fibers have caught much attention for their high fracture strength and toughness. The tension tests of the melt-extracted $Cu_{48}Zr_{48}Al_4$ amorphous microwires were carried out by Sun et al. [19], indicating that the fracture strength ranged from 1724 MPa to 1937 MPa, and the average strength and standard deviation were 1836 MPa and 56 MPa, respectively. In contrast, the strength value is highly improved after the glass coat is added. It should be noted that the glass coats produced by the glass-coated drawing fiber equipment show circular morphology and a smooth surface, which effectively impede the rapid expansion of the shear band in the amorphous phase. As a result, the abrupt failure of one dominant shear band and highly localized deformation can be put off. This can explain the phenomenon of the present glass-coated FeCuNbSiB amorphous fibers exhibiting much higher strength of 2500–3000 MPa.

As for the low-coercivity effect, the microstructure of fibers, combined with the morphology, play an important role. In coarse crystal fibers which have exactly the similar particle sizes to that of the domain wall width, the anisotropies act a pivotal part and impede the magnetic domain wall in rotation or mobility. As for the amorphous or nanocrystalline structure, however, the ferromagnetic exchange interactions start to dominate and thus average out the magneto-crystalline anisotropy which randomly fluctuates on a scale much smaller than the domain wall width [20]. Consequently, there is no net anisotropy effect on the magnetization process, leading to a smaller coercive force value.

In practical application, the glass-coated fibers show such properties as anticorrosion and high-temperature resistance while, when the glass is removed, better soft magnetic property can be obtained. That is to say, the coated-glass also makes great contribution to the lower coercivity, which can be explained by the residual inner stress. The typical characteristic of the glass-coated fibers is that there are great internal stress including radial compressive stress between glass and inner metal core, because of the different coefficient of thermal expansion and the axial tensile stress during the drawing preparation process. The removal of the glass reduces part of the stress in fibers, and thus the distribution of the magnetic domain is changed. With the decrease of radial stress, the radial magnetic domain in the inner core decreases, leaving only the axial domain, while the circular domain in the outer shell increase greatly, as shown in Figure 14. As a result, the magnetic anisotropy is reduced, which in turn leads to a lower coercivity.

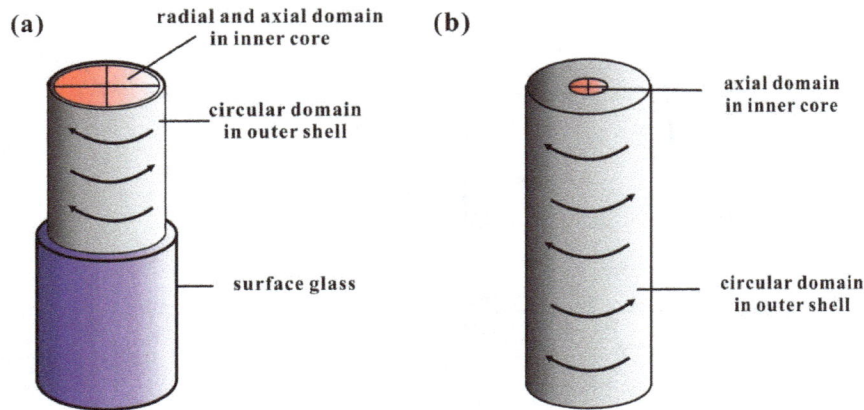

Figure 14. Magnetic domain distribution of (**a**) as-cast and (**b**) glass-removed $Fe_{73.5}Cu_1Nb_3Si_{13.5}B_9$ fibers.

A direct comparison of the tensile strength and saturated magnetization of the FeCuNbSiB alloy system with a different shape is presented in Figure 15. Although having excellent magnetic properties such as high saturation magnetization, the ribbons and the films cannot show high tensile strength, compared with the fibers. In all of the three kinds of fibers, the glass-coated FeCuNbSiB fibers exhibit the most outstanding properties with the combination of relatively high saturation magnetization and high tensile strength, especially for the amorphous one. The saturated magnetization of the amorphous fibers is higher than that of the as-cast and glass-removed fibers and they have excellent comprehensive performance. Consequently, the glass-coated FeCuNbSiB amorphous fibers are of a good choice for the components requirement both for great magnetic and tensile properties.

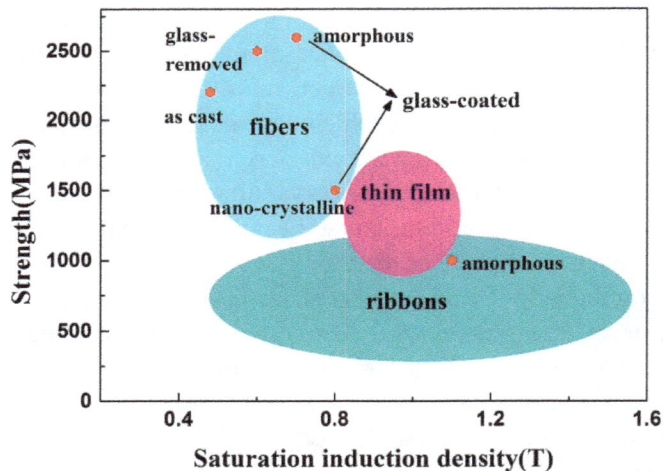

Figure 15. The map of tensile strength and saturated magnetization of the FeCuNbSiB alloy system with different shapes and structures.

5. Conclusions

The mechanical and magnetic properties of glass-coated $Fe_{76.5-x}Cu_1Nb_xSi_{13.5}B_9$ ($x = 0, 1, 2, 3, 3.5$) fibers prepared by a modified-Taylor method were studied. The main findings of the present work can be summarized as follows:

1.   The continuous glass-coated FeCuNbSiB fibers with similar diameters were successfully fabricated by the glass-coated drawing fiber equipment which is developed in our group.

The fibers show circular morphology and smooth surface and the thinnest glass layer is about 0.5 µm.

2. The tensile stress-strain curves show that the glass-coated fibers have a typical brittle fracture. The strength of the amorphous fibers is higher than that of the crystalline ones prepared in this test.

3. Weibull statistical analysis has been introduced to characterize the strength reliability of the fibers. The Weibull modulus m of the amorphous fibers reaches above 20, far higher than that of the crystalline ones.

4. The magnetic behavior is greatly influenced by the structure and composition of the fibers. The coercivity of the amorphous or nanocrystalline structure is far lower than that of coarse crystal ones. Compared with all the other kinds of fibers, the glass-coated FeCuNbSiB fibers with the intrinsic amorphous nature have the best comprehensive performance, combining the great magnetic and tensile properties.

Acknowledgments: Yong Zhang would like thank the financial support from National Natural Science Foundation of China (NSFC), granted No. 51471025 and 51671020.

Author Contributions: Haiying Hao, Yangyong Zhao and Yong Zhang conceived and designed the experiments; Haiying Hao and Yangyong Zhao performed the experiments; Haiying Hao and Ruixuan Li analyzed the data; Haiying Hao and Ruixuan Li wrote the paper.

Conflicts of Interest: The authors declare no conflict of interest.

References

1. McHenry, M.E.; Willard, M.A.; Laughlin, D.E. Amorphous and nanocrystalline materials for applications as soft magnets. *Prog. Mater. Sci.* **1999**, *44*, 291–433. [CrossRef]

2. Gotō, T.; Yoshino, A. Mechanical properties and microstructure of high toughness Fe-base filaments produced by glass-coated melt spinning. *J. Mater. Sci.* **1986**, *21*, 1809–1813. [CrossRef]

3. Chiriac, H. Preparation and characterization of glass covered magnetic wire. *Mater. Sci. Eng. A* **2001**, *304–306*, 166–171. [CrossRef]

4. Vázquez, M.; Zhukov, A.P. Magnetic properties of glass coated amorphous and nanocrystalline microwires. *J. Magn. Magn. Mater.* **1996**, *160*, 223–228. [CrossRef]

5. Chin, T.S.; Lin, C.Y.; Lee, M.C.; Huang, R.T.; Huang, S.M. Bulk nano-crystalline alloys. *Mater. Today* **2009**, *12*, 34–39. [CrossRef]

6. Zhukova, V.; Kaloshkin, S.; Zhukov, A.; Gonzalez, J. DSC studies of finemet-type glass-coated microwires. *J. Magn. Mater.* **2002**, *249*, 108–112. [CrossRef]

7. Chiriac, H.; Óvári, T.A. Amorphous glass-covered magnetic wires: preparation, properties, applications. *Prog. Mater. Sci.* **1996**, *40*, 333–407. [CrossRef]

8. Hu, Z.; Liu, X.; Wang, Z.; Xie, J. The rapid solidification prepration and application of glass-coated metallic fibers. *Mater. Rev.* **2004**, *18*, 8–11. (In Chinese)

9. Larin, V.S.; Torcunov, A.V.; Zhukov, A.; Gonzalez, J.; Vazquez, M.; Panina, L. Preparation and properties of glass-coated microwires. *J. Magnet. Magnet. Mater.* **2002**, *249*, 39–45. [CrossRef]

10. Donald, I.W.; Metcalfe, B.L. The preparation, properties and applications of some glass-coated metal filaments prepared by the Taylor-wire process. *J. Mater. Sci.* **1996**, *31*, 1139–1149. [CrossRef]

11. Zhao, Y.; Li, H.; Wang, Y.; Zhang, Y.; Liaw, P.K. Shape Memory and Superelasticity in Amorphous/Nanocrystalline Cu-15.0 Atomic Percent (at. %) Sn Wires. *Adv. Eng. Mater.* **2014**, *16*, 40–44. [CrossRef]

12. Zhang, Y.; Li, M.; Wang, Y.D.; Lin, J.P.; Dahmen, K.A.; Wang, Z.L.; Liaw, P.K. Superelasticity and serration behavior in small-sized NiMnGa alloys. *Adv. Eng. Mater.* **2014**, *16*, 955–960. [CrossRef]

13. Wang, C.D.; Wang, F.M.; Zhang, Z.H.; Xie, J.X. The influence of process parameters on the size, structure and mechanical properties of glass-coated Fe-based fibers. *J. Univ. Sci. Technol. Beijing* **2009**, *31*, 1436–1441. (In Chinese)

14. Herzer, G. Nanocrystalline soft magnetic materials. *J. Magnet. Magnet. Mater.* **1996**, *157/158*, 133–136. [CrossRef]

15. Gotō, T. Fe–B and Fe–Si–B System Alloy Filaments Produced by Glass-Coated Melt Spinning. *Trans. Japan Inst. Metals* **1980**, *21*, 219–225. [CrossRef]

16. Zberg, B.; Arata, E.R.; Uggowitzer, P.J.; Löffler, F. Tensile properties of glassy MgZnCa wires and reliability analysis using Weibull statistics. *Acta Mater.* **2009**, *57*, 3223–3231. [CrossRef]

17. Wu, Y.; Wu, H.H.; Hui, X.D.; Chen, G.L.; Lu, Z.P. Effects of drawing on the tensile fracture strength and its reliability of small-sized metallic glasses. *Acta Mater.* **2010**, *58*, 2564–2576. [CrossRef]

18. Liao, W.; Hu, J.; Zhang, Y. Micro forming and deformation behaviors of $Zr_{50.5}Cu_{27.45}Ni_{13.05}Al_9$ amorphous wires. *Intermetallics* **2012**, *20*, 82–86. [CrossRef]

19. Sun, H.; Ning, Z.; Wang, G.; Liang, W.; Shen, H.; Sun, J.; Xue, X. Tensile strength reliability analysis of $Cu_{48}Zr_{48}Al_4$ amorphous microwires. *Metals* **2016**, *6*. [CrossRef]

20. Herzer, G. Modern soft magnets: Amorphous and nanocrystalline materials. *Acta Mater.* **2013**, *61*, 718–734. [CrossRef]

Effect of Heavy Ion Irradiation Dosage on the Hardness of SA508-IV Reactor Pressure Vessel Steel

Xue Bai [1,2,3,*], **Sujun Wu** [1,3,*], **Peter K. Liaw** [2], **Lin Shao** [4] and **Jonathan Gigax** [4]

[1]   School of Materials Science and Engineering, Beihang University, Beijing 100191, China
[2]   Department of Materials Science and Engineering, The University of Tennessee, Knoxville, TN 37996, USA; pliaw@utk.edu
[3]   Beijing Key Laboratory of Advanced Nuclear Materials and Physics, Beihang University, Beijing 100191, China
[4]   Department of Nuclear Engineering, Texas A&M University, College Station, TX 77843, USA; lshao@tamu.edu (L.S.); gigaxj@tamu.edu (J.G.)
[*]   Correspondence: 20051702bx@163.com (X.B.); wusj@buaa.edu.cn (S.W.)

Academic Editor: Hugo F. Lopez

Abstract: Specimens of the SA508-IV reactor pressure vessel (RPV) steel, containing 3.26 wt. % Ni and just 0.041 wt. % Cu, were irradiated at 290 °C to different displacement per atom (dpa) with 3.5 MeV Fe ions (Fe^{2+}). Microstructure observation and nano-indentation hardness measurements were carried out. The Continuous Stiffness Measurement (CSM) of nano-indentation was used to obtain the indentation depth profile of nano-hardness. The curves showed a maximum nano-hardness and a plateau damage near the surface of the irradiated samples, attributed to different hardening mechanisms. The Nix-Gao model was employed to analyze the nano-indentation test results. It was found that the curves of nano-hardness versus the reciprocal of indentation depth are bilinear. The nano-hardness value corresponding to the inflection point of the bilinear curve may be used as a parameter to describe the ion irradiation effect. The obvious entanglement of the dislocations was observed in the 30 dpa sample. The maximum nano-hardness values show a good linear relationship with the square root of the dpa.

Keywords: SA508-IV; displacement per atom (dpa); ion irradiation hardening; inflection point; reactor pressure vessel (RPV) steel; nano-hardness

1. Introduction

With the increasing demands for high capacity and long design life of nuclear reactors, the performance of existing reactor pressure vessel (RPV) steels is becoming less adequate. It is important that new RPV steels should possess higher yield strength and fracture toughness, lower ductile to brittle transition temperature, and excellent neutron irradiation embrittlement resistance. The SA508-IV RPV steel studied in this work is a new generation RPV steel, with higher nickel content and lower copper content, compared with SA508-III. Ni is well known for improving the low-temperature cleavage toughness by decreasing the energy barrier of kink formation [1,2]. To determine the performance of this steel under nuclear environments, it is necessary to investigate the effect of irradiation doses on the microstructure evolution and the mechanical properties of this new type of RPV steel. The microstructure of this SA508-IV steel is mainly martensite structure with retained austenite, which could guarantee strength and toughness.

Neutron irradiation embrittlement of RPV steels causes the most severe damage during operation of nuclear power stations. Due to the long-term damage production and the radioactivity of the neutron irradiation, the real neutron irradiation experiments are limited and very difficult to carry out.

Instead, high-energy heavy ion irradiation has been extensively used to study the irradiation response of the candidate RPV steel because of the simplicity of use, easier control of irradiation parameters, reduction of cost, rapid damage production, and absence of induced radioactivity [3–6]. Ion irradiation is one of the best methods for investigation of irradiation embrittlement [7], and provides a rapid and flexible way to achieve high doses without the hazards induced by activation of the materials [8]. Furthermore, the displacement cascades induced by ion irradiation are similar to those induced by neutron irradiation [9]. On the other hand, the disadvantages of ion irradiation are inhomogeneous damage profiles and shallow irradiation depth compared to neutrons [10].

The purpose of this paper is to fully characterize the relationship between the measured nano-indentation test results and the irradiation doses. The classic Nix-Gao model was used to analyze the test results. Considering the limitation of ion-irradiation depth and sample size, the nano-indentation test is a quite useful method for the study of the mechanical properties of irradiated materials, which uses the recorded depth into the specimen of an indenter along with the applied load to determine the area of contact and obtain the hardness of the test specimen [11]. The technique Continuous Stiffness Measurement (CSM) offers a significant improvement in nano-indentation testing [12]. Compared with the loading-unloading method used in the past, the CSM has the great advantage of obtaining the depth profile with only a single indent. The irradiation hardening should be related to the irradiated microstructural features, which commonly belong to two groups: solute atom clusters and matrix damage [13].

2. Experimental Procedures

2.1. Testing Materials and Irradiation Conditions

The material used in this study is a new generation RPV steel SA508-IV, designed with high Ni (3.26 wt. %) and low Cu (0.041 wt. %) content. The detailed chemical compositions are shown in Table 1. The rolled steel underwent a heat treatment process of holding for 1.5 h at 920 °C, followed by water quenching (WQ) to room temperature (RT), re-heating to 650 °C for 30 h and then water cooling to RT, called the as-received state. Heavy ion irradiations were conducted with 3.5 MeV Fe ions (Fe^{2+}) using the ion accelerator in the Texas A&M University (College Station, TX, USA) on these as-received state samples, at 290 ± 5 °C. The ion accelerator is shown in Figure 1. The Stopping and Ranges of Ions in Matters (SRIM) was used for the irradiation damage calculation. The irradiation damage (D, in dpa) could be calculated by the equation [14],

$$D = \frac{F \cdot N_d^{1ion} \cdot M_{mol}}{\rho \cdot d \cdot N_A} \tag{1}$$

in which, F is the ion fluence, in ion/cm^2; M_{mol} is the target's molar mass, in g/mol; N_A is the Avogadro's constant.

Table 1. Chemical composition of SA508-IV Ni-Cr-Mo alloy steel in wt. %.

C	Si	Mn	P	S	Ni	Cr	Cu	Mo	Al	Fe
0.15	0.36	0.34	0.011	0.008	3.26	1.66	0.041	0.46	0.005	Balance

The displacement damage and ion distribution versus depth calculated by SRIM is shown in Figure 2. It can be seen that the peak damage depth sits at around 1000 nm below the ion irradiated surface. The ion distribution reaches the peak value at around 1200 nm depth.

Figure 1. The picture of the ion accelerator.

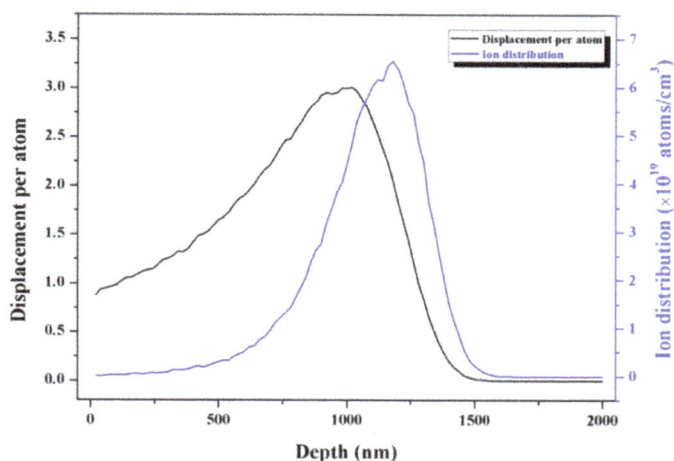

Figure 2. Depth profile of displacement per atom (dpa) and ion distribution in 3.5 MeV Fe^{2+} ion irradiation.

The corresponding dose calculation was done in pure Fe with the displacement energy $E_d = 40$ eV. The dose and dose rate at a depth of 300 nm were used for damage parameters. The applied dose rate was fixed at 1.74×10^{-3} dpa/s at the peak or 0.7×10^{-3} dpa/s at the depth of 300 nm. The doses reached 1, 2, 3, 10, 20, 30 dpa, respectively. A 6 mm × 6 mm ion beam was used in order to ensure full coverage on the samples. The parameters of ion-irradiation are shown in Table 2.

From Table 2, it can be seen that 30 dpa of ion-irradiation needs 420 min. Although the 30 dpa dosage of neutron irradiation needs hundreds of years in the neutron reactor in service, this RPV steel SA508-IV is designed for the fourth-generation nuclear reactor where the neutron flux inside will be much higher, especially in the fast reactor, therefore, it is essential to study the performance of SA-508-IV RPV steel under high dosage.

Table 2. The parameters of ion-irradiation.

dpa	Fluence	Duration (min)
1	2.46×10^{15}	14.01
2	4.92×10^{15}	28.02
3	7.38×10^{15}	42.03
10	2.46×10^{16}	140.1
20	4.92×10^{16}	280.2
30	7.38×10^{16}	420.3

2.2. Microstructure Observation

After ion-irradiation, the samples were cut from the sample center along the irradiation direction. Standard grinding and polishing was applied, followed by etching with 5 vol % nital. The microstructure of the etched samples was examined using the Quanta 200F Field Emission Environmental Scanning Electron Microscopy (FEESEM) (FEI, Hillsboro, OR, USA).

Thin foils for TEM observation were prepared perpendicular to the irradiated surface using the Zeiss Auriga Crossbeam focused ion beam (FIB) system (Zeiss, Jena, Germany). To remove defects appearing under Ga implantation at the FIB preparation, the specimens were polished by low-energy ion sputtering of Ar ions accelerated at 150 V using an ultralow-energy Ar ion beam sputtering system. After sufficiently thinning, the samples were observed using the F20 Field Emission Transmission Electron Microscopy (FE-TEM) (FEI, Hillsboro, OR, USA).

2.3. Nano-Indentation Tests

The nano-indentation hardness at room temperature was measured using a Nano Indenter XP (MTS, Eden Prairie, MN, USA) with a 2 µm diamond Berkovitch indenter (three-sided pyramid) under a maximum load of 500 mN. The Continuous Stiffness Measure (CSM) was used to study the mechanical properties. The depth (h) profiles of nano-hardness (H) up to a depth of about 2000 nm were obtained using this CSM method in the study. Five indentations were made on each of the specimens.

3. Results and Discussion

3.1. Microstructure

In the as-received state samples, the coarse carbides are mainly $M_{23}C_6$ types, which were observed in our previous work [15]. High density of fine M_6C carbides formed after the ion irradiation, which were supposed to form in the high-nickel RPV steels after irradiation [16,17]. In our recent study, the Cu-rich clusters (CRCs) preferred to form under low dose irradiation (3 dpa) but disappeared after high dose (30 dpa), whereas Mo atoms started to segregate onto Cr/Mn precipitates to form Cr/Mn/Mo until irradiated at high dose (30 dpa) [18].

The microstructures of samples after different ion-irradiation doses, 3 dpa, 10 dpa, 20 dpa, and 30 dpa, are shown in Figure 3a–d, respectively, from which it can be seen that there are no significant differences between the martensitic microstructure of the four differently dosed ion-irradiated samples. Thus it could be speculated that ion irradiation affects the density or size of the nano-clusters/precipitates much more than grain size or phases structure.

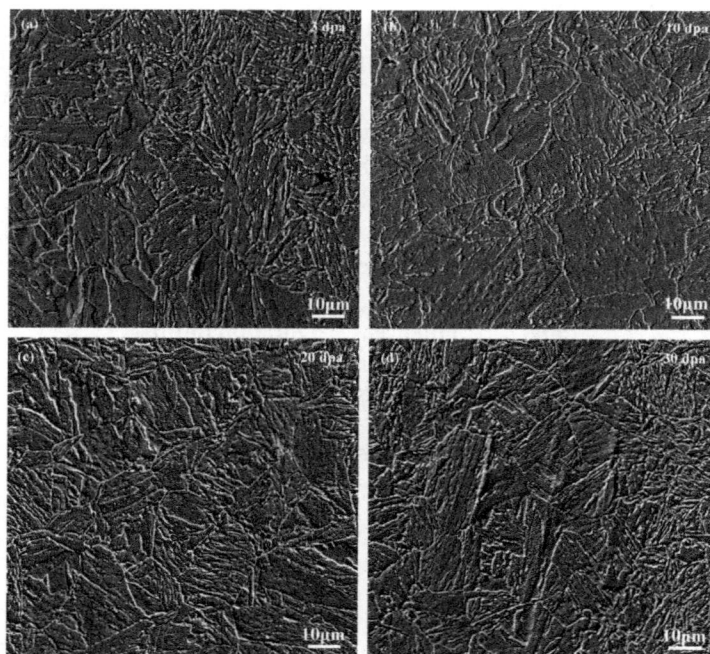

Figure 3. Microstructure of samples after different ion-irradiation doses, (**a**) 3 dpa; (**b**) 10 dpa; (**c**) 20 dpa; and (**d**) 30 dpa.

3.2. Nano-Hardness

The indentation depth profiles of nano-hardness corresponding to different dpa are plotted in Figure 4 which demonstrates that the nano-hardness curves of the irradiated samples are related to the irradiation doses with the non-irradiated at the bottom and the 30 dpa irradiated at the top. It can be seen that the nano-hardness curve of each state sample can be divided into three regions. The nano-hardness value goes up with increase of depth initially, and reaches the highest value near the surface (region A), and then goes down with the further penetration into the samples (region B and C in Figure 4). The variation in region A is known as the reverse indentation size effect (RISE), whereas the phenomenon in the broader regions B and C (the decrease in hardness with further depth) is called the indentation size effect (ISE) [19,20].

In the region B, there are plateau damage profiles in the low dose irradiation samples (1–3 dpa). With the increase of the dosage, the plateau stage weakens (10 dpa), and then disappears (20 dpa), but shows up again in the 30 dpa sample. Considering the CRCs in the 3 dpa sample which disappear in the 30 dpa sample, demonstrated in our previous study [18], the formation of CRCs could be related to the appearance of the plateau stage (90–180 nm depth) in the nano-hardness curves of the low irradiation dose samples. The plateau damage profile represents the damage saturation. Therefore, it could be concluded that the hardening effect after low doses of ion irradiation is because of the formation of CRCs, and the nano-hardness profiles reach the plateau stage due to the saturation of CRCs. Furthermore, there is also an obvious plateau profile (110–230 nm depth region) in the 30 dpa sample. The hardening effect under higher dosage is probably caused by the formation of the interstitial-type dislocation loops and the Cr/Mn/Mo precipitates. The plateau stage of the 30 dpa curve can be considered to result from loop coalescence [21]. As for the depth of the plateau stage, it could be seen in Figure 4 that the plateau stage in the 30 dpa sample is deeper than that in the low dose samples. This could be explained as accumulation of irradiation defects and a move-forward towards the inner of the samples with increasing irradiation dose.

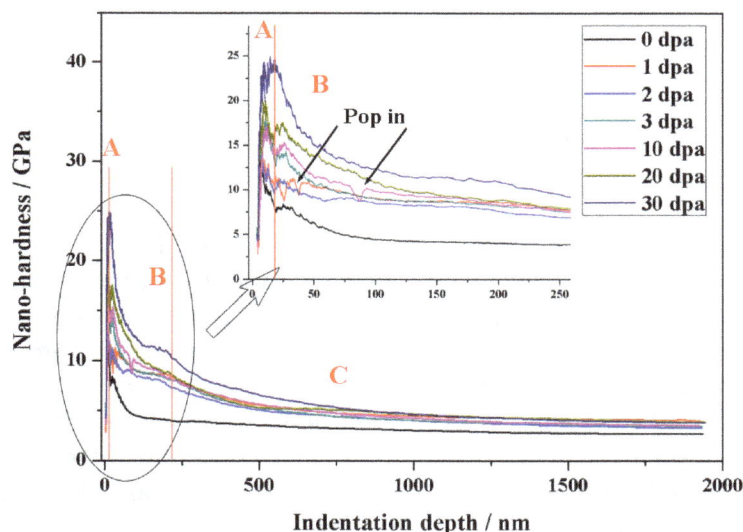

Figure 4. Depth profiles of nano-hardness of SA508-IV after different ion-irradiation doses.

According to the Nix and Gao equation [22] for the ISE stage, which describes the relationship between nano-hardness (H) and depth (h),

$$H = H_0 \left(1 + \left(\frac{l^*}{h}\right)\right)^{0.5} \tag{2}$$

where H is the hardness for a given depth of indentation h; H_0 is the hardness at the limit of infinite depth; l^* is a characteristic length that depends on the shape of the indenter, the shear modulus and H_0.

The curves of the ISE stage (region B and C) in Figure 4 could be replotted in Figure 5. The circled area in Figure 5a is enlarged and shown in Figure 5b. There is a difference between the irradiated samples and the non-irradiated sample (0 dpa). The slope of the non-irradiated curve remains constant in the ISE stage, while the slopes of the irradiated ones change at the critical indentation depth, forming the bilinear curves. According to Equation (2), $H_0^2 \times l^*$ is equal to the slope of the curves in Figure 5. It could be found that there are not many differences between the H_0^{irr} and H_0^{unirr}, thus the ion irradiation affects l^* dramatically. As stated in Equation (2), the l^* depends on the shape of the indenter, the H_0 and the shear modulus. The effect of ion irradiation is therefore mainly on the shear modulus of the materials. The C region in Figure 4 corresponds to the initial linear part of the curves in Figure 5, which is not significantly affected by the ion-irradiation. The following increased part of the irradiated curves in Figure 5 represents the region B (Figure 4), which is the area influenced mostly by the ion irradiation. The bilinear profile of the irradiated samples in Figure 5b is formed due to the soft substrate effect (SSE) of the deeper area just beneath the irradiated depth area [20,22].

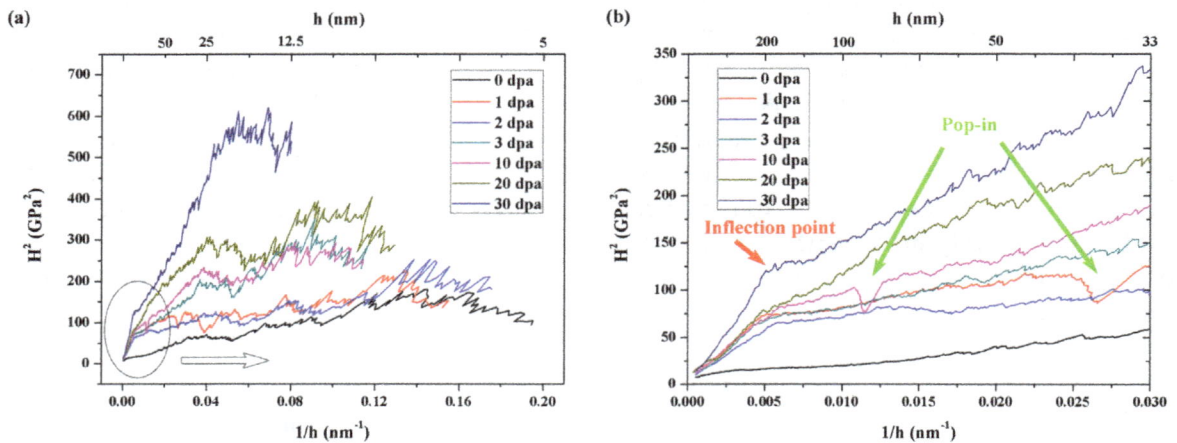

Figure 5. H^2 versus $1/h$ profile for the indentation size effect (ISE) region according to the Nix-Gao model, the circled area in (a) is enlarged in (b).

From Equation (2) it could be found that the slope of the second line just after the inflection point of the bilinear curves in Figure 5b is equal to $H_0^2 \times l^*$. Thus the l^* of the irradiated stage could be obtained by the slope divided by H_0^2 for each state sample. The relationship between l^* and dpa is plotted in Figure 6. In Figure 6, it could be seen that the increase rate of l^* is higher under low dosage than that under high dosage, which means the l^* value becomes constant when the ion irradiation dose is high enough for the irradiation effect on the nano-hardness to reach the saturation state.

Figure 4 shows that each nano-hardness curve has a maximum value which varies with the irradiation doses. The relationship between the maximum nano-hardness and dpa is presented in Figure 7. The dose dependence of the maximum nano-hardness in Figure 7a is similar to that of the characteristic length l^* as shown in Figure 6 which demonstrates a rapid increase at low doses and a slowdown of the rate of increase at higher doses, implying that the irradiation effect on nano-hardness would become saturated eventually after high dosage. Zhang et al. [3] proposed a power-law dependence of the nano-hardness on dpa. In this work, the maximum nano-hardness versus the square root of dpa in Figure 7b shows a linear increase relationship. The linear fitness result is,

$$H_{max} = 12.26 + 2.09\sqrt{dpa} \qquad (3)$$

where the constant 12.26 is the peak value of the nano-hardness curve for the non-irradiated sample, the coefficient 2.09 can be considered as the ion-irradiation hardening parameter, which could be used to evaluate the performance of the RPV steels under irradiation.

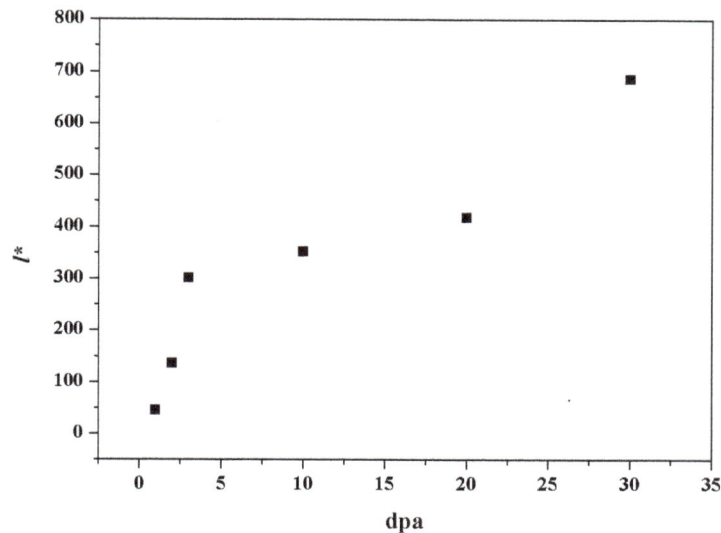

Figure 6. The relationship between the l^* and dpa.

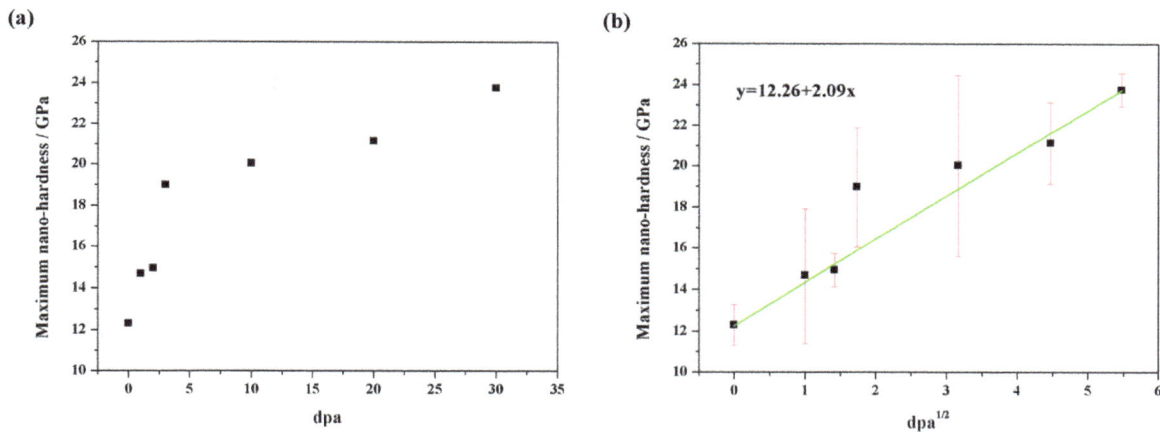

Figure 7. The relationship between the maximum nano-hardness and (**a**) dpa; (**b**) the square root of dpa.

It can be seen in Figure 7 that the 3 dpa point remains higher up on the linear fitting line. Since the precipitates or clusters are formed in different stage of the Fe^{2+} ion irradiation process, the hardening mechanisms should also be different from low dose to high dose. The CRCs are known to form at low doses and to disappear at high doses [18,23–25], and Bohmert et al. have demonstrated that the increase in hardness has good correlation with the square root of the volume fraction of the solute atom clusters [26]. Thus, it could be speculated that the first hardening stage could be attributed to the formation of the CRCs. From the plateau damage profile of the 3 dpa in Figure 4 and the higher maximum nano-hardness point of 3 dpa in Figure 7, it can be concluded that the formation of CRCs becomes saturated when the dose of irradiation reaches 3 dpa. When the dose increases to 10 or 20 dpa, the hardening mechanism changes from the CRCs to the precipitates, such as the Cr/Mn/Mo, and different dislocation structures. The large scatter of 1 and 10 dpa data found from Figure 7b may be because of the variety of the microstructures or the surface condition.

The TEM observation results of the dislocations in the 3 dpa and 30 dpa samples are shown in Figure 8. It can be seen that the dislocations in the 3 dpa irradiated sample (Figure 8a) are mainly parallel and along the directions of ion irradiation. In contrast, in Figure 8b, the dislocations are found to be entangled with one other. Y. Takayama et al. [27] found that nickel element addition enhances the irradiation hardening of RPV steel during the irradiation process. Ni and Mn elements are reported to have the tendency to promote the formation of interstitial loops during irradiation [28–30] and the high dose of irradiation could generate high density of self-interstitial atoms by the cascade effect which results in the formation of interstitial-type dislocation loops [31,32]. Therefore, it can be considered that the entanglement of the dislocations and the interstitial-type dislocation loops are one of the main factors responsible for the hardening effect after high dose irradiation.

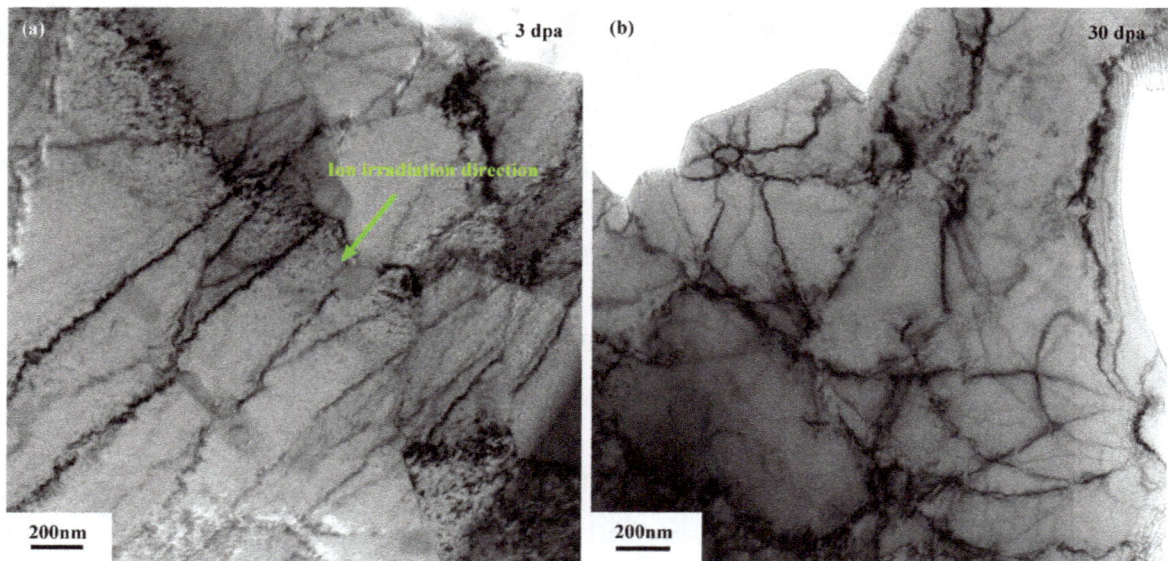

Figure 8. Dislocation density of different dose ion-irradiation samples, (**a**) 3 dpa; (**b**) 30 dpa.

In Figure 4, it can be seen that there are multiple discontinuous displacement bursts occurring clearly in the nano-hardness versus depth curves of some ion-irradiated samples. Such a non-uniform displacement response named "pop-in" is marked by arrows in Figures 4 and 5. The pop-ins on the 1 dpa curve occur around 25–50 nm depth, whereas the one on the 10 dpa curve stays around 85 nm depth. Generally, the occurrence of pop-in phenomenon in metals can be attributed to several factors such as dislocation nucleation and propagation, crack formation or phase transformation [33–36].

The pop-ins on nano-hardness curves of the SA508-IV steel samples in this study probably occurred due to the transformation of the retained austenite to martensite. During the measurement of the nano-hardness along the depth of the sample, the indenter may encounter the retained austenite. Because the retained austenite is relatively soft, the corresponding nano-hardness is lower which results in the drop of the nano-hardness on the curve. On the other hand, the plastic deformation of the retained austenite may induce transformation of the austenite to martensite which is much harder than the retained austenite. Thus the nano-hardness in the hardness-depth curve will increase, therefore, forming the pop-ins on the curves. Whether there are pop-ins on the curves depends on the distribution of the retained austenite. If there are retained austenite grains along the measuring depth of the sample, pop-ins would be generated, otherwise, there would be no pop-ins.

Attention should be paid to the inflection points of the bilinear curves (the crossing points of the two-line curves) in Figure 5b, corresponding to the cut-off point between B region and C region (Figure 4). The indentation depths at the inflection points are similar for the irradiated samples, but the nano-hardness value increases with the increase of the irradiation dosage. As mentioned above, the bilinear profile is attributed to the SSE of the non-irradiated area just beneath the irradiated area. Therefore, the nano-hardness value (H_{inf}) and the indentation depth ($h^{irr/unirr}$) corresponding to the inflection point, can be considered as a parameter representing the irradiation mostly affected layer on the surface of the material, and below this layer, the effect of irradiation can be neglected.

The nano-hardness profiles in Figure 4 manifest when the measuring depth is over 1500 nm, the nano-hardness value remains almost constant. Therefore, to measure the nano-hardness as a property of a material, the indentation depth must be greater than 1500 nm. The nano-hardness value at a depth of ~2000 nm is taken as the material nano-hardness property in normal practice. To characterize the effect of ion irradiation on the nano-hardness of a material, however, different parameters should be applied since the effect of ion irradiation is limited within the surface thin layer of the material. As discussed above, the maximum nano-hardness value H_{max} or the hardness value corresponding to the inflection point (H_{inf}) could be used to describe the irradiation effect on the nano-hardness of materials.

4. Conclusions

The heavy ion irradiation process was demonstrated to produce a similar cascade effect to neutron irradiation. Micro-mechanical properties were obtained by nano-hardness tests. Based on the above analysis, conclusions can be drawn as follows:

(1) The CSM nano-hardness profiles of the irradiated SA508-IV steel samples exhibit a maximum value and a plateau.
(2) The maximum nano-hardness (H_{max}) increases rapidly at low doses and slowly at higher doses. The relationship between H_{max} and dpa follows a $1/2$-power law.
(3) The inflection point of the bilinear curve is the point separating the B region and C region on the CSM nano-hardness profile. Beyond the inflection point, the ion irradiation effect on the materials is insignificant.
(4) The characteristic length l^* which is affected by ion-irradiation was obtained by fitting the bilinear curves, and also had a relationship with the displacement per atom (dpa).

Acknowledgments: This work is financially supported by the National Science and Technology Key Specific Project: Life Management Technology of Nuclear Power Plant of China (Grant No. 2011ZX06004-002). The authors would like to express thanks for the help with ion-irradiation experiments conducted in the Department of Nuclear Engineering in Texas A&M University (USA).

Author Contributions: Xue Bai designed the research, conducted the experiments, and then finished the draft of this paper, Sujun Wu and Peter K. Liaw supervised the project and help revised the manuscript, Lin Shao and Jonathan Gigax helped with the ion-irradiation experiments, all authors reviewed the manuscript.

Conflicts of Interest: The authors declare no conflicts of interest.

References

1. Chen, Y.T.; Atteridge, D.G.; Gerberich, W.W. Plastic flow of Fe-binary alloys-I. A description at low temperatures. *Acta Metall.* **1981**, *29*, 1171–1185. [CrossRef]
2. Gerberich, W.W.; Chen, Y.T.; Atteridge, D.G.; Johnson, T. Plastic flow of Fe-binary alloys-II. Application of the description to the ductile-brittle transition. *Acta Metall.* **1981**, *29*, 1187–1201. [CrossRef]
3. Zhang, H.; Zhang, C.; Yang, Y.; Meng, Y.; Jang, J.; Kimura, A. Irradiation hardening of ODS ferritic steels under helium implantation and heavy-ion irradiation. *J. Nucl. Mater.* **2014**, *455*, 349–353. [CrossRef]
4. Liu, P.P.; Bai, J.W.; Ke, D.; Wan, F.R.; Wang, Y.B.; Wang, Y.M.; Ohnuki, S.; Zhan, Q. Effects of deuterium implantation and subsequent electron irradiation on the microstructure of Fe-10Cr model alloy. *J. Nucl. Mater.* **2012**, *423*, 47–52. [CrossRef]
5. Klueh, R.L.; Gelles, D.S.; Jitsukawa, S.; Kimura, A.; Odette, G.R.; van der Schaaf, B.; Victoria, M. Ferritic/martensitic steels-overview of recent results. *J. Nucl. Mater.* **2002**, *307*, 455–465. [CrossRef]

6. Zhang, C.H.; Sun, Y.M.; Song, Y.; Shibayama, T.; Jin, Y.F.; Zhou, L.H. Defect production in silicon carbide irradiated with Ne and Xe ions with energy of 2.3 MeV/amu. *Nucl. Instrum. Methods Phys. Res. B* **2007**, *256*, 243–247. [CrossRef]

7. Fujii, K.; Fukuya, K.; Hojo, T. Effects of dose rate change under irradiation on hardening and microstructural evolution in A533B steel. *J. Nucl. Sci. Technol.* **2013**, *50*, 160–168. [CrossRef]

8. Zhang, Z.W.; Liu, C.T.; Wang, X.L.; Miller, M.K.; Ma, D.; Chen, G.; Williams, J.R.; Chin, B.A. Effects of proton irradiation on nanocluster precipitation in ferritic steel containing fcc alloying additions. *Acta Mater.* **2012**, *60*, 3034–3046. [CrossRef]

9. Was, G.S.; Busby, J.T.; Allen, T.; Kenik, E.A.; Jenssen, A.; Bruemmer, S.M.; Gan, J.; Edwards, A.D.; Scott, P.M.; Andresen, P.L. Emulation of neutron irradiation effects with protons: Validation of principle. *J. Nucl. Mater.* **2002**, *300*, 198–216. [CrossRef]

10. Liu, P.P.; Wan, F.R.; Zhan, Q. A model to evaluate the nano-indentation hardness of ion-irradiated materials. *Nucl. Instrum. Methods Phys. Res. B* **2015**, *342*, 13–18. [CrossRef]

11. Fischer-Cripps, A.C. *Nanoindentation Testing, Nanoindentation*; Springer: Berlin, Germany, 2011.

12. Li, X.D.; Bhushan, B. A review of nanoindentation continuous stiffness measurement technique and its applications. *Mater. Charact.* **2002**, *48*, 11–36. [CrossRef]

13. Fujii, K.; Fukuya, K.; Hojo, T. Concomitant formation of different nature clusters and hardening in reactor pressure vessel steels irradiated by heavy ions. *J. Nucl. Mater.* **2013**, *443*, 378–385. [CrossRef]

14. Hengstler-Eger, R.M.; Baldo, P.; Beck, L.; Dorner, J.; Ertl, K.; Hoffmann, P.B.; Hugenschmidt, C.; Kirk, M.A.; Petry, W.; Pikart, P.; et al. Heavy ion irradiation induced dislocation loops in AREVA's M5 alloy. *J. Nucl. Mater.* **2012**, *423*, 170–182. [CrossRef]

15. Bai, X.; Wu, S.; Liaw, P.K. Influence of thermo-mechanical embrittlement processing on microstructure and mechanical behavior of a pressure vessel steel. *Mater. Des.* **2016**, *89*, 759–769. [CrossRef]

16. Klueh, R.L.; Hashimoto, N.; Sokolov, M.A.; Shiba, K.; Jitsukawa, S. Mechanical properties of neutron-irradiated nickel-containing martensitic steels: I. Experimental study. *J. Nucl. Mater.* **2006**, *357*, 156–168. [CrossRef]

17. Klueh, R.L.; Hashimoto, N.; Sokalov, M.A.; Maziasz, P.J.; Shiba, K.; Jitsukawa, S. Mechianical properties of neutron-irradiated nickel-containing martensitic steels: II. Review and analysis of helium-effects studies. *J. Nucl. Mater.* **2006**, *357*, 169–182. [CrossRef]

18. Bai, X.; Wu, S.; Liaw, P.K.; Shao, L.; Gigax, J.; Wei, L. Effect of ion irradiation on microstructure and nano-hardness of the SA508-IV reactor pressure vessel steel. *Sci. Rep.* submitted for publication. 2017.

19. Pharr, G.M.; Herbert, E.G.; Gao, Y. The indentation size effect: A critical examination of experimental observations and mechanistic interpretations. *Annu. Rev. Mater. Res.* **2010**, *40*, 271–292. [CrossRef]

20. Kasada, R.; Takayama, Y.; Yabuuci, K.; Kimura, A. A new approach to evaluate irradiation hardening of ion-irradiated ferritic alloys by nano-indentation techniques. *Fusion Eng. Des.* **2011**, *86*, 2658–2661. [CrossRef]

21. Jiang, J.; Wu, Y.C.; Liu, X.B.; Wang, R.S.; Nagai, Y.; Inoue, K.; Shimizu, Y.; Toyama, T. Microstructural evolution of RPV steels under proton and ion irradiation studied by positron annihilation spectroscopy. *J. Nucl. Mater.* **2015**, *458*, 326–334. [CrossRef]

22. Nix, W.D.; Gao, H. Indentation size effects in crystalline materials: A law for strain gradient plasticity. *J. Mech. Phys. Solids* **1998**, *46*, 411–425. [CrossRef]

23. Yabuuchi, K.; Yano, H.; Kasada, R.; Kishimoto, H.; Kimura, A. Dose dependence of irradiation hardening of binary ferritic alloys irradiated with Fe^{3+} ions. *J. Nucl. Mater.* **2011**, *417*, 988–991. [CrossRef]

24. Akamatsu, M.; van Duysen, J.C.; Pareige, P.; Auger, P. Experimental evidence of several contributions to the radiation damage in ferritic alloys. *J. Nucl. Mater.* **1995**, *225*, 192–195. [CrossRef]

25. Tobita, T.; Suzuki, M.; Iwase, A.; Aizawa, K. Hardening of Fe-Cu alloys at elevated temperatures by electron and neutron irradiaions. *J. Nucl. Mater.* **2001**, *299*, 267–270. [CrossRef]

26. Bohmert, J.; Viehrig, H.-W.; Ulbricht, A. Correlation between irradiation-induced changes of microstructural parameters and mechanical properties of RPV steels. *J. Nucl. Mater.* **2004**, *334*, 71–78. [CrossRef]

27. Takayama, Y.; Kasada, R.; Sakamoto, Y.; Yabuuchi, K.; Kimura, A.; Ando, M.; Hamaguchi, D.; Tanigawa, H. Nanoindentation hardness and its extrapolation to bulk-equivalent hardness of F82H steels after single- and dual-ion beam irradiation. *J. Nucl. Mater.* **2013**, *442*, S23–S27. [CrossRef]

28. Nishiyama, Y.; Onizawa, K.; Suzuki, M.; Anderegg, J.W.; Nagai, Y.; Toyama, T.; Hasegawa, M.; Kameda, J. Effects of neutron-irradiation-induced intergranular phosphorus segregation and hardening on embrittlement in reactor pressure vessel steels. *Acta Mater.* **2008**, *56*, 4510–4521. [CrossRef]

29. Shibamoto, H.; Kitao, K.; Matsui, H.; Hasegawa, M.; Yamaguchi, S.; Kimura, A. *Effects of Radiation on Materials, 20th International Symposium, ASTM STP 1405*; Rosinski, S.T., Grossbeck, M.L., Allen, T.R., Kumar, A.S., Eds.; American Society for Testing and Materials: West Conshohocken, PA, USA, 2001; p. 722.

30. Watanabe, H.; Arase, S.; Yamamoto, T.; Wells, P.; Onishi, T.; Odette, G.R. Hardening and microstructural evolution of A533b steels irradiated with Fe ions and electrons. *J. Nucl. Mater.* **2016**, *471*, 243–250. [CrossRef]

31. Liu, P.P.; Zhao, M.Z.; Zhu, Y.M.; Bai, J.W.; Wan, F.R.; Zhan, Q. Effects of carbide precipitate on the mechanical properties and irradiation behavior of the low activation martensitic steel. *J. Alloy. Compd.* **2013**, *579*, 599–605. [CrossRef]

32. Radiguet, B.; Pareige, P.; Barbu, A. Irradiation induced clustering in low copper or copper free ferritic model alloys. *Nucl. Instrum. Methods Phys. Res. B* **2009**, *267*, 1496–1499. [CrossRef]

33. Qian, X.; Han, L.; Zhang, W.; Gu, J. Nano-indentation investigation on the mechanical stability of individual austenite in high-carbon steel. *Mater. Charact.* **2015**, *110*, 86–93.

34. Kim, B.; Trang, T.T.T.; Kim, N.J. Deformation behavior of ferrite-austenite duplex high nitrogen steel. *Met. Mater. Int.* **2014**, *20*, 35–39. [CrossRef]

35. Mistra, R.D.K.; Zhang, Z.; Jia, Z.; Venkat Surya, P.K.C.; Somani, M.C.; Karjalainen, L.P. Nanomechanical insights into the deformation behavior of austenitic alloys with different stacking fault energies and austenitic stability. *Mater. Sci. Eng. A* **2011**, *528*, 6958–6963. [CrossRef]

36. Bahr, D.F.; Kramer, D.E.; Gerberich, W.W. Non-linear deformation mechanisms during nano-indentation. *Acta Mater.* **1998**, *46*, 3605–3617. [CrossRef]

14

A Study on the Aging Behavior of Al6061 Composites Reinforced with Y_2O_3 and TiC

Chun-Liang Chen * and Chen-Han Lin

Department of Materials Science and Engineering, National Dong Hwa University, Hualien 97401, Taiwan; 610222013@ems.ndhu.edu.tw
* Correspondence: chunliang@mail.ndhu.edu.tw

Academic Editor: Hugo F. Lopez

Abstract: The reinforcement particles play important roles in determining microstructural development and properties of Al6061 composites. In the present work, the aging behavior of Al6061 reinforced with Y_2O_3 and TiC particles produced via mechanical alloying are investigated. The results indicate that the peak-aged Al6061 alloy without reinforcement demonstrates the highest hardness, which corresponds to the formation of the Mg–Si precipitates. However, precipitation formation is not observed in the case of the Al6061 composites, which can be attributed to the fact that the Mg–Si clusters and GP zones are inhibited by the presence of the reinforcement particles. The solute elements segregate in the complex oxides or carbides and contribute to only a slight increase in hardness.

Keywords: aging behavior; mechanical alloying; Al6061; composite; precipitation

1. Introduction

Al6061 alloys exhibit high corrosion resistance, excellent formability, and a high strength-to-weight ratio [1–3]. They are promising candidates for structural materials in the automotive, aircraft, military, and marine industries. Al6061 metal matrix composites (MMCs) have recently attracted considerable interest due to higher specific strength at high temperature, good fatigue performance, high stiffness, and excellent wear resistance [4–9]. Reinforcement particles play an important role in determining the properties of materials. It has been reported that the presence of nanoparticles enhances strength and plasticity by interacting with dislocations, while simultaneously retarding grain growth [10,11]. In a previous work [12], we have studied Al6061 composites reinforced with Y_2O_3 and TiC. The formation of intermetallic compounds and complex oxide dispersoids associated with reinforcement particles–matrix interaction was investigated, and the results demonstrated that a complicated crystal structure of the $Y_2(Al,Si)_2O_7$ phase was found in the Al6061-Y_2O_3 composite. The addition of Y_2O_3 can provide nucleation sites to facilitate the formation of intermetallic compounds and complex oxide particles [12]. In the case of the Al6061-TiC composite, the TiC particles can act as a milling agent that helps refine the matrix particles and greatly enhances the reactivity of the mixed components, facilitating the formation of iron-rich intermetallic compounds [12]. The Al6061 composites demonstrate a pronounced increase in hardness (about 50%) compared with the Al6061 alloy [12]. In the present work, we further investigate the effect of the aging behavior of the Al6061 and Al6061 composites. It is important to understand the role of the reinforcement particles and how they influence precipitation. During aging treatment, metastable Mg–Si precipitations are formed and the Mg:Si atomic ratio can be altered at different stages of the precipitation evolution, which can be correlated to the mechanical performance of Al6061 alloys. Although there are some studies on the aging behavior of Al-based composite materials [13–16], there is a limited amount of research work reporting the effect of reinforcement particles on the formation of precipitates of Al6061-Y_2O_3 and Al6061-TiC composites during aging treatment. Therefore, in this paper, the features of non-reinforced

Al6061 and reinforced-Al6061 from the supersaturated to the peak time state have been clarified using TEM electron microscopy. The interface between reinforcements and the Al6061 matrix and how they control an accelerated aging kinetic in the Al6061 composites will also be discussed.

2. Materials and Methods

An Al6061 alloy was used as the composite matrix with a particle size of 44–62 μm. Reinforcement particulates used were the yttrium oxide (Y_2O_3) powder of 99.99% purity with a particle size in the range 20–50 nm, and titanium carbide (TiC) powder of 99.50% purity with a particle size range of 40–50 μm. The three different model materials were fabricated by mechanical alloying (Al6061 alloy, Al6061 composite reinforced with 2 wt % Y_2O_3, and Al6061 composite reinforced with 2 wt % TiC). The mechanical alloying process was carried out using a planetary ball mill (Retsch PM 100, RETSCH, Germany) with a speed of 350 rpm under an argon atmosphere. Milling experiments were performed by using a hardened stainless steel grinding medium with a ball-to-powder ratio of 10:1. The process control agent (PCA) stearic acid was used to inhibit agglomeration of powders during ball milling. The mechanically alloyed powders were consolidated into green compacts with a pressure of 350 MPa and were then further sintered in a mixed hydrogen–argon atmosphere at 600 °C for 2 h. The Al6061 alloy and Al6061 composites were solution treated at 550 °C in a furnace for 2 h and were water quenched at room temperature. All specimens were then aged at room temperature (T4 treatment). Artificial aging (T6 treatment) was carried out at 160 °C in an oil bath. The age-hardening curves of the Al6061 composite were characterized using Vickers hardness measurements performed at room temperature using a load of 1 kg for 15 s. The microstructure of the Al6061 composite specimens at different aging stages (solution treatment, peak aging, and over-aging) was examined using a Hitachi-4700 SEM (HITACHI, Tokyo, Japan). Precipitate formation of the peak-aged Al6061 composites was then further investigated on a FEI Tecnai F20 G2 Field Emission Gun TEM (FEI, Hillsboro, OR, USA).

3. Results and Discussion

3.1. Aging Hardness

In order to understand the aging behavior of the model materials, Vickers hardness was measured in the Al6061 alloy (non-reinforcement) and the Al6061 composites (reinforcements of Y_2O_3 and TiC) aged for different times at 160 °C. The age hardening curves of the materials with T6 treatment are shown in Figure 1. The peak aging hardness of the Al6061 alloy was achieved after aging up to 18 h (Figure 1A). This shows about a 66% increase in hardness from the supersaturated solid solution status to the peak aging (123.3 HV). In comparison with the Al6061 composites, the age hardening curve of the Al6061-Y_2O_3 composite demonstrates that the peak hardness (96.4 HV) was obtained after aging for 13 h as shown in Figure 1B. In addition, the aging peak of the Al6061-TiC composite sample was revealed at 8 h of aging (75.8 HV) (Figure 1C). The hardness increment of the Al6061-Y_2O_3 and the Al6061-TiC composites at the aging peak is about 22% and 15%, respectively, which are much smaller than that of the Al6061 alloy. However, the hardness of the Al6061 alloy decreased much faster than that of Al6061 composites in the over-aging period. On the other hand, it can be clearly seen that the time required to achieve the aging peak for the Al6061 composites is shorter than that of the Al6061 alloy. These results imply that the formation of precipitates can significantly influence the aging behavior of the model materials. The aging hardness profiles of the materials with T4 treatment are shown in Figure 2. The Al6061 alloy has a significant increase in hardness of about 20% after the solution heat treatment for 24 h (Figure 2A). However, the aging hardening curves of the Al6061-Y_2O_3 and Al6061-TiC composites with T4 treatment only demonstrate a minor increase in hardness about 5%–7% after the solution heat treatment for 12 h (Figure 2B,C). The results of age hardening also imply that the addition of reinforcement particles plays a major role in accelerating the aging process and affecting the formation of precipitates. The precipitation mechanism will be further discussed and investigated in more detail in the following TEM section.

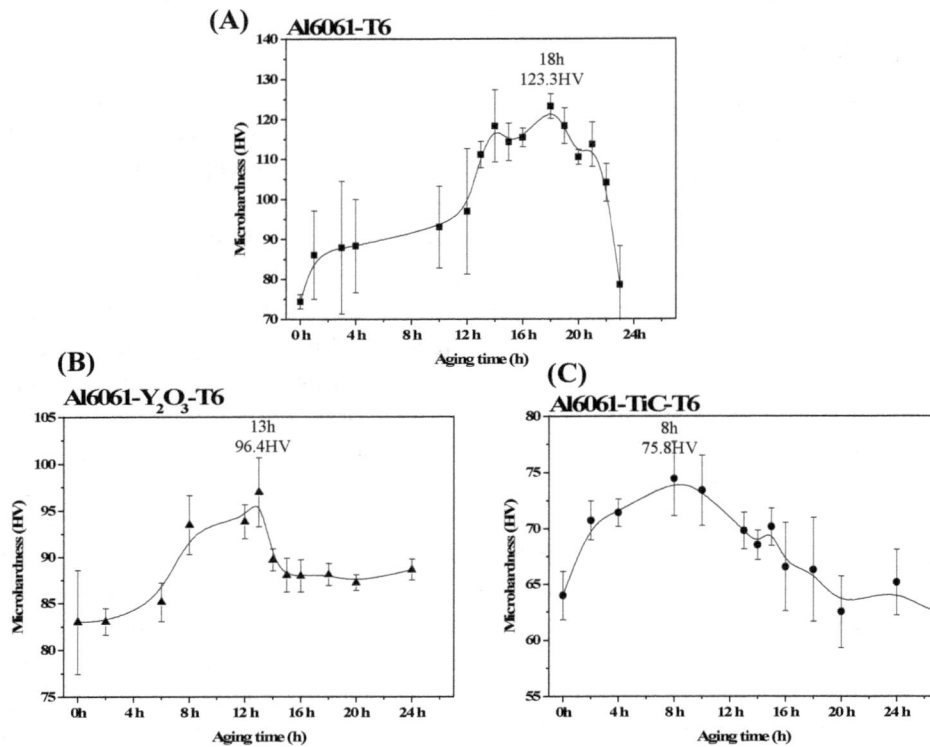

Figure 1. Artificial aging hardening curves of (**A**) Al6061 alloy; (**B**) Al6061-Y_2O_3; and (**C**) Al6061-TiC composites at different aging times.

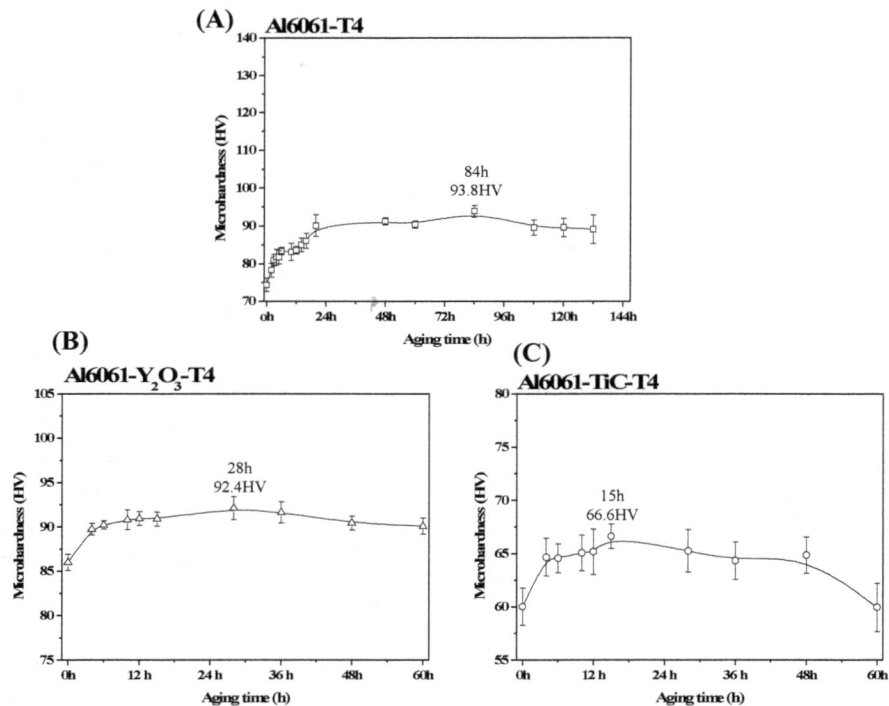

Figure 2. Natural aging hardening curves of (**A**) Al6061 alloy; (**B**) Al6061-Y_2O_3; and (**C**) Al6061-TiC composites at different aging times.

3.2. Evolution of Microstructure

3.2.1. SEM Observation

The microstructure of the Al6061 alloy and Al6061 composites at the different stages of T6 treatment was further studied. Figure 3 shows the SEM images of the Al6061, Al6061-Y_2O_3, and Al6061-TiC samples after solid solution treatment and quenching. In this case, the materials are quenched to keep the material in a supersaturated solid solution state. The Al6061 alloy sample without added reinforcements is shown in Figure 3A; however, a large number of reinforcement particles can be seen uniformly distributed throughout the Al6061-Y_2O_3 and Al6061-TiC composites, as shown in Figure 3B,C. Microstructures of the Al6061 alloy and Al6061 composites at the peak aging stage are shown in Figure 4. A clear matrix was also observed in the Al6061 alloy, and the reinforcement particles were again uniformly distributed in the Al6061-Y_2O_3 and Al6061-TiC composites. It should be noted that, at the peak aging stage, different types of needle/rod-shaped precipitates of Mg_xSi_y would form within the Al matrix. However, the formation of precipitates would be only nano-sized, and these cannot be easily observed by SEM; therefore, the precipitation evolution of the Al6061 model materials during aging will be further investigated by TEM analysis. The microstructure of over-aged Al6061 samples can also be seen in Figure 5, and this demonstrates that there are a few intermetallic compounds that formed in the microstructure of the Al6061 alloy (Figure 5A). Moreover, there seems to be a coarsening of the reinforcement particles of the Al6061 composites as shown in Figure 5B,C. In the case of the Al6061-Y_2O_3 composite sample, the distribution of the reinforcement particles became non-uniform in the matrix, as shown in Figure 5B. It should be noted that the crystal structure of an initial Y_2O_3 reinforcement particle in the Al6061 composite can be modified during processing. In an earlier work [12], it was proposed that high-energy ball milling can facilitate a solid-state reaction between the Y_2O_3 particles and the Al6061 matrix, and a subsequent sintering processing promotes the formation of the complex Al–Si–Y–O-based oxide particle, which has been identified as the $Y_2(Al,Si)_2O_7$ phase. In the case of the present study, the reinforcement particles played a crucial role in altering the diffusion of the solute elements in the Al6061 composites where a particles–matrix interaction can result in microstructural changes corresponding to different aging behaviors.

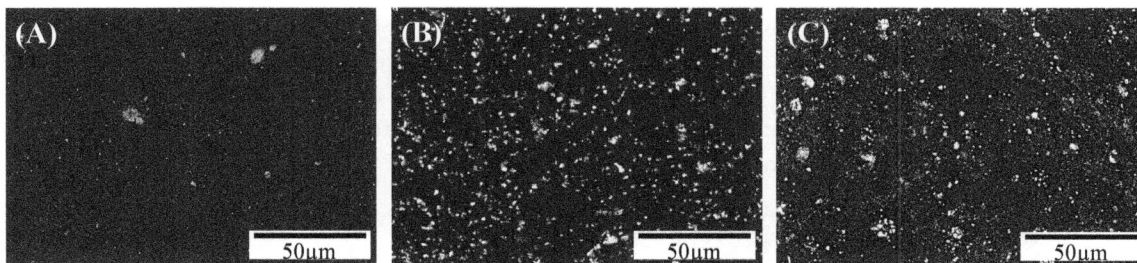

Figure 3. SEM images of (**A**) Al6061 alloy; (**B**) Al6061-Y_2O_3; and (**C**) Al6061-TiC composites after solid solution treatment and quenching.

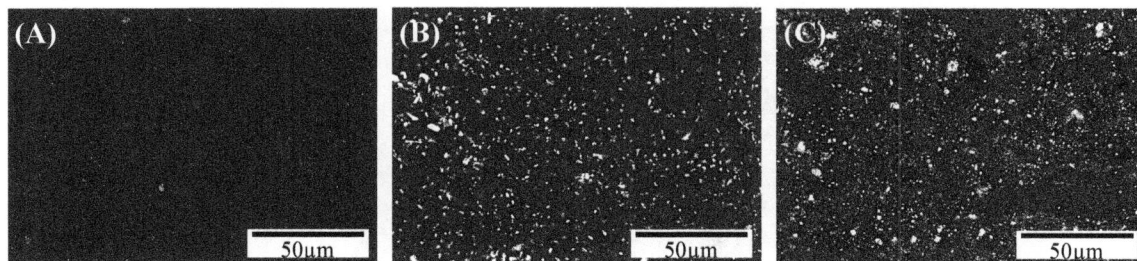

Figure 4. SEM images of (**A**) Al6061 alloy; (**B**) Al6061-Y_2O_3; and (**C**) Al6061-TiC composites at peak aging condition.

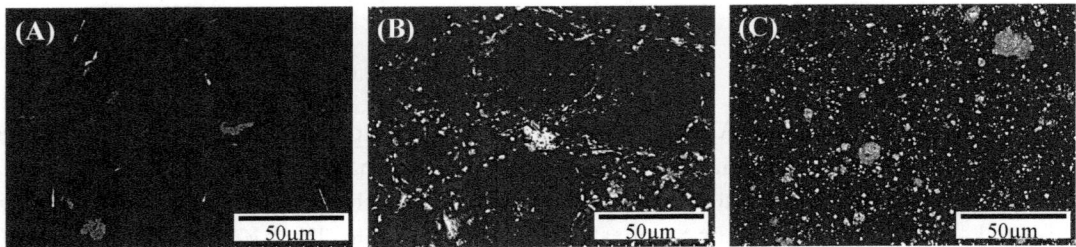

Figure 5. SEM images of (**A**) Al6061 alloy; (**B**) Al6061-Y$_2$O$_3$; and (**C**) Al6061-TiC composites at over-aging condition.

3.2.2. TEM Investigation

TEM was further used to understand the precipitation evolution of the Al6061 model materials during aging treatment. Figure 6A shows a TEM image of the peak-aged Al6061 alloy sample. A large number of needle-like precipitates were observed in the alloy. These nano-precipitates lined up along the <100> direction of the Al matrix, which correspond to the selected area diffraction (SAD) pattern in Figure 6C. They are most likely to be considered β"(Mg$_5$Si$_6$) and β'(Mg$_9$Si$_5$) phases, which are normally observed in Al6061 alloys at peak aged conditions. The evolution of aging on the precipitation behavior of the Al6061 alloy can be described in the following stages. After solution treatment and quench, the Al matrix experienced a high supersaturation of solute atoms and vacancies. This promoted the formation of clusters of Mg and Si atoms from the supersaturated matrix. These are known as GP zones. During aging treatment, the Mg:Si ratio can be altered due to the poor solubility and high diffusibility of Si in Al [17]. Mg can diffuse into the GP zones and increase the Mg:Si ratio as aging time increases. This results in metastable coherent or semi-coherent β"(Mg$_5$Si$_6$) and β'(Mg$_9$Si$_5$) precipitates in the peak-aged condition, which finally transform to stable incoherent β(Mg$_2$Si) precipitates after over-aging [18].

The microstructure of the over-aged Al6061 alloy is shown in Figure 6B. Fine precipitates were observed inside the grains and lie along the <100> directions of the Al matrix (the SAD pattern in Figure 6D), which might correspond to semi-coherent β'(Mg$_9$Si$_5$) precipitates. However, precipitate particles grow large at the grain boundary and are believed to be stable incoherent β(Mg$_2$Si) precipitates. In addition, a precipitation-free zone (PFZ) was clearly found in the over-aged Al6061 alloy sample, as shown in Figure 6B. This can be attributed to the solute depletion from areas adjacent to the boundaries. The decrease in Mg concentration within the PFZ becomes noticeable with increased aging time, corresponding to the growth of grain boundary precipitates.

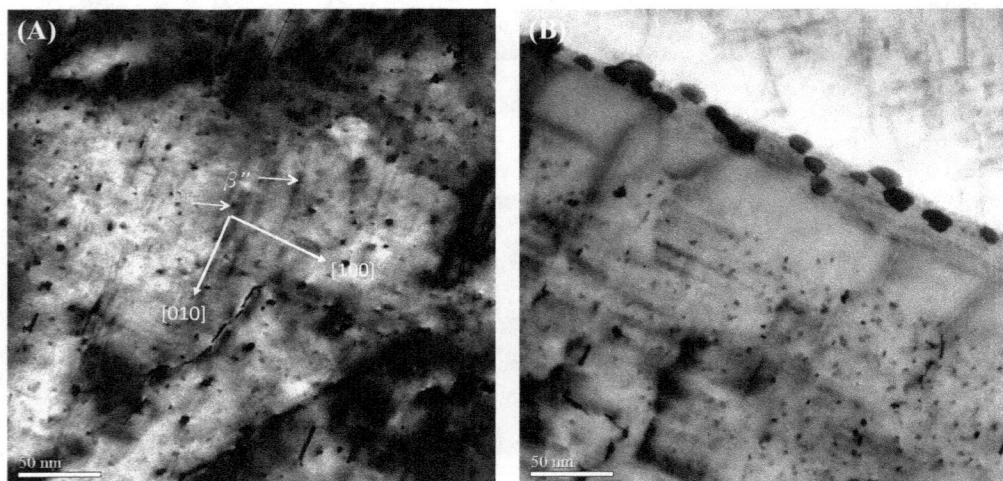

Figure 6. *Cont.*

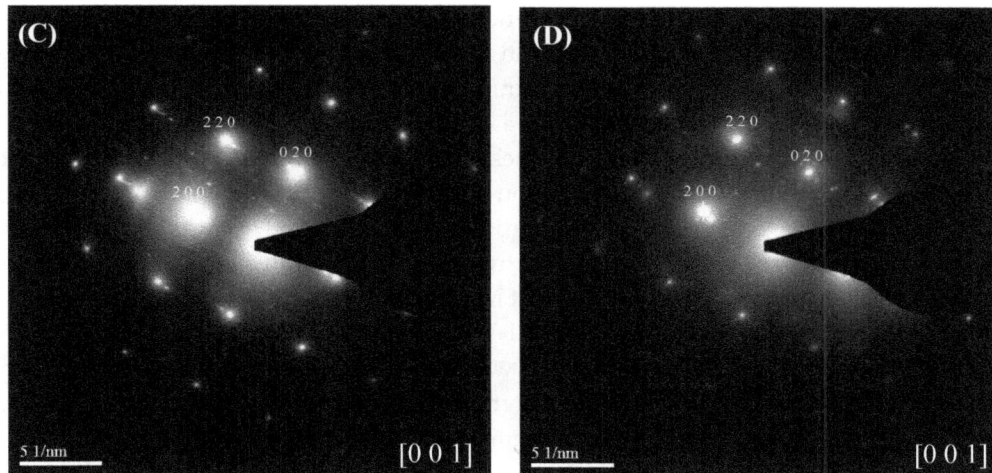

Figure 6. TEM images of (**A**) the peak-aged and (**B**) over-aged Al6061 alloys; (**C,D**) the SAD patterns of (**A**) and (**B**) along the <100> direction of the Al matrix.

Figure 7A,B shows the microstructure of the peak-aged Al6061-Y_2O_3 and Al6061-TiC composites, respectively. There was almost no precipitation observed in these two alloy samples. In the Al6061-Y_2O_3 sample, a number of dark particles were observed (see the arrows of Figure 7A), and they contain a high level of Al, Mg, Si, Y, and O elements measured by EDS. In early work, it has been reported that Y_2O_3 particles in the Al6061 alloy can act as nucleation sites to facilitate the formation of the complex Al–Si–Y–O oxide particle [12]. The aging process could further promote and accelerate the diffusion of Mg atoms; therefore, a high concentration of Mg was found in the complex oxides. In the case of the Al6061-TiC sample, TiC particles containing a small amount of Al, Mg, and Si elements were observed in the Al6061 matrix, as indicated by the arrows in Figure 7B.

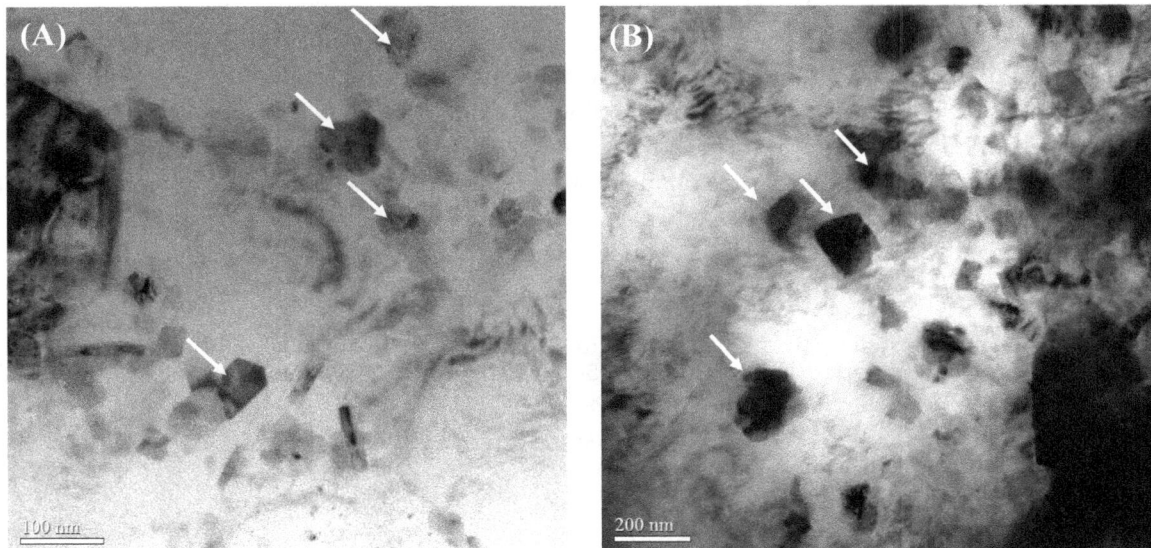

Figure 7. TEM images of (**A**) the peak-aged Al6061-Y_2O_3 and (**B**) the peak-aged Al6061-TiC composites.

A comparison of the results from the three model materials, the Al6061 has significantly different microstructural characteristics and aging behaviors. The Al6061 alloy was artificially aged at 160 °C for a prolonged period of time; as a result, the precipitation starts with the formation of Mg or Si atomic clusters. However, mechanisms of precipitation hardening can be considerably different in the case of the Al6061 composites. The presence of reinforcement particles in the aged-Al6061 composites can effectively impede the formation of GP zones. The mechanical alloying process can facilitate a solid-state reaction between reinforcement particles and the solute elements of Mg and Si in the Al6061 matrix and generate a large number of vacancies and dislocations in the crystal structure of the milled powders. These vacancies at the interface of oxide or carbide particles and Al matrix might provide nucleation sites for the solutes where Mg and Si could be segregated. The subsequent sintering processing promotes the formation of complex oxides or carbides. As the Al6061 composites are subjected to aging treatments, the solutes are difficult to diffuse back into the Al matrix. Consequently, the formation of Mg–Si clusters and GP zones is inhibited.

In early work, the Al6061-Y_2O_3 and Al6061-TiC composites show a pronounced increase in hardness compared with the Al6061 alloy sample without aging treatments [12]. In the present work, the aging precipitation behavior of the model materials can be associated to the age hardening curve. The hardness of the non-reinforced Al6061 alloy significantly increased after T6 treatment, which can be attributed to the formation of fine precipitates as shown in Figure 6. However, the Al6061 composites reinforced with Y_2O_3 and TiC particles have only a slight increase in hardness, and the time to peak aging was accelerated. These results can be ascribed to the fact that the Mg–Si precipitation is inhibited by the presence of the reinforcement particles. The amount of Mg in the Al matrix is insufficient to form Mg–Si clusters and subsequent β″ and β′ phase formation. Instead, the solute elements can then segregate in the reinforcement particles and the formation of complex oxides and carbides is encouraged, which can only contribute to a slight increase in hardness.

4. Conclusions

The aging behavior of the Al6061 alloy, as well as Al6061 composites fabricated by mechanical alloying under T6 and T4 treatments, has been studied. It is important to understand the effects of the presence of reinforcement on the age hardening, precipitation formation, and microstructural change of the Al6061 composites. The results show that the peak aging hardness of the Al6061 alloy shows a significant increase in hardness of about 66% after solution treatment. However, the Al6061 composites reinforced with Y_2O_3 and TiC particles demonstrate only a slight increase in hardness (only 15% to 22%) under T6 treatment. Age hardening of the materials can be associated with the formation of precipitation. TEM investigation shows that a large number of the needle-like precipitates were observed in the peak-aged Al6061 alloy, which line up along the <100> direction of the Al matrix. However, precipitates are hard to form in the peak-aged Al6061 composites. This result could be attributed to the fact that the Mg–Si precipitation can be inhibited by the presence of the reinforcement particles. The amount of Mg in the Al matrix of the two composite materials is insufficient to form Mg–Si clusters and GP zones. The solute elements segregate in the complex oxides or carbides and only contribute to a slight increase in hardness.

Acknowledgments: The authors would like to gratefully acknowledge financial support from the National Science Council of Taiwan under the grant MOST 105-2221-E-259-010. The authors would also like to thank Jian-Yih Wang, National Dong Hwa University, for Al6061 material support.

Author Contributions: C.L.C. conceived and designed the experiments; C.H.L. performed the experiments; C.L.C. and C.H.L. analyzed the data; C.L.C. wrote the paper.

Conflicts of Interest: The authors declare no conflict of interest.

References

1. Liu, Y.Q.; Cong, H.T.; Wang, W.; Sun, C.H.; Cheng, H.M. AlN nanoparticle-reinforced nanocrystalline Al matrix composites: Fabrication and mechanical properties. *Mater. Sci. Eng. A* **2009**, *505*, 151–156. [CrossRef]

2. Ma, Z.Y.; Li, Y.L.; Liang, Y.; Zheng, F.; Bi, J.; Tjong, S.C. Nanometric Si_3N_4 particulate-reinforced aluminum composite. *Mater. Sci. Eng. A* **1996**, *219*, 229–231. [CrossRef]

3. Gupta, M.; Surappa, M.K.; Qin, S. Development of Aluminium Based Silicon Carbide Particulate Metal Matrix Composite. *J. Mater. Process. Technol.* **1997**, *67*, 94–99. [CrossRef]

4. Rahimian, M.; Parvin, N.; Ehsani, N.; Mater, A. Investigation of particle size and amount of alumina on microstructure and mechanical properties of Al matrix composite made by powder metallurgy. *Mater. Sci. Eng. A* **2010**, *527*, 1031–1038. [CrossRef]

5. Ahamed, H.; Senthilkumar, V. Hybrid Aluminium Metal Matrix Composites and Reinforcement Materials: A Review. *Mater. Des.* **2012**, *37*, 182–192. [CrossRef]

6. Nemati, N.; Khosroshahi, R.; Emamy, M.; Zolriasatein, A. Sintering and Hardness Behavior of Fe-Al_2O_3 Metal Matrix Nanocomposites Prepared by Powder Metallurgy. *Mater. Des.* **2011**, *32*, 3718–3729. [CrossRef]

7. Fogagnolo, J.B.; Robert, M.H.; Torralba, J.M. Mechanically alloyed AlN particle-reinforced Al-6061 matrix composites: Powder processing, consolidation and mechanical strength and hardness of the as-extruded materials. *Mater. Sci. Eng. A* **2006**, *426*, 229–231. [CrossRef]

8. Krizik, P.; Balog, M.; Matko, I.; Sr, P.S.; Cavojsky, M.; Simancik, F. The effect of a particle–matrix interface on the Young's modulus of Al–SiC composites. *J. Compos. Mater.* **2015**, 1–10. [CrossRef]

9. Moghadam, A.D.; Schultz, B.F.; Ferguson, J.B.; Omrani, E.; Rohatgi, P.K.; Gupta, N. Functional Metal Matrix Composites: Self-lubricating, Self-healing, and Nanocomposites-An Outlook. *JOM* **2014**, *66*, 872–881. [CrossRef]

10. Zhang, Z.; Topping, T.; Li, Y.; Vogt, R.; Zhou, Y.; Haines, C.; Paras, J.; Kapoor, D.; Schoenung, J.M.; Lavernia, E.J. Mechanical behavior of ultrafine-grained Al composites reinforced with B4C nanoparticles. *Scr. Mater.* **2011**, *65*, 652–655. [CrossRef]

11. Li, Y.; Zhang, Z.; Vogt, R.; Schoenung, J.M.; Lavernia, E.J. Boundaries and interfaces in ultrafine grain composites. *Acta Mater.* **2011**, *59*, 7206–7218. [CrossRef]

12. Chen, C.-L.; Lin, C.-H. Effect of Y_2O_3 and TiC Reinforcement Particles on Intermetallic Formation and Hardness of Al6061 Composites via Mechanical Alloying and Sintering. *Metall. Mater. Trans. A* **2015**, *46*, 3687–3695. [CrossRef]

13. Li, B.; Zeng, B.L.H.; Fan, W.; Bai, Z. Effect of aging on interface characteristics of Al–Mg–Si/SiC composites. *J. Alloy. Compd.* **2015**, *649*, 495–499. [CrossRef]

14. Maisonnette, D.; Suery, M.; Nelias, D.; Chaudet, P.; Epicier, T. Effects of heat treatments on the microstructure and mechanical properties of A 6061 aluminium alloy. *Mater. Sci. Eng. A* **2011**, *528*, 2718–2724. [CrossRef]

15. Rao, P.N.; Singh, D.; Brokmeier, H.-G.; Jayaganthan, R. Effect of ageing on tensile behavior of ultrafine grained Al6061 alloy. *Mater. Sci. Eng. A* **2015**, *641*, 391–401. [CrossRef]

16. Dong, R.; Yang, W.; Yu, Z.; Wu, P.; Hussain, M.; Jiang, L.; Wu, G. Aging behavior of 6061Al matrix composite reinforced with high content SiC nanowires. *J. Alloy. Compd.* **2015**, *649*, 1037–1042. [CrossRef]

17. Lang, P.; Karadeniz, E.P.; Falahati, A.; Kozeschnik, E. Simulation of the effect of composition on the precipitation in 6xxx Al alloys during continuous heating DSC. *J. Alloy. Compd.* **2014**, *612*, 443–449. [CrossRef]

18. Vissers, R.; van Huis, M.A.; Jansen, J.; Zandbergen, H.W.; Marioara, C.D.; Andersen, S.J. The crystal structure of the β' phase in Al–Mg–Si alloys. *Acta Mater.* **2007**, *55*, 3815–3823. [CrossRef]

15

Effect of Initial Oriented Columnar Grains on the Texture Evolution and Magnetostriction in Fe–Ga Rolled Sheets

Jiheng Li [1,2], Chao Yuan [3], Qingli Qi [1], Xiaoqian Bao [1] and Xuexu Gao [1,*]

[1]   State Key Laboratory for Advanced Metals and Materials, University of Science and Technology Beijing, Beijing 100083, China; lijh@ustb.edu.cn (J.L.); qiql@xs.ustb.edu.cn (Q.Q.); bxq118@ustb.edu.cn (X.B.)

[2]   Department of Chemistry-Ångström Laboratory, Uppsala University, Uppsala 75121, Sweden

[3]   Grirem Advanced Materials Co., Ltd., Beijing 100088, China; chaoyuan916@sina.com

*   Correspondence: gaox@skl.ustb.edu.cn

Academic Editor: Hugo F. Lopez

Abstract: The effects of initial oriented columnar grains on the texture evolution and magnetostriction in $(Fe_{83}Ga_{17})_{99.9}(NbC)_{0.1}$ rolled sheets were investigated. The recrystallization texture evolution exhibited the heredity of initial orientations, concerning the formation of cube and Goss textures in the primary recrystallized sheet for the columnar-grained sample. Moreover, the growth advantage of Goss grains was more obvious than that of cube grains during the secondary recrystallization process. Because of the combined effect of this and Nb-rich precipitates as inhibitors, a sharp Goss texture and very large Goss grains were achieved in the secondary recrystallized sheet for the columnar-grained sample. For comparison, the secondary recrystallization in the equiaxed-grained sample was not fully developed although there were Nb-rich precipitates as inhibitors. We think this could be ascribed to the large particle size and premature coarsening of precipitates. Magnetostriction of the secondary recrystallized columnar-grained sheet was up to 232 ppm owing to the ideal Goss texture and quite large grain size. As for the equiaxed-grained sample, the magnetostriction was only 163 ppm in the secondary recrystallized sheet.

Keywords: magnetostriction; Fe–Ga alloy; texture; columnar-grains; rolling

---

1. Introduction

Magnetostrictive Fe–Ga alloys (Galfenol) have received increasing attention—particularly, a new kind of magnetostrictive smart material for actuator, sensor, and energy harvesting applications [1,2]. These interests stem from the fact that, unlike existing smart material systems, Galfenol is the first kind of material that shows a good combination of magnetostrictive properties and mechanical properties. The addition of Ga increases the magnetostrictive capability of Fe over tenfold up to 400 ppm ($\times 10^{-6}$) along the <100> direction in single crystal material [3]. Mechanically [4], Fe–Ga alloys are robust, as opposed to materials such as PZT, Ni–Mn–Ga, or Terfenol-D. In addition, Fe–Ga alloys have high permeability [5], a high Curie temperature [6], and highly thermal-stable ferromagnetism [7]. These make Fe–Ga alloys unique.

The high conductivity of Fe–Ga alloy requires its use in the form of thin sheets to avoid eddy current losses when it is used in high frequency. In order to produce thin sheets with reasonable robustness and magnetostrictive properties, efforts have been made to produce the textured sheets by rolling and secondary crystallization processes [8–12]. Up to now, the secondary recrystallization, also named abnormal grain growth (AGG), of Goss-oriented ({110}<001>) grains has been considered as the most effective way to achieve the sharp <001> orientation along the rolling direction in the rolled Fe–Ga alloy sheets [13–18]. Previous studies have reported the achievement of sharp Goss orientation, by the combined effects of NbC particles as inhibitors and sulfur-induced surface energy [15–17]. Presently, most studies are about the rolling and recrystallization behaviors of the Fe–Ga polycrystalline alloys with equiaxed grains, and mainly focus on the influence of the final annealing process on abnormal grain growth. Recently, we have prepared Goss-oriented Fe–Ga alloy sheets using <001> oriented columnar-grained alloys for rolling [18]. On one hand, as Fe–Ga polycrystalline alloys exhibit grain boundary embrittlement, the rollability of these alloys with columnar grains can be improved by suppressing the fracture along the transverse boundaries when the long axes of columnar grains are arranged along the rolling direction. On the other hand, the anisotropic feature of columnar grains is different from that of the equiaxed grains with a random texture, and the role of the restriction at the special columnar grain boundaries on orientation formation cannot be ignored during the rolling and recrystallization process. In this work, effects of the initial oriented columnar grains on texture evolution and magnetostriction in the rolled $(Fe_{83}Ga_{17})_{99.9}(NbC)_{0.1}$ alloy sheets are investigated. It proves that without a sulfur-induced surface energy effect, the use of initial oriented columnar grains for rolling can improve the secondary recrystallization, and a sharp Goss texture and quite large Goss grains are achieved in the secondary recrystallized alloy sheets, resulting in a high magnetostriction.

2. Materials and Methods

The alloys with nominal composition $(Fe_{83}Ga_{17})_{99.9}(NbC)_{0.1}$ were prepared from Fe (99.9%, weight percent), Ga (99.99%, weight percent), and master alloys of Nb–Fe and Fe–C. The columnar-grained rod was produced by directional solidification at a growth rate of $720 \ mm \cdot h^{-1}$. A detailed description of the directional solidification process could be found in Ref. [19]. The as-cast ingot was prepared by induction melting, and then hot forged to reduce casting defects. The slabs with a thickness of ~18 mm were cut by electrical discharge machining from the directionally solidified and the hot-forged samples, respectively. The long axes of columnar grains were arranged along the rolling direction when the directionally solidified sample was used for rolling, as shown in Figure 1. The slabs were hot-rolled at 1150 °C to ~2.1 mm, followed by warm rolling at 500 °C to ~1.1 mm. After an intermediate annealing at 850 °C for 5 min, further cold rolling was undertaken to make a final thickness of ~0.3 mm. The as-rolled $(Fe_{83}Ga_{17})_{99.9}(NbC)_{0.1}$ sheets, 12 mm × 16 mm cut by electrical discharge machining, were enclosed in quartz ampoules using 0.3 atm Ar as protecting gas. The sheets enclosed in the ampoules were primarily annealed at 850 °C for 6 min. After the primary annealing, samples were rapidly heated from 850 to 900 °C at a rate of 10 °C/min in the furnace, and they were then slowly heated from 900 to 1080 °C at a controlled rate of 0.25 °C/min, and they were finally cooled by air to room temperature without dwell at 1080 °C.

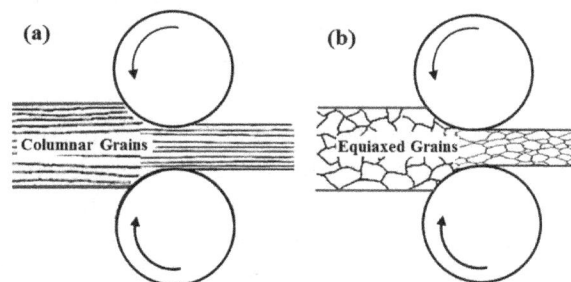

Figure 1. Schematic diagrams of the rolling method: (**a**) rolling of the columnar-grained sample; (**b**) rolling of the equiaxed-grained sample.

Microstructures and phases were characterized by optical microscopy (Carl Zeiss AG, Heidenheim, Germany) and X-ray diffraction (XRD) (Rigaku Corporation, Tokyo, Japan) respectively. Precipitates were examined by transmission electron microscopy (TEM) (Technai F30, FEI, Hillsboro, OR, USA), and energy dispersive X-rays spectroscopy (EDS) (Technai F30, FEI, Hillsboro, OR, USA) was used to identify the composition of precipitates. The texture was analyzed using electron back-scattering diffraction (EBSD). The EBSD was carried out on a SUPRA™ 55 field emission scanning electron microscope (Zeiss Supra 55, Oberkochen, Germany). The EBSD patterns were captured and analyzed to obtain the inverse pole figure (IPF) and the orientation distribution function (ODF). The magnetostriction was measured by strain gauge, and the gauges were positioned along the rolling direction. For the magnetostriction measurement ($\lambda_{//}$ and $\lambda_{\perp}$), a magnetic field parallel and perpendicular to the rolling direction (RD) was applied, respectively.

3. Results and Discussion

Figure 2 shows the microstructures, phases, and orientations of the directionally solidified and as-cast $(Fe_{83}Ga_{17})_{99.9}(NbC)_{0.1}$ alloys. In the specimen fabricated by directional solidification, some columnar grains with a width above 1000 μm are distributed homogeneously, as shown in the longitudinal optical photograph (Figure 2a). The grain boundary is parallel to the drawing direction (grain growth direction), because the grain morphology of the directionally solidified specimen is related to the direction of heat dissipation. The XRD pattern of the directionally solidified sample captured from the cross section of the rod is shown in Figure 2b. It demonstrates that the α-Fe phase (A2) is the dominant phase in the directionally solidified sample. The (200) peak dominates the pattern, rather than the (110) peak, indicating the preferred <100> orientation along the grain growth direction. Moreover, a very weak peak corresponding to NbC at ~40.29° is observed. Additionally, an unexpected small peak around 30.7° corresponding to the DO_3 phase (long-range order structure) appears, which could be attributed to the low temperature gradient (about 55 K/cm) and relatively slow cooling rate during the directional solidification process. In order to further detect the orientation information, the EBSD pattern on the cross section of the directionally solidified rod was captured. IPF shows that a strong <100> orientation was achieved, consistent with the dominant (200) diffraction peak in the XRD pattern, as seen in Figure 2c. By contrast, in Figure 2d, many equiaxed grains can be observed in the as-cast alloy. The main phase of A2 is visible in the XRD pattern of the as-cast alloy, as indicated in Figure 2e. The EBSD pattern of the as-cast alloy was captured on the same surface of plate sample used for XRD measurement. These equiaxed grains are without an obvious preferred orientation although a substantially weak <110> fiber texture can be seen in Figure 2f. This may be due to the fact that the dominant diffraction peak is usually (110) peak in random oriented Fe–Ga alloys, as shown in Figure 2e. The directionally solidified and the as-cast alloys are called the columnar-grained (CG) and the equiaxed-grained (EG) specimens, respectively.

To analyze the texture evolution during the rolling process, EBSD patterns were captured on the RD-ND (normal direction) section of the rolled sheets. IPF maps for the hot-rolled sheets and the warm-rolled sheets before and after intermediate annealing are shown in Figure 3. In the IPF maps, the red, blue, and green colors represent the crystal directions of <001>, <111>, and <101> arranged along the RD, respectively. All samples show the through-thickness structures and texture gradients on the lateral face. The <001> oriented columnar grains in the CG sample are elongated along the rolling direction during the rolling process, as shown in Figure 3a,b. An increase in the area of near green and blue colors indicates that the orientation begins to deviate from <001>. The deformation microstructure consists of the shear-deformed surface regions due to the friction between roll and sheet and a homogeneously deformed center region in the EG sample, as shown in Figure 3d,e. The grains in the surface region are markedly refined by the shearing stress, while the <110> fiber textured grains (green color shown in Figure 3d,e) are slightly refined under the compressive stress in the center region. In addition, there are many scanning blind spots shown by the white color in the surface region due to the many deformation defects by shearing, as shown in Figure 3a,b,d,e. Figure 3c,f show that the intermediate annealing, at 850 °C for 5 min, leads to the partial recrystallization of the warm-rolled sheets. The grain sizes are very inhomogeneous in these intermediate annealed sheets. In Figure 3c, a marked rotation from the <001> to <111> can be observed on both sides of the intermediate annealed CG sheet (indicated by black arrow) due to the recrystallization in the strong shear deformation region.

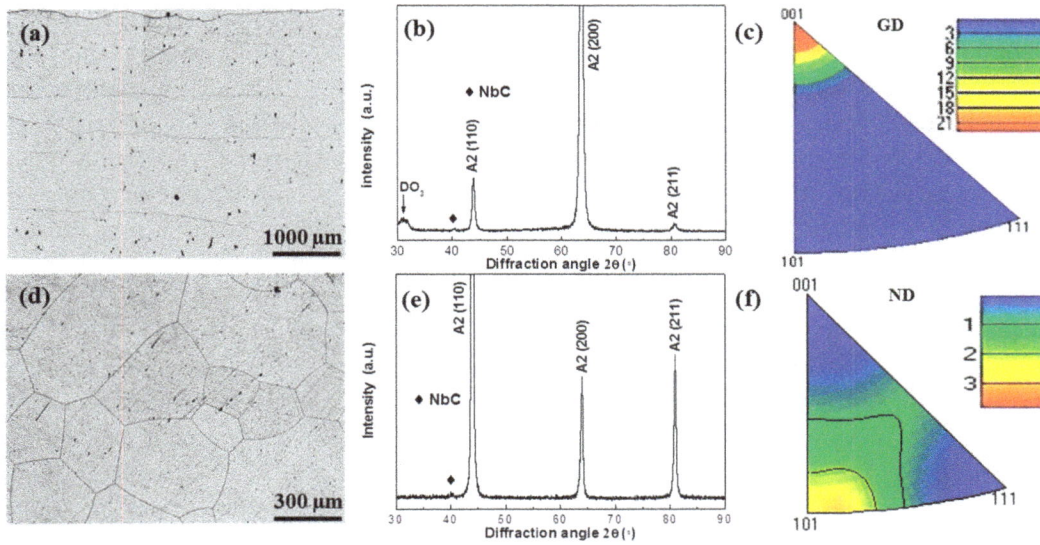

Figure 2. (**a**) Optical microstructure, (**b**) XRD pattern, and (**c**) IPF of the directionally solidified alloy. (**d**) Optical microstructure, (**e**) XRD pattern, and (**f**) IPF of the as-cast alloy. GD: growth direction of the directionally solidified rod; ND: normal direction of the plate sample used for XRD measurement.

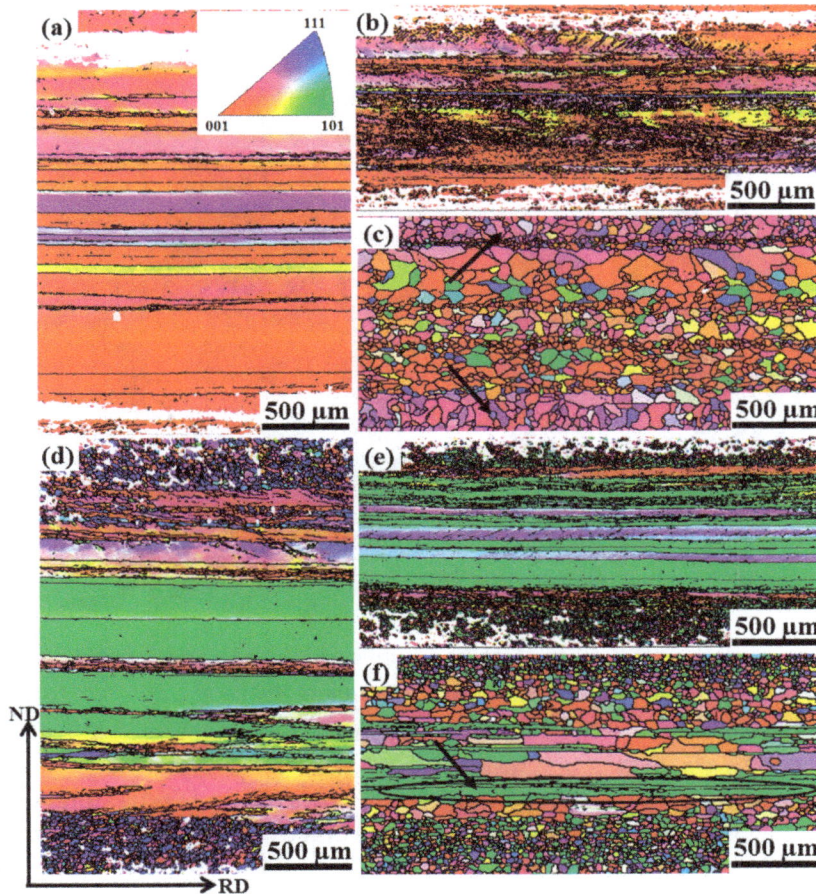

Figure 3. IPF maps of (**a**) the hot-rolled sheet and warm-rolled sheet (**b**) before and (**c**) after intermediate annealing for the CG sample; IPF maps of (**d**) the hot-rolled sheet and warm-rolled sheet (**e**) before and (**f**) after intermediate annealing for the EG sample. Grain boundaries are high angle boundaries ($\omega \geq 15°$).

Figure 4 presents the ODF ($\phi_2 = 45°$) plots corresponding to Figure 3. It shows that the major texture components in the hot and warm deformed CG samples are a strong Goss texture and a weaker near-cube texture {100}<001> (or {100}<021>), as seen in Figure 4a,b. In general, the initial texture has less influence on the formation of shear texture components, which is accordant with the study results of Shimizu et al. [20]: that any initial orientation can induce a sharp Goss texture after the hot rolling of a single crystal silicon–iron alloy. The near-cube texture likely comes from the slight rotations of the initial cube grains. The rotation from the <001> to <111> on both sides of the intermediate annealed CG sheet, which is reflected in the decrease in intensity of the Goss texture and the appearance of the {110}<113> texture, as shown in Figure 4c. In contrast, the rolling textures are dominated by {111}<uvw> and {100}<011> (also named by the 45° rotated cube texture) in the hot- and warm-rolled EG sample, as shown in Figure 4d,e. In this case, the change of grain orientation takes place as a consequence of shear on specific favorably oriented crystal planes and directions, and slip preferentially occurs on {hkl}<111> slip systems where {hkl} could be {110} or {112}. As is well known, this is common in body-centered cubic (BCC) metals. In addition, the texture of {100}<011>, which is a typical component in BCC metals at high reduction, is retained in the warm-rolled EG sample, which is attributed to the fact that the {100}<011> oriented grains do not lead to shear strain during deformation. After an intermediate annealing, orientations of the deformed {111}<112> textured grains change due to the recrystallization in the strong shear deformation region on both sides of the intermediate annealed EG sheet, while the retained strong {100}<011> texture could be attributed to the poor recrystallization behavior of the deformed grains (indicated by the black arrow in Figure 3f), as shown in Figure 4f.

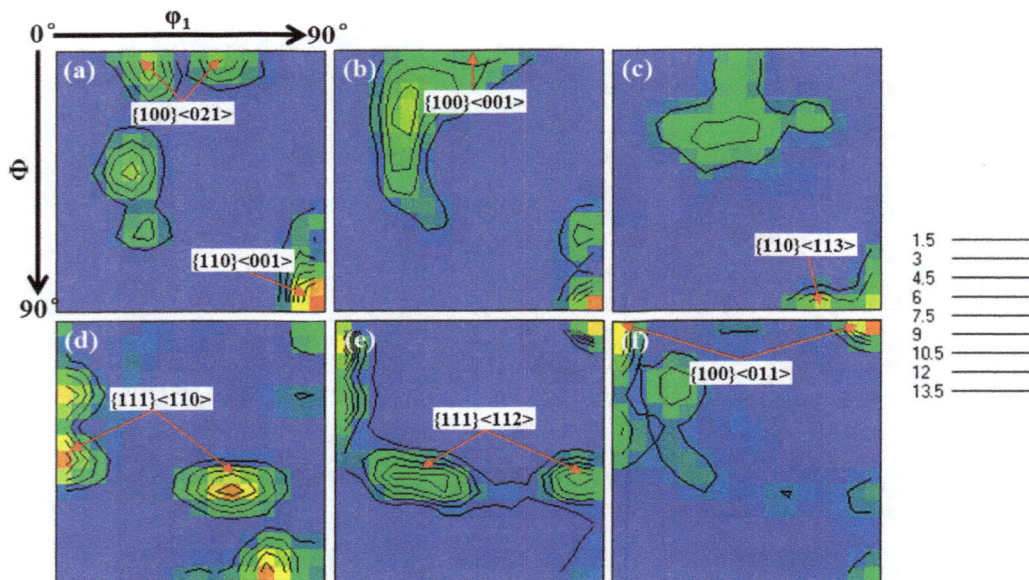

Figure 4. ODF plots at $\phi_2 = 45°$ section of (**a**) the hot-rolled sheet and warm-rolled sheet (**b**) before and (**c**) after intermediate annealing for the CG sample; ODF plots at $\phi_2 = 45°$ section of (**d**) the hot-rolled sheet and warm-rolled sheet (**e**) before and (**f**) after intermediate annealing for the EG sample.

Figure 5 displays the changes of magnetostriction in the CG and EG samples during the rolling process. The observed magnetostriction values are an average of three similarly treated samples, and the error bars show the standard deviation. As for the bulk samples, the magnetostriction of $\lambda_{//}$ is only measured for the as-cut CG and EG slabs, while the magnetostrictions of $\lambda_{//}$ and $\lambda_{\perp}$ are measured for the rolled and annealed sheets. The magnetostrictive value decreases sharply from 215 ppm in the as-cut CG slab to 24 ppm in the cold-rolled sheet, which is mainly ascribed to the deviation of orientation from the <001> direction and many severely deformed grains. Magnetostriction of the as-cut EG slab, which has no preferred orientation, is relatively small, and that of the rolled sheets decreases during the rolling process. The change of magnetostriction in the CG samples is markedly larger than in the EG samples, which could be ascribed to the non-uniform deformation of large columnar grains during the rolling process. This is likely another effect of the CG samples' having a different favorable initial texture, which changes upon rolling. Overall, after a heavy rolling reduction, magnetostriction of the cold-rolled CG and EG sheets are both very low.

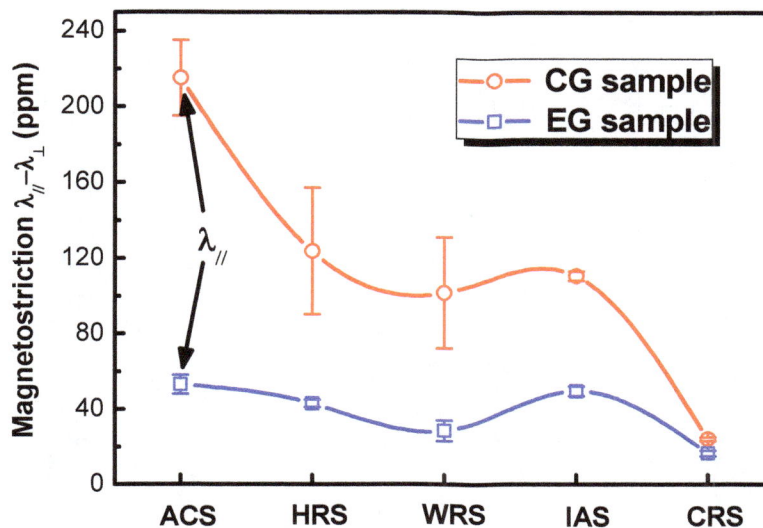

Figure 5. Magnetostriction of the CG and EG samples during rolling process. ACS, HRS, WRS, IAS, and CRS denote the sample of as-cut slabs, hot-rolled sheets, warm-rolled sheets, intermediate annealed sheets, and cold-rolled sheets, respectively.

The warm-rolled sheets with intermediate annealing are further cold-rolled to a final thickness of 0.3 mm, and the primary annealing is carried out at 850 °C for 6 min. The IPF maps and ODF plots of the cold-rolled sheets and the primary recrystallized sheets are shown in Figure 6. The figure shows obvious differences in the cold rolling texture and the primary recrystallization texture between CG and EG samples owing to their different initial orientation and grain morphology. The dominant texture is a sharp {223}<362> (near {113}<361>) with minor components of {100}<023>, {111}<112>, and {110}<001> in the cold-rolled CG sample, as shown in Figure 6a,b. With a heavy rolling reduction, most initial cube grains gradually rotate to {223}<362>. The rotation to {113}<361> is a typical path of cube grains during rolling, whereas a part of the initial Goss grains rotates to {111}<112> [21]. In addition, the shear deformation texture of {110}<001> (red color indicated by the black arrow in Figure 6a) remains in the surface and subsurface regions of the cold-rolled CG sample. After the primary recrystallization, in addition to a weak Goss texture, the CG sample shows predominantly a γ-fiber texture ({111}<110> and {111}<112>) and a near-cube texture, as seen in Figure 6c,d. Hu [22] has pointed out that recrystallization proceeds easily in the deformed γ-fiber- and cube-textured grains, whereas {100}<011> and {100}<023> grains can hardly recrystallize. As such, the primary recrystallized texture in the CG sample originates from the recrystallization of the deformed γ-fiber, cube, and Goss grains. The deformed {100}<023> grains are so stable in resisting recrystallization that they could only be consumed slowly by other oriented grains.

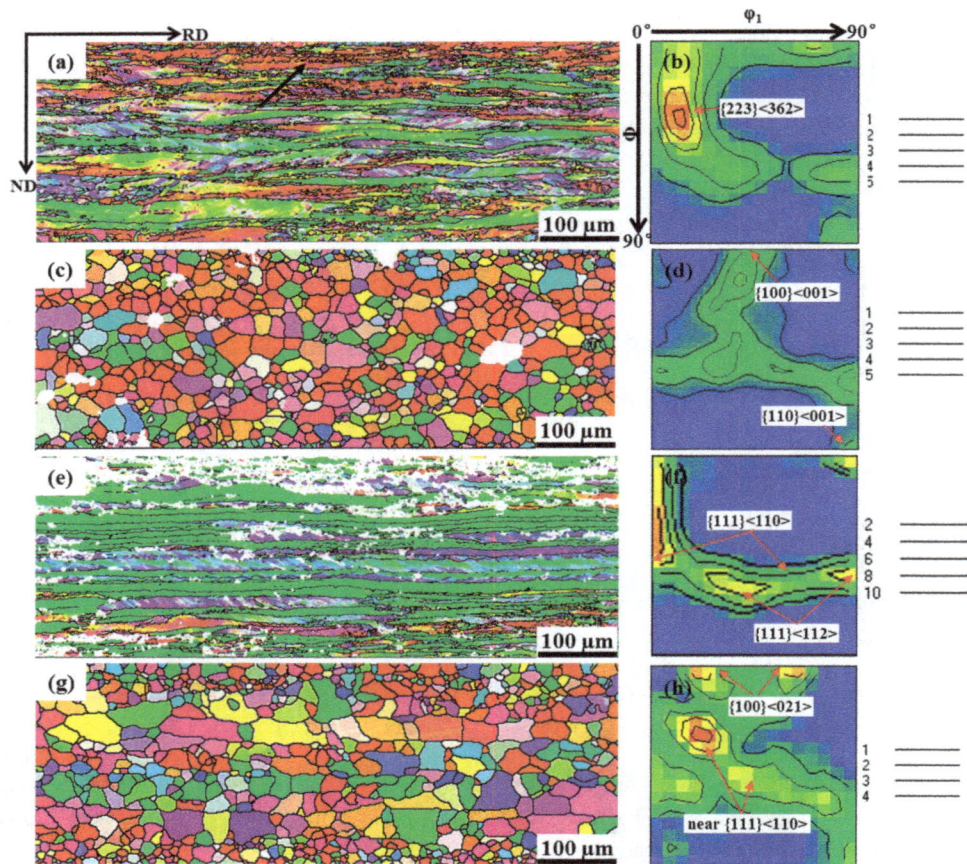

Figure 6. (**a**) IPF map and (**b**) ODF plot ($\phi_2 = 45°$) of the cold-rolled sheet for the CG sample; (**c**) IPF map and (**d**) ODF plot ($\phi_2 = 45°$) of the primary recrystallized sheet for the CG sample; (**e**) IPF map and (**f**) ODF plot ($\phi_2 = 45°$) of the cold-rolled sheet for the EG sample; (**g**) IPF map; (**h**) ODF plot ($\phi_2 = 45°$) of the primary recrystallized sheet for the EG sample. Grain boundaries are high angle boundaries ($\omega \geq 15°$).

By contrast, strong γ-fiber and {100}<011> textures are obtained in the cold-rolled EG sample, as shown in Figure 6e,f. Figure 6h indicates that near {111}<110> and {100}<021> textures and weak {110}<001> textures are formed in the primary recrystallized EG sample. For the equiaxed crystals in BCC metals, following the start of three main slip system families of {110}<111>, {112}<111>, and {123}<111>, the crystal orientation tends to {100}-{112}<110>, and {111}<110> or {111}<112>, and the α-fiber and γ-fiber textures form easily after recrystallization. Table 1 shows the area fractions of differently oriented grains in the primary recrystallized sheets. It can be seen that, with a deviation from the ideal texture within 20°, the area fractions of {100}<001>, {100}<011>, {111}<110>, and {111}<112> textures in the CG sample are higher than that in the EG sample. Especially, the area fraction of {100}<001> texture with a deviation within 20° is up to 16.8% in the CG sample, which is nearly twice that of in the EG sample. Moreover, the slightly higher area fraction of Goss texture is also visible in the CG sample. A part of the primary Goss-oriented grains may come from the initial cube grains. Because, in addition to the rotation to {223}<362>, the cube grains also rotate to Goss orientation by shearing during the rolling process, and the recrystallization grains have the same orientation as the sub-band, which serves as the point of origin of the nucleus in deformed Goss grains. This observation agrees with the results obtained in the rolled single crystal silicon–iron alloy [23]. Overall, the primary recrystallization texture in the CG sample presents the heredity of initial orientations, which is reflected in the formation of cube and Goss textures in the primary recrystallized sheet. This could be explained by the theory of oriented nucleation.

Table 1. Area fractions of different oriented grains in the primary recrystallized sheets (0°~20° is deviation degree from the ideal texture).

Texture Component	CG Sample		EG Sample	
	0°~15°	15°~20°	0°~15°	15°~20°
{110}<001>	3.56	5.19	3.43	3.77
{100}<001>	6.89	9.91	4.29	5.11
{100}<011>	2.44	1.91	0.44	1.15
{111}<112>	8.10	8.80	7.10	8.70
{111}<110>	7.03	4.27	4.21	4.47

As is well-known, the primary recrystallized microstructure and texture are important for the final formation of a sharp Goss texture by abnormal grain growth. The primary recrystallized sheets go through a continuous heating process from 900 to 1080 °C to induce the abnormal grain growth. IPF maps and ODF plots for the secondary recrystallized sheets are shown in Figure 7. In IPF maps, the red, blue, and green colors represent the crystal directions of <001>, <111>, and <101> arranged along the ND, respectively. Figure 7a demonstrates that most of the grains in the CG sample are very large, up to several centimeters, although some small grains still exist. We think this is attributed to the abnormal growth of Goss grains. As a result, a sharp Goss texture is achieved in the secondary recrystallized CG sample, as shown in Figure 7b. By contrast, it can be seen from Figure 7c that the size of the {110} textured grains is not very large in the secondary recrystallized EG sample, and many grains with other orientations remain. This indicates that, although the abnormal grain growth also takes place in the EG sample during the continuous heating process, the secondary recrystallization is not complete. Correspondingly, Figure 7d displays a texture of {110}<113> in addition to a strong Goss texture.

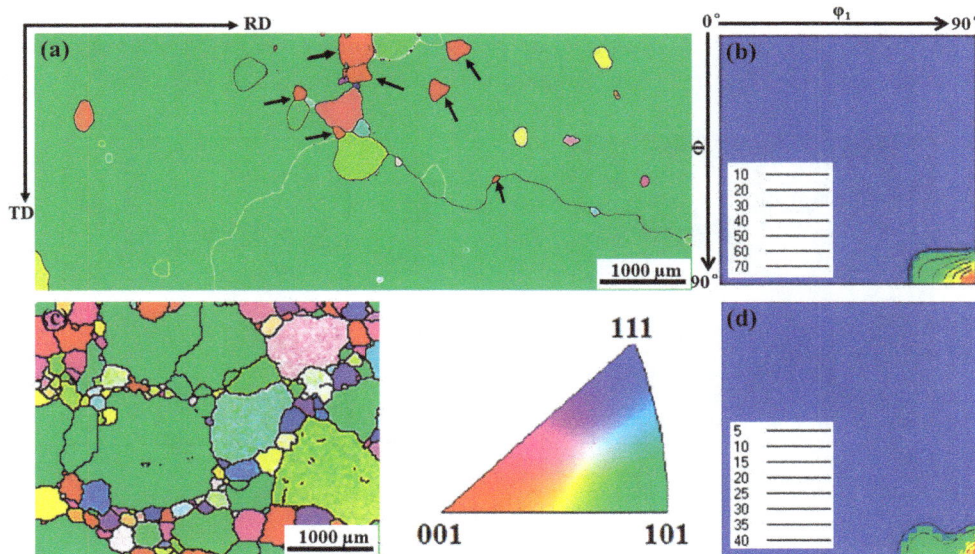

Figure 7. (a) IPF map and (b) ODF plot ($\phi_2 = 45°$) of the secondary recrystallized sheet for the CG sample; (c) IPF map and (d) ODF plot ($\phi_2 = 45°$) of the secondary recrystallized sheet for the EG sample. The low angle boundaries ($\omega < 15°$) are shown in white; the high angle boundaries ($\omega \geq 15°$) are shown in black.

In Goss-oriented silicon steels, precipitates as inhibitors are used to inhibit the grain growth of primary recrystallization, and an inhibitor is one of the basic conditions for the occurrence of secondary recrystallization. The abnormal grain growth accompanied by precipitated particles of sizes smaller than 0.2 μm is commonly observed in grain-oriented silicon steels. In addition, various mechanisms for the development of secondary recrystallization have already been proposed [24–26]. The coincidence site lattice (CSL) model and the high energy grain boundary (HEGB) model are frequently used to quantify the grain boundary characteristics in Goss-textured silicon steels during the abnormal grain growth process. Theoretical interpretations for the development of secondary recrystallization in the oriented silicon steels are also applicable for Fe–Ga alloys, because they have the same BCC structure. Figure 8 shows the TEM image of Nb-rich precipitates in the primary recrystallized EG sample. It can be found that the sizes of most Nb-rich precipitates are larger than 0.2 μm in the primary recrystallized EG sample. Furthermore, Nb-rich precipitates would be further coarsened during the continuous heating process, resulting in a decrease in inhibition force. Therefore, the large particle size and premature coarsening of precipitates would be the reason for the incomplete secondary recrystallization in the EG sample. In addition, there are more Goss-oriented grains in the primary recrystallized CG sample, which could provide more nuclei for the secondary recrystallization. On the other hand, although the orientation of most initial Goss grains rotated to {111}<112> during the rolling process, Goss grains easily increase in size during the recrystallization annealing process due to the high mobility boundaries between them and the deformed matrix. The {111}<112> oriented grains are rotated by about 35° with respect to the {110}<001> orientation, which produces boundaries close to the Σ9 coincident orientation. Thus, higher diffusivity can be obtained on the boundaries between {111}<112> and Goss grains [27]. Moreover, Figure 7a suggests that, in the late stage of secondary recrystallization, Goss grains grow competitively with cube grains (shown by black arrows), and the growth advantage of Goss grains is more obvious than that of cube grains. This is consistent with the result obtained in silicon steel by Zhang et al. [28]. Goss grains annex other textured grains, and several adjacent Goss grains then meet and merge into a large Goss grain, resulting in a sharp Goss texture finally. This growth pattern is perhaps the reason that the secondary recrystallization could be quickly completed from the beginning to the end, and very large Goss grains can be obtained. In addition, the heredity of initial oriented columnar grains, which leads to strong cubic and Goss orientations in the primary recrystallized sheet, possibly compensate to some extent for the adverse effect of the insufficient inhibition of inhibitors on the development of secondary recrystallization.

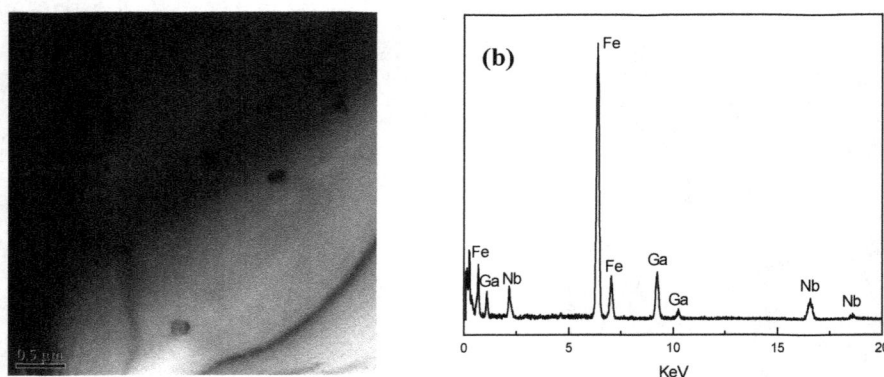

Figure 8. (a) TEM image and (b) EDS profile of Nb-rich precipitates in the primary recrystallized EG sample.

Figure 9 shows magnetostrictive curves of the secondary recrystallized sheets. After the secondary recrystallization annealing, the magnetostriction $(\lambda_{//}-\lambda_{\perp})$ of the CG sample has an increase up to 232 ppm due to the ideal Goss texture and the improvement in the crystal perfection after recrystallization. For comparison, the magnetostriction is only 163 ppm in the secondary recrystallized EG sample. Moreover, there is little hysteresis in the magnetostriction vs. magnetic field, resulting in an almost reversible magnetostriction.

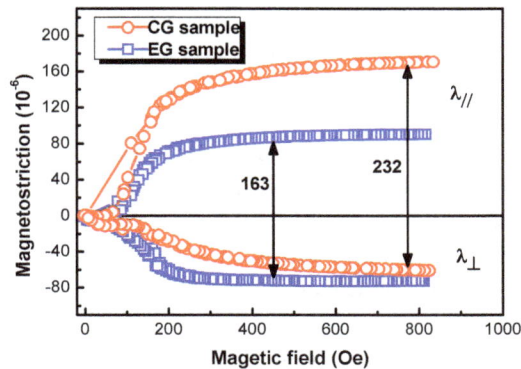

Figure 9. Magnetostrictive curves of the secondary recrystallized sheets.

4. Conclusions

(1) The dominant texture in the cold-rolled CG sample is a sharp {223}<362> with minor components of weak {100}<023>, {111}<112>, and {110}<001>, whereas a strong γ-fiber and {100}<011> textures can be obtained in the cold-rolled EG sample. During the rolling process of the CG sample, most initial cube grains rotate gradually to {223}<362> or to Goss orientation by shearing; meanwhile, the orientation of initial Goss grains rotates to {111}<112>. By contrast, there is a conventional rotation path from α-fiber to γ-fiber during the rolling process of the EG sample.

(2) The recrystallization texture evolution in the CG sample presents the heredity of initial orientations, concerning the formation of cube and Goss textures in the primary recrystallized sheet. Moreover, the growth advantage of Goss grains is higher than that of cube grains during the secondary recrystallization process. By the combined effect of this and Nb-rich precipitates as inhibitors, a sharp Goss texture and some very large Goss grains are achieved in the secondary recrystallized CG sample. In contrast, the development of secondary recrystallization in the EG sample was not fully completed, which could be attributed to the large particle size and premature coarsening of precipitates.

(3) Magnetostriction of the secondary recrystallized CG sample is up to 232 ppm due to the ideal Goss texture and very large grain size. As for the EG sample, the magnetostriction is only 163 ppm in the secondary recrystallized sheet.

Acknowledgments: This study was financially supported by the National Natural Science Foundation of China (No.51271019, 51501006), and partly supported by a scholarship from the China Scholarship Council. The authors thank Ya Hu for her critical reading of the manuscript and helpful suggestions.

Author Contributions: Jiheng Li and Chao Yuan conceived of and designed the experiments; Jiheng Li, Chao Yuan, and Qingli Qi performed the experiments; Jiheng Li, Xiaoqian Bao, and Xuexu Gao analyzed the data; Jiheng Li wrote the paper.

Conflicts of Interest: The authors declare no conflict of interest.

References

1. Guruswamy, S.; Srisukhumbowornchai, N.; Clark, A.E.; Restorf, J.B.; Wun-Fogle, M. Strong, ductile, and low-field-magnetostrictive alloys based on Fe–Ga. *Scr. Mater.* **2000**, *43*, 239–244. [CrossRef]
2. Clark, A.E.; Restorf, J.B.; Wun-Fogle, M.; Lograsso, T.A.; Schlagel, D.L. Magnetostrictive Properties of Body-Centered Cubic Fe–Ga and Fe–Ga-Al Alloys. *IEEE Trans. Magn.* **2000**, *36*, 3238–3240. [CrossRef]
3. Clark, A.E.; Hathaway, K.B.; Wun-Fogle, M.; Restorf, J.B.; Lograsso, T.A.; Keppens, V.M.; Petculescu, G.; Taylor, R.A. Extraordinary magnetoelasticity and lattice softening in bcc Fe–Ga alloys. *J. Appl. Phys.* **2003**, *93*, 8621–8623. [CrossRef]
4. Kellogg, R.A.; Russell, A.M.; Lograsso, T.A.; Flatau, A.B.; Clark, A.E.; Wun-Fogle, M. Tensile properties of magnetostrictive iron-gallium alloys. *Acta Mater.* **2004**, *52*, 5043–5050. [CrossRef]
5. Clark, A.E.; Wun-Fogle, M.; Restorf, J.B.; Lograsso, T.A.; Cullen, J.R. Effect of Quenching on the Magnetostriction of $Fe_{1-x}Ga_x$ (0.13 < x < 0.21). *IEEE Trans. Magn.* **2001**, *37*, 2678–2680.

6.   Clark, A.E.; Wun-Fogle, M.; Restorf, J.B.; Lograsso, T.A. Magnetostrictive Properties of Galfenol Alloys under Compressive Stress. *Mater. Trans.* **2002**, *43*, 881–886. [CrossRef]

7.   Kellogg, R.A.; Flatau, A.B.; Clark, A.E.; Wun-Fogle, M.; Lograsso, T.A. Temperature and stress dependencies of the magnetic and magnetostrictive properties of $Fe_{0.81}Ga_{0.19}$. *J. Appl. Phys.* **2002**, *91*, 7821–7823. [CrossRef]

8.   Srisukhumbowornchai, N.; Guruswamy, S. Crystallographic textures in rolled and annealed Fe–Ga and Fe-Al alloys. *Metall. Mater. Trans. A* **2004**, *35*, 2963–2970. [CrossRef]

9.   Li, J.H.; Gao, X.X.; Zhu, J.; Bao, X.Q.; Xia, T.; Zhang, M.C. Ductility, texture and large magnetostriction of Fe–Ga-based sheets. *Scr. Mater.* **2010**, *63*, 246–249. [CrossRef]

10. Summers, E.M.; Meloy, R.; Na, S.M. Magnetostriction and texture relationships in annealed galfenol alloys. *J. Appl. Phys.* **2009**. [CrossRef]

11. He, Z.H.; Sha, Y.H.; Fu, Q.; Lei, F.; Zhang, F.; Zuo, L. Secondary recrystallization and magnetostriction in binary $Fe_{81}Ga_{19}$ thin sheets. *J. Appl. Phys.* **2016**, *119*, 123904–123906. [CrossRef]

12. Sun, A.L.; Liu, J.H.; Jiang, C.B. Recrystallization, texture evolution, and magnetostriction behavior of rolled $(Fe_{81}Ga_{19})_{98}B_2$ sheets during low-to-high temperature heat treatments. *J. Mater. Sci.* **2014**, *49*, 4565–4675. [CrossRef]

13. Na, S.M.; Yoo, J.H.; Flatau, A.B. Abnormal (110) grain growth and magnetostriction in recrystallized Galfenol with dispersed niobium carbide. *IEEE Trans. Magn.* **2009**, *45*, 4132–4135.

14. Meloy, R.; Summers, E.M. Magnetic property-texture relationships in galfenol rolled sheet stacks. *J. Appl. Phys.* **2011**. [CrossRef]

15. Na, S.M.; Flatau, A.B. Single grain growth and large magnetostriction in secondarily recrystallized Fe–Ga thin sheet with sharp Goss (011) [100] orientation. *Scr. Mater.* **2012**, *66*, 307–310. [CrossRef]

16. Na, S.M.; Flatau, A.B. Surface-energy-induced selective growth of (001) grains in magnetostrictive ternary Fe–Ga-based alloys. *Smart Mater. Struct.* **2012**, *21*, 055024. [CrossRef]

17. Na, S.M.; Flatau, A.B. Global Goss grain growth and grain boundary characteristics in magnetostrictive Galfenol sheets. *Smart Mater. Struct.* **2013**, *22*, 125026. [CrossRef]

18. Yuan, C.; Li, J.H.; Zhang, W.L.; Bao, X.Q.; Gao, X.X. Sharp Goss orientation and large magnetostriction in the rolled columnar-grained Fe–Ga alloys. *J. Magn. Magn. Mater.* **2015**, *374*, 459–462. [CrossRef]

19. Li, J.H.; Yuan, C.; Zhang, W.L.; Bao, X.Q.; Zhu, J.; Gao, X.X. Retaining the <001> orientation from initial columnar grains and magnetostriction in binary Fe–Ga alloy sheets. *Mater. Trans.* **2015**, *56*, 1940–1944. [CrossRef]

20. Shimizu, Y.; Ito, Y.; Iida, Y. Formation of the Goss Orientation near the Surface of 3pct Silicon Steel during Hot Rollin. *Metall. Trans. A* **1986**, *17*, 1323–1334. [CrossRef]

21. Cheng, L.; Zhang, N.; Yang, P.; Mao, W.M. Retaining {100} texture from initial columnar grains in electrical steels. *Scr. Mater.* **2012**, *67*, 899–902. [CrossRef]

22. Hu, H. Annealing of Silicon–iron single crystals. In *Recovery and Recrystallization of Metals*; Himmel, L., Ed.; Interscience: New York, NY, USA, 1963; Volume 14, pp. 311–362.

23. Walter, J.L.; Koch, E.F. Substructures and recrystallization of deformed (100) [001]-oriented crystals of high-purity silicon–iron. *Acta Mater.* **1963**, *11*, 923–938. [CrossRef]

24. Homma, H.; Hutchinson, B. Orientation dependence of secondary recrystallization in silicon–iron. *Acta Mater.* **2003**, *51*, 3795–3805. [CrossRef]

25. Hayakawa, Y.; Szpunar, J.A. A new model of Goss texture development during secondary recrystallization of electrical steel. *Acta Mater.* **1997**, *45*, 4713–4720. [CrossRef]

26. Rajmohan, N.; Szpunar, J.A.; Hayakawa, Y. A role of fractions of mobile grain boundaries in secondary recrystallization of Fe-Si steels. *Acta Mater.* **1999**, *47*, 2999–3008. [CrossRef]

27. Matsuo, M. Texture control in the production of grain oriented silicon steels. *ISIJ Int.* **1989**, *29*, 809–827. [CrossRef]

28. Zhang, N.; Yang, P.; Mao, W.M. Influence of columnar grains on the recrystallization texture evolution in Fe-3%Si electrical steel. *Acta Metall. Sin.* **2012**, *48*, 307–314. (In Chinese) [CrossRef]

16

Properties of Resistance Spot-Welded TWIP Steels

Havva Kazdal Zeytin [1], Hayriye Ertek Emre [2],* and Ramazan Kaçar [2]

[1]   TUBITAK MAM, Gebze, Kocaeli 41470, Turkey; havva.zeytin@mam.gov.tr
[2]   Department of Manufacturing Engineering, Karabuk University, Karabuk 78050, Turkey; rkacar@karabuk.edu.tr
*   Correspondence: hayriyeertek@karabuk.edu.tr

Academic Editors: Halil Ibrahim Kurt, Adem Kurt and Necip Fazil Yilmaz

Abstract: High manganese TWIP (twinning-induced plasticity) steels are particularly attractive for automotive applications because of their exceptional properties of strength combined with an excellent ductility. However, the microstructure and properties of TWIP steels are affected by excessive thermal cycles, such as welding and heat treatment. This paper deals with characterization and understanding the effect of welding current and time on the mechanical properties and microstructure of the resistance spot welded TWIP steel. For this purpose, weld nugget diameter was evaluated and the hardness, tensile shear strength of the weldment, and failure mode of samples were also determined. It has been found that the tensile shear strength of the samples increased with increasing welding current and welding time without expulsion, which reduces the strength of the weldment. Tensile shear samples failed by a partial interfacial fracture mode for low-heat input welds. The pullout fractures were observed with a sufficient heat input without expulsion.

Keywords: TWIP steel; welding parameters; resistance spot weld; mechanical properties; microstructure

1. Introduction

There is an increasing demand for high strength steel sheets in the automotive industry in order to improve the fuel efficiency, occupant's safety, and reduction of auto body weight [1]. More commonly used advanced high strength steels (AHSS) are dual phase (DP), transformation induced plasticity (TRIP), and twinning-induced plasticity (TWIP). TRIP and TWIP are highly-deformable steels that give them certain amounts of plasticity for machine pressing with a useful hardening characteristic in the event of an accident. These attractive properties stem from a fully-austenitic structure and sufficient principal glide plane symmetry, known as the "twinning-induced plasticity" (TWIP) effect [2]. Several alloying concepts for TWIP steels have been developed and published and the most popular ones are based on the Fe-Mn-C [3,4] or Fe-Mn-Al-Si [5,6] systems.

Resistance spot welding is an assembly process widely used in the automotive industry for joining steel sheet components. Generally, there are three measures for the quality evaluation of resistance spot welds, including physical weld attributes, mechanical properties, and failure mode [1,7,8]. Due to high alloying content of AHSS, which influences the heat generation during resistance spot welding, high alloying content increase the bulk resistivity of the metal, which leads to excessive heat at the interface and, therefore, suitable current range shifts to the lower current side. The failure mode of spot welded AHSS's are often the interfacial mode [9].

Spena et al. [10] carried out a preliminary study on the effects of the main important process parameters (welding time, welding current, and clamping force) on the mechanical and microstructural properties of resistance spot-welded TWIP sheets (with 22.4% Mn content) of 1.4 mm thickness with using the L-9(3^3) orthogonally array. They noted that the tensile shear strength of the welded joints mainly increases as the welding current increases without any expulsion. Their study show that there is an almost linear relationship between the tensile shear strength and the weld spot size that can be defined for the samples that failed with an interfacial fracture [10]. Saha et al. [11] conducted a study on resistance spot welded high-Mn (18% Mn) TWIP steels with 1.4 mm thickness. They confirm that due to high chemical composition in high-Mn steels; expulsion occurred earlier and high welding current increases the cracking tendency because of increasing nugget pressure and the tensile stress in heat affected zone (HAZ) during cooling [11]. Deformation behavior of high-manganese TWIP steels has been widely studied in relation to microstructure and texture evolution by microscopy analysis scanning electron microscope (SEM) and transmission electron microscopy (TEM), X-ray diffraction (XRD) measurements [2,12–21]. However, weldability of such steel has been poorly investigated. Due to insufficient knowledge available regarding the resistance spot weldability of TWIP steel, it is still not widely accepted by automobile industries [9].

In this work, the effects of welding parameters, such as welding current and welding time, on the mechanical properties and metallurgical characterization of spot-welded high Mn and Al (32% Mn, 3.16% Al) TWIP steel in 1 mm thickness is investigated in detail. For this purpose, the microstructure of welded samples are evaluated and tensile shear load bearing capacity, failure mode, and hardness distribution of weldment are determined for weld quality.

2. Materials and Methods

In this study, high-Mn-Al (TWIP) steel which was manufactured experimentally, having a chemical composition of 0.024% C, 32% Mn, 2.36% Si and 3.16% Al, Bal. % Fe with minor alloying elements (<0.01% B) was used. The bulk specimens were cut into samples with dimensions of 100 mm × 25 mm × 1 mm and were cleaned with ethanol to remove dirt and surface oil prior to joining (Figure 1).

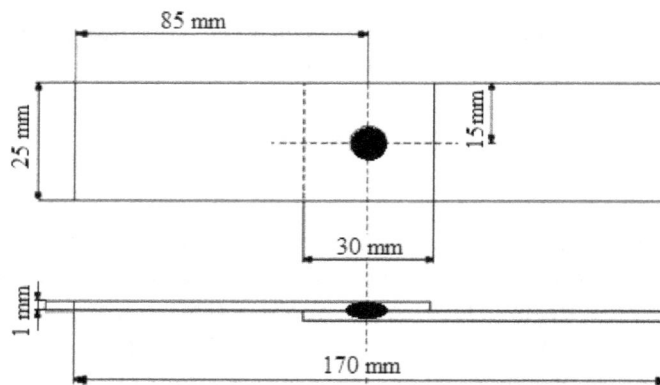

Figure 1. Geometry of the resistance spot-welded tensile shear test specimen.

Test samples were resistance spot welded using a pneumatic, phase-shift-controlled AC spot welding machine with a capacity of 60 kVA. The water-cooled F16 type Cu-Cr alloy spherical head electrode was employed for joining process. A set of samples were spot welded at a constant welding time with various welding currents. However, the other set of samples was welded at a selected welding current of 7 kA which gave the optimum tensile strength for various welding times. The welding parameters were given in Table 1.

Tensile shear load bearing capacity of weldment was determined by subjecting the specimens to tensile testing at ambient temperature. Three tensile shear test samples were prepared and tested for each of the weld variables by using a Shimadzu testing machine ((Shimadzu Sanjo Works, Kyoto, Japan). The information about thetensile shear test sample has been given in Figure 1. The tensile shear test was performed at a crosshead speed of 5 mm/min.

A welded sample was cross-sectioned through the center of the weld nugget for metallographic evaluation and hardness measurements. The test sample was mounted, ground, polished, and finally electrically etched in a solution containing 10% HNO_3 and 90% water for 2 s and then etched in a solution containing 3% HNO_3 and 97% ethanol for 2 s (3% nital solution). Optical examination of specimens was carried out by using a Nikon DIC microscope (Nikon Instruments, Karfo-Karacasulu Dis. Tic. A.S., Istanbul, Turkey) and Zeiss Ultra Plus type SEM microscope (Leibniz Institute for Solid State and Materials Research, Helmholtzstraße, Dresden).

Table 1. Welding parameters.

Welding Current (kA)	Electrode Force (kN)	Weld Time (Cycle)	Holding Time (Cycle)	Squeeze Time (Cycle)	Clamping Time (Cycle)
3 5 7 9	6	20	20	25	15
7	6	5 10 15 20 25 30	20	25	15

Note: (1 cycle = 0.02 s).

The Vickers microhardness measurement was carried out on a diagonal length to the weld nugget, heat-affected zone, and base metal with a load of 500 g and a loading time of 10 s (Figure 2).

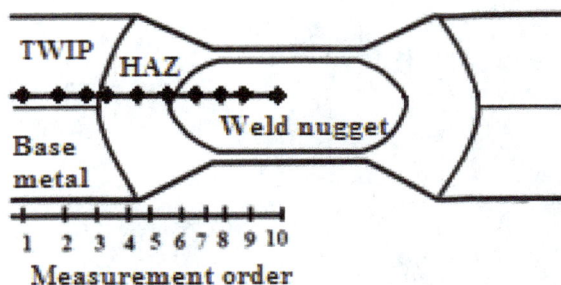

Figure 2. Schematic illustration of hardness measurements.

3. Results and Discussions

3.1. Fracture Characteristics of Sample

The failure of the resistance spot welded sample occurs generally in three modes: pull-out failure (PF), partial interfacial failure (PIF), and interfacial failure (IF) [22–24]. These main failure mechanisms and failure zones are: (as identified and reported by Choa et al. [25]) (i) strain localization in the base metal/subcritical HAZ; and (ii) ductile shear at the interface in the weld nugget. In general, the maximum tensile shear strength is obtained when the nugget tears from the sheet [26]. An effect of welding parameters on the failure mode of tensile shear test samples is shown (Figure 3a,b).

As seen in Figure 3a,b, the tensile shear tested sample failed in PIF mode from HAZ through the weld nugget made with a low heat input (at 3 kA welding current and from five to 15 cycles welding time). The detailed fracture surface of the sample that failed in PIF mode is shown in Figure 4a–d. As seen in Figure 4a,b, the crack initiated at the steel sheet/steel sheet interface and propagated along the weld nugget circumference in PIF failing mode [11]. Figure 4c,d shows the PIF mode with ductile characteristics in the weld nugget periphery. In the case of lack of defects, cracks normally initiated at the edge of weld nugget or in the HAZ (Figure 4e,f). The primary cause of weakening of the weldment may be identified as the second phase or precipitate phases in the HAZ, which may weaken the transition zone from weld nugget and HAZ and grain growth occurred in HAZ adjacent to the weld nugget.

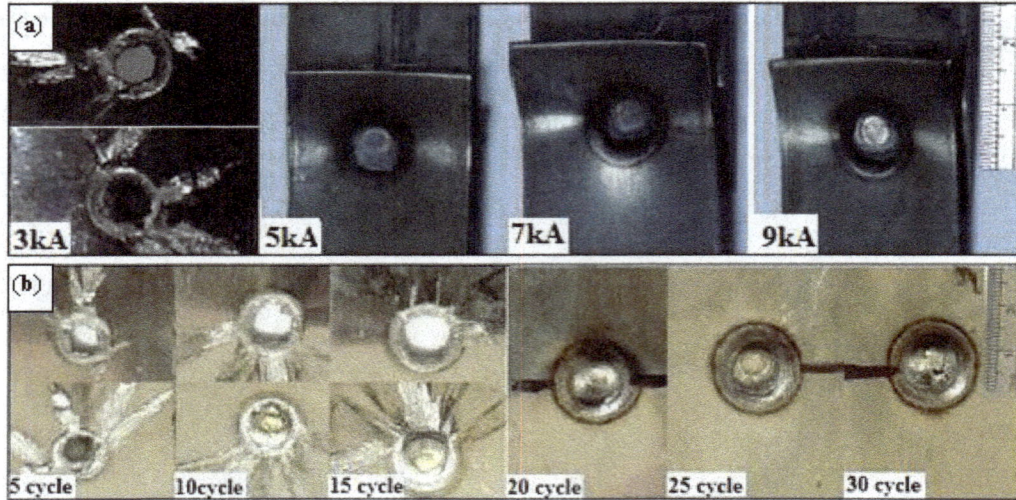

Figure 3. Failure mode of welded samples: (**a**) constant welding time for various welding currents; and (**b**) constant welding current for various welding times.

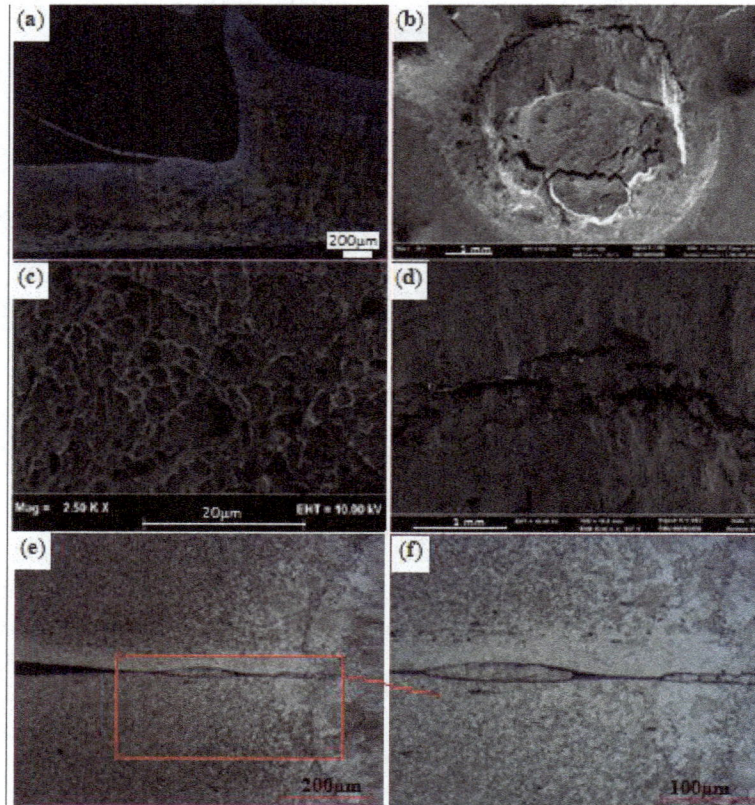

Figure 4. PIF mode of a spot-welded sample: (**a**) PIF; (**b**) macrograph of weldment; (**c**) fracture in weld nugget; (**d**) HAZ; (**e**) microstructure of HAZ; (**f**) high magnification of HAZ microstructure

Transition from PIF to PF mode depends on the physical characteristics, strength of the crack initiation zone, and especially the weld nugget sizes [11,27,28]. In this study, the transition from PIF to PF failure mode was observed for tensile shear tested samples which were joined with higher than 5 kA welding current and 15 cycles welding time. The desired PF mode occurred along the circumference of the weld nugget in the samples which were joined with from 5 kA to 9 kA welding current and 20–25 cycles welding time. The PF mode can be attributed to the strain concentration and grain growth in HAZ, which can encourage the starting of failure from this region [24]. Circumferential failure can be seen under pure opening loading conditions (with a loading angle of 0°) as defined in Lin et al. [28]. The strength of the region where failure occurred is important for the tensile shear load bearing capacity of weldment [29]. Expulsion of molten metal was observed for all welding parameters. However, it is believed that the low amount of expulsion did not affect the enlargement of the nugget size and tensile shear strength of the weldment up to the 7 kA welding current and 25-cycle welding parameters. A high amount of expulsion of molten metal from the fusion zone was observed in high welding parameters (over 7 kA welding current and 25 cycles of welding time) with respect to the high input or electrode pressure.

Due to excessive heat input and consequently higher electrode/sheet interface temperature, the amount of plastic deformation in the sheet surface under the applied electrode force is amplified, resulting in excessive electrode indentation [30]. It is believed that the excessive electrode indentation induced a surface crack which is shown in Figure 3b (30 cycles of welding time).

3.2. Effect of Welding Parameters on Weld Nugget Size, Strength, and Hardness

The most important factors that affect spot weld quality are weld nugget size, strength and ductility, surface appearance, weld penetration, and internal discontinuities [23]. The relationship between welding parameters and weld nugget size, as well as tensile shear strength, is determined and shown graphically in Figure 5a,b.

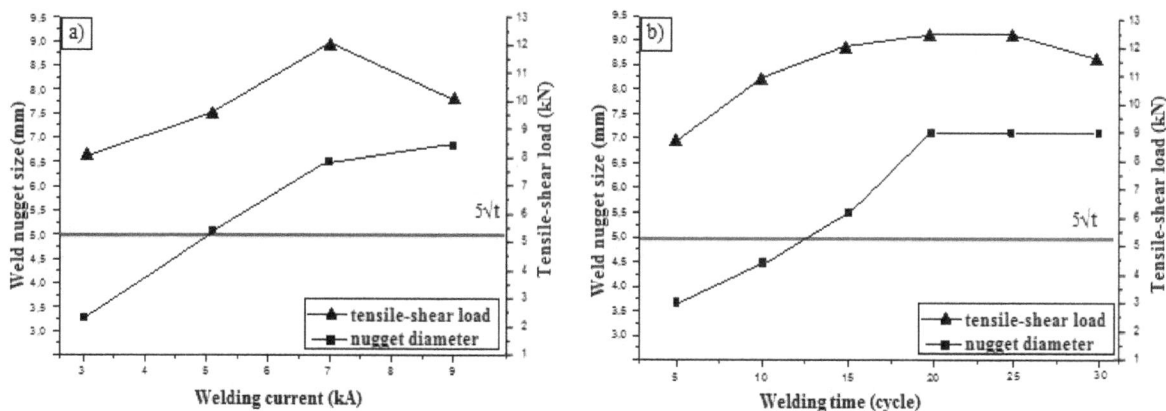

Figure 5. Effect of (**a**) welding current; and (**b**) welding time on weld nugget size and tensile shear strength of samples.

The quality and approximate strength of the weldment can be estimated by measuring the nugget diameter of fusion [23,26]. The weld nugget size depends on materials type and electrode geometry, which is associated with welding parameters such as welding current and welding time. The diameter of the fused zone must meet the requirements of the appropriate specifications or the design criteria. Normally, the spot weld that is reliably reproduced under normal production conditions should have a minimum weld nugget size of the 3.5–4 times the thickness of the thinnest outside part of the joint [26]. The measured weld nugget size is determined approximately 5–7 times of the thickness of the thinnest outside part of the joint which were welded with over 5 kA welding current (Figure 5a) and 15 cycles of welding time (Figure 5b).

Normally, the weldable current and time ranges is determined using the equation $4\sqrt{t}$ (where t is the thickness of used steel) condition. However, in AHSS steel welds, in order to ensure the PF mode, a larger weld nugget diameter than the value recommended $4\sqrt{t}$ is required [31]. Kumar Pal and Bhowmick [32] showed that the average weld nugget diameter should be equal to or larger than $4\sqrt{t}$ for PF mode in dual-phase steels for sheet thicknesses less than 1.5 mm. In this study, the recommended nugget diameter is found to be $5\sqrt{t}$ to obtain the desired PF mode for TWIP spot weldment. The results also confirm that weld nugget size clearly increased with increasing heat input up to a critical level. Increasing welding current and welding time causes enhancement of heat input which, in turn, increases the extent of the nugget size of the weldment [9,33–35].

As seen in Figure 5, the tensile shear strength of a sample that was joined with 7 kA welding current for 20 and 25 cycles of welding time, reached the maximum value. With an increase of welding current from 3 kA to 7 kA, the strength of the weldment increased from 8 kN to 12 kN due to the enlargement in weld nugget size, which is determined to be 3.8 mm and 7.2 mm, respectively. Since excessive grain growth in HAZ and high heat generation at the sheet interface led to early expulsion, the strength of the welded sample, which was joined at 9 kA, declined to 10 kN. Due to high expulsion, the strength of the weldment also decreased for a higher welding time (over 25 cycles). Although the weld nugget diameter, which is 7.8 mm, did not change much by increasing welding times from 15 to 35 cycles, the nugget cross-section (0.8 mm) thickness slightly decreased. In fact, when the heat input is relatively sufficient, the contact surfaces of two TWIP sheets are completely melted to form the weld nugget. The contact surface area increases with increasing welding parameters, such as welding current and welding time, so the weld nugget size also increases. On the other hand, due to the expulsion of molten metal from the fusion zone, the thickness of the nugget cross-section decreases. Figure 5b indicates that the tensile shear strength of the welded sample increases with increasing welding parameters up to 7 kA welding current and 25 cycles of welding time. However, the strength of the weldment decreases by excessive heat input (over 7 kA welding current and 25 cycles of welding time) due to high expulsion of molten metal from the fusion zone, and decreasing cross-section thickness of the weldment. Spena et al. [10] indicate that there are several industrial standards in which the recommend minimum tensile shear strength of a spot weld for a specific sheet metal are given. The acceptable tensile shear strength (TSS) of a RSW-welded joint, computed as a function of the tensile strength (UTS) and thickness (t) of the base metal, is:

$$TSS = [(-6.36 \times 10^{-7} \times UTS + 6.58 \times 10^{-4} \times UTS + 1.674) \times S \times 4 \times t^{1.5}]/1000 \qquad (1)$$

For the examined TWIP sheet (1 mm thickness, 697 MPa tensile strength), the acceptable tensile shear strength is 9766 N. Therefore, except 3 kA and 5 kA welding current for 20 cycles of welding time and 7 kA welding current for five cycles of welding time, all of the combinations of the welding process parameters in this study provide acceptable tensile strength for the automotive industry. However, because of high expulsion caused an excessive thinning in the weld nugget cross-section height, greater than 25 cycles of welding time for 7 kA welding current, should not be recommended for this sample. In addition, cracks along the periphery of the weld nugget also decreases the strength of the weldment.

Based on the facts mentioned above, it could be concluded that welding current and welding time alters the weld nugget size, which is the main controlling factor for shear strength of the weldment. As also mentioned earlier, tensile shear strength of a spot-welded sample is primarily affected by the strength of the failure region [28,36]. Some researchers were also reported that nugget size increases with increasing heat input, hence, the strength of weldment increases [10,29].

Conclusively, the optimum welding parameters resulted in maximum joint strength was established at 7 kA welding current for 20–25 cycles of welding time. Test results clearly point out that heat input is related to the welding current and welding time has an obvious effect on the tensile shear strength of the weldment, and there is a critical heat input level related to welding parameters with which the mechanical properties of the welded joint attain their optimum values.

The results also point out that an effect of twinning-induced plasticity in the austenitic structure is completely lost in the weld nugget due to melting. It is partially lost in HAZ because of recrystallization and grain growth which decreases the failure location strength.

Hardness measurement was also performed diagonally in the weld nugget, HAZ, and base metal of the weldment and the results are shown graphically in Figure 6.

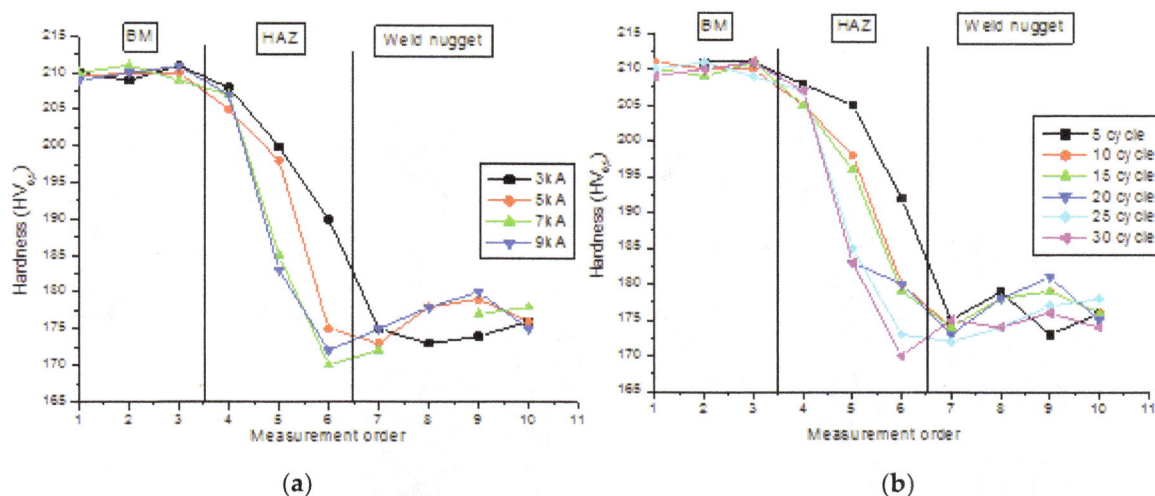

Figure 6. The hardness of weldment versus: (**a**) welding current; and (**b**) welding time.

As seen in Figure 6a,b, the weld nugget and HAZ present considerably lower hardness as compared with base metal. Hardness results also indicate that twinning-induced strength in austenitic structure is completely lost due to melting, but it is partially lost in HAZ due to recrystallization and grain growth. The lower hardness in the weld nugget can also be explained by a lower amount of carbon. Weld nugget hardness for all test samples was found to be similar due to the nature of high-manganese austenitic (TWIP) steel structure that cannot be hardened by heat treatment.

The maximum hardness in HAZ is determined in samples which were welded by using 3 kA welding current for 20 cycles of welding time or 7 kA welding current for five cycles welding time, respectively (Figure 6a,b). The hardness in HAZ decreases by increasing the welding parameters. This could be attributed to high heat input that causes enlargement of HAZ, grain growth, and partial loss of the twinning-induced plasticity.

3.3. Microstructure Evaluation of Weldment

The microstructure of the weldment is shown in Figure 7a–d.

The weld nugget is decorated by a columnar austenite grain which consists of different dendritic structures with directional solidification toward the center of fusion zone (Figure 7a,b). In other words, the equiaxed austenite grains transformed to columnar dendritic morphology in the fusion zone due to the weld thermal cycle (Figure 7e,f). It is well known that microstructure change from the fusion zone towards the base metal depends on the highest temperature reached at each region. It was reported that the dendrites in the fusion zone initially grow as primary arms and, depending upon the cooling rate, composition, and agitation, secondary arms grow outward from the primary arms. The tertiary arms grow outward from the secondary arms. The equiaxed dendrites, which were only present at the center of the fusion zone and totally absent toward approaching fusion line, resulted because of the slow cooling rate at the center of the fusion zone [11].

The microstructure of HAZ is shown in Figure 7c,d, in which grain coarsening occurred due to the thermal cycle of the welding process. Since the HAZ has been heated up to temperature approaching the solidus temperature of the alloy, the grain growth is present and many of the second-phase particles, which might also be present in the base metal, may dissolve. This can lead to a super-saturation of the austenite matrix during cooling [9]. The microstructure of the TWIP steel is decorated with equiaxed austenite grains as seen in Figure 7d.

In contrast to base metal chemical composition, point EDS analysis carried out in HAZ adjacent to the weld nugget in which Mn and Al segregation is attracting attention (Figure 8a–c). It was reported that alloys were segregated according to the condition of equilibrium segregation coefficients; the elements, which have segregation coefficients less than unity, were segregated between the two dendritic cores (interdendritic zone) during solidification [10,37].

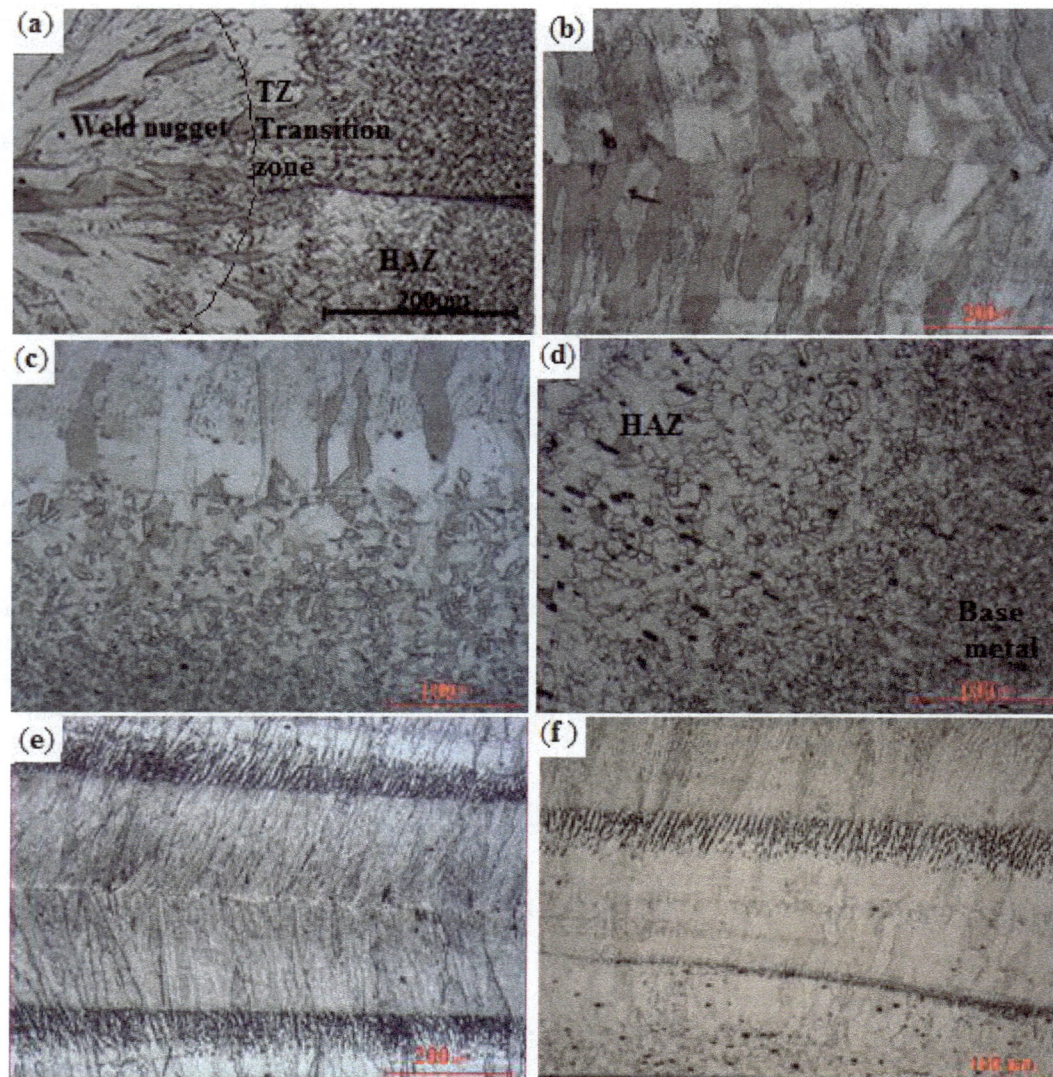

Figure 7. The microstructure of a spot-welded sample: (**a**) three distinct zones of weldment; (**b**) fusion zone; (**c**) transition microstructure from the weld nugget to HAZ; (**d**) microstructure from HAZ to base metal; and (**e**,**f**) columnar dendrites in the fusion zone.

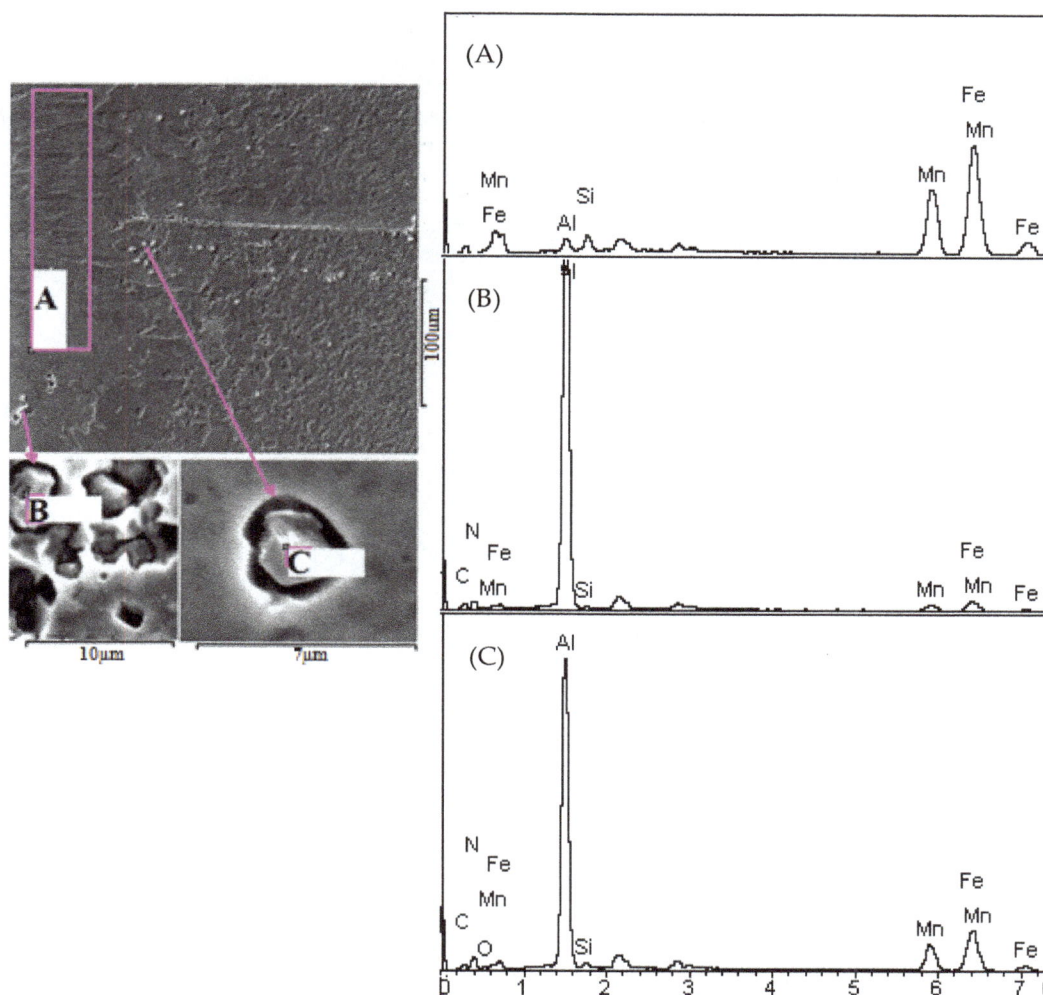

Figure 8. EDS analysis results of welded samples (**A**) EDS analysis of weld nugget (**B**) EDS analysis of particles in weld nugget, (**C**) EDS analysis of particles in HAZ.

The formed particles observed in the transition zone from the weld nugget to HAZ, at which point EDS analysis was carried out (Figure 8b,c), respectively. Results indicate that this formed particles containing high amounts of aluminum. It is thought that alloying element segregation in the transition zone from weld nugget to HAZ can cause the formation of second-phase particles. Saha et al. [11] investigated the element segregation in fusion zone in one of their studies. They reported that Mn and C were segregated in interdendritic areas, while Al was in dendritic areas. This segregated C and Mn combine with each other and form MC (metal carbides). It is worth noting that second-phase particles, such as carbide and nitride particles, can inhibit grain growth in steels by hindering the movement of grain boundaries [37,38]. These particles, if not dissolved during welding, tend to inhibit grain growth in the HAZ. Investigation into formed particles should be conducted.

It was observed that the grain size in the weld fusion zone was quite larger than in HAZ and BM. HAZ consists of larger austenitic grains, as compared to those of the base metal, due to the grain coarsening caused by the heat generated during spot welding.

Optical microscopy examination was also carried out to identify internal defects in weld nugget. The appearance, location, and orientation of internal defects are shown in Figure 9a–c.

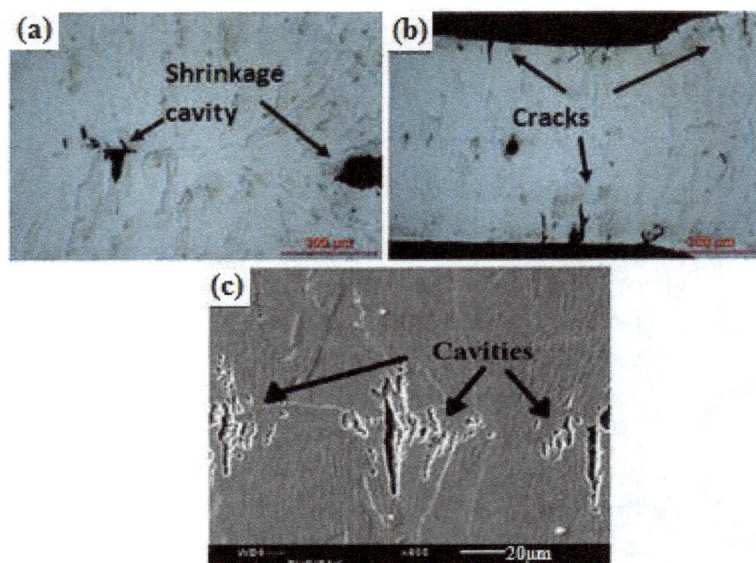

Figure 9. Defects in weld nugget: (**a**) shrinkage cavity; (**b**) cracks; and (**c**) cavities.

As seen in Figure 9a–c, internal defects, such as cavities, are located in the center of the nugget, and cracks are located at the periphery of weld. It is worth noting that the amount of cavities increased with increasing heat input. These internal defects in spot welds are generally caused by low electrode force, high heat input, or any other conditions that produce excessive weld heat. It was reported that these discontinuities will have no detrimental effect on the static or fatigue strength of a weld if they are located entirely in the central portion of the weld nugget because the stresses are essentially zero in the central portion of the weld nugget [25].

The cracks are formed at the periphery of weld nugget due to high heat input (upper than 7 kA welding current and 25 cycles of welding time) followed by a high cooling rate. An increase in the welding current and welding time increases the cracking tendency because of the solidification mode, aluminum segregation, high electrode pressure, and residual stress in the weld nugget during cooling. These cracks can be detrimental for mechanical properties of the weldment where the load stresses are highly concentrated [26].

4. Conclusions

The conclusions derived from this study can be given as follows:

- The tensile shear strength of welded samples increased up to 7 kA welding current for 20 cycles of welding time. The results indicate that, except for 3 kA and 5 kA welding current for 20 cycles of welding time and 7 kA welding current for five cycles of welding time, all of the combinations of the welding process parameters in this study provide acceptable tensile strength for the automotive industry. Over the critical heat input level (greater than 25 cycles of welding time for 7 kA welding current) the strength of the welded sample decreases due to expulsion or decreasing in cross-section thickness of the nugget.
- The failure in the tensile shear test sample occurred in PIF mode for lower welding parameters (3 kA welding current, up to 15 cycles of welding time). PIF mode was present in the ductile characteristics in the weld nugget. Over these welding parameters (except expulsion) PF mode was started by a crack in HAZ and then the crack propagation occurred by tearing from the sheet. It is thought that the primary cause of weakening in HAZ could be the grain growth mechanism.
- The fully austenitic solidification present in the weld nugget was due to a high amount of manganese and a low amount of carbon in chemical composition. Since the fusion zone

microstructure has been fully austenitic, the weld thermal cycle has not changed the structure. The ferritic or martensitic transformation has not been detected through metallographic investigation. The formed particles observed in transition zone from the weld nugget to HAZ, at which point EDS analysis was carried out. Results indicate that this formed particle contains high amounts of Al and Mn that can be cause the formation of second-phase particles.

- The hardness of the weld nugget and HAZ were found to be lower than those of the base metal due to the nature of the weld thermal cycles, the chemical composition of TWIP steel, and grain coarsening. The hardness results also indicate that the high strength caused by twinning-induced plasticity is almost lost due to the weld thermal cycle.

- Due to electrode force, high heat input, or any other parameters that produce excessive weld heat, the shrinkage cavities and cracks were observed in the fusion zone of the weldment. It is believed that the cracks at the periphery of a weld nugget where the load stresses are highly concentrated were formed due to the solidification mode, interdendritic aluminum segregation, high electrode pressure, and residual stress in the weld nugget.

- In conclusion, the optimum welding parameters that guarantee acceptable tensile shear strength and fracture mode (PF) for the automotive industry were obtained at 7 kA welding current for 20 and 25 cycles of welding time in the examined range. The results indicate that the acceptable welding parameter area is very narrow for resistance spot-welded TWIP steels, because of cracks and cavities in the weld nugget, surface cracks in the HAZ that causes PIF mode, unacceptable weld nugget geometry in adequate welding parameters because of low heat input, and high metal expulsion reducing the partial thickness due to high heat input.

Author Contributions: H.K.Z. produced and provided TWIP steels. R.K. conceived and designed the welding; H.E.E. performed the experiments; H.E.E., H.K.Z and R.K. analyzed the data; R.K. contributed materials and analysis tools; H.E.E. wrote the paper.

Conflicts of Interest: The authors declare no conflict of interest.

References

1. Poranvari, M.; Mousavizadeh, S.M.; Marashi, S.P.H.; Goodarzi, M.; Ghorbani, M. Influence of fusion zone size and failure mode on mechanical performance of dissimilar resistance spot welds of AISI 1008 low carbon steel and DP600 advanced high strength steel. *Mater. Des.* **2011**, *32*, 1390–1398. [CrossRef]

2. Beal, C. Mechanical Behavior of New Automotive High Manganese TWIP Steel in the Presence of Liquid Zinc. Ph.D. Thesis, L'Institut National des Sciences Appliquées de Lyon, Lyon, France, 2011; pp. 1–11.

3. Bouaziz, O.; Allain, S.; Scott, C.P.; Cugy, P.; Barbier, D. High manganese austenitic twinning induced plasticity steels. A review of the microstructure properties relationships. *Curr. Opin. Solid State Mater. Sci.* **2011**, *15*, 141–168. [CrossRef]

4. Frommayer, G.; Brüx, U.; Neumann, P. Supra-ductile and high-strength manganese-TRIP/TWIP steels for high energy absorption purposes. *ISIJ Int.* **2003**, *43*, 438–446. [CrossRef]

5. Grässel, O.; Krüger, L.; Frommeyer, G.; Meyer, L.W. High strength Fe-Mn-(Al, Si) TRIP/TWIP steels development—Properties—Application. *Int. J. Plast.* **2000**, *16*, 1391–1409. [CrossRef]

6. Roncery, L.M.; Weber, S.; Theisen, W. Welding of twinning-induced plasticity steels. *Scr. Mater.* **2012**, *66*, 997–1001. [CrossRef]

7. Sawhill, J.M.; Baker, J.C. Spot weldability of high-strength sheet steels. *Weld. J.* **1952**, *31*, 931–943.

8. Pouranvari, M.; Abedi, A.; Marashi, P.; Goodarzi, M. Effect of expulsion on peak load and energy absorption of low carbon resistance spot welds. *Sci. Technol. Weld. Join.* **2008**, *13*, 39–43. [CrossRef]

9. Chandra Saha, D.; Han, S.; Chin, K.G.; Choi, I.; Park, Y.D. Weldability evaluation and microstructure analysis of resistance-spot-welded High-Mn steel in automotive application. *Steel Res. Int.* **2012**, *83*, 1–6.

10. Spena, P.R.; Maddis, D.M.; Lombardi, F.; Rossini, M. Investigation on resistance spot welding of TWIP steel sheets. *Steel Res. Int.* **2015**, *86*, 1480–1489. [CrossRef]

11. Saha, D.C.; Cho, Y.; Park, Y.D. Metallographic and fracture characteristics of resistance spot welded TWIP steels. *Sci. Technol. Weld. Join.* **2013**, *18*, 711–720. [CrossRef]

12. Choi, H.C.; Ha, T.K.; Shin, H.C.; Chang, Y.W. The formation kinetics of deformation twin and deformation induced ε-martensite in an austenite FE-C-Mn steel. *Scr. Mater.* **1999**, *40*, 1171–1177. [CrossRef]

13. Dai, Q.; Yang, R.; Chen, K. Deformation behavior of Fe-Mn-Cr-N austenitic steel. *Mater. Charact.* **1999**, *42*, 21–26. [CrossRef]

14. Vercammen, S.; Blanpain, B.; de cooman, B.C.; Wollants, P. Cold rolling behavior of an austenitic Fe-30Mn-3Al-3Si TWIP-steel: The importance of deformation twinning. *Acta Mater.* **2004**, *52*, 2005–2012. [CrossRef]

15. Yang, P.; Xie, Q.; Meng, L.; Ding, H.; Tang, Z. Dependence of deformation twinning on grain orientation in a high manganese steel. *Scr. Mater.* **2006**, *55*, 629–631. [CrossRef]

16. Ueji, R.; Tsuchida, N.; Terada, D.; Tsuji, N.; Tanaka, Y.; Takumera, A.; Kunishige, K. Tensile properties and twinning behavior of high manganese austenitic steel with fine-grained structure. *Scr. Mater.* **2008**, *59*, 963–966. [CrossRef]

17. Barbier, D.; Gey, N.; Allain, S.; Bozzolo, N.; Humbert, M. Analysis of the tensile behavior of a TWIP steel based on the texture and microstructure evolutions. *Mater. Sci. Eng. A* **2009**, *500*, 196–206. [CrossRef]

18. Bracke, L.; Verbeken, K.; Kestens, L.; Penning, J. Microstructure and texture evolution during cold rolling and annealing of high Mn TWIP steel. *Acta Mater.* **2009**, *57*, 1512–1524. [CrossRef]

19. Idrissi, H.; Ryelandt, L.; Veron, M.; Schryvers, D.; Jacques, P.J. Is there a relationship between the stacking fault character and the activated mode of plasticity of Fe-Mn based austenitic steels. *Scr. Mater.* **2009**, *60*, 941–944. [CrossRef]

20. Dai, Y.J.; Tang, D.; Mi, Z.L.; Li, J.C. Microstructure characteristics of an Fe-Mn-C TWIP steel after deformation. *J. Iron Steel Res. Int.* **2010**, *17*, 53–59. [CrossRef]

21. Idrissi, H.; Renard, K.; Ryelandt, L.; Schryvers, D.; Jacques, P.J. On the mechanism of twin formation in Fe-Mn-C TWIP steels. *Acta Mater.* **2010**, *58*, 2464–2476. [CrossRef]

22. Gutierrez-Urrutia, I.; Zaefferer, S.; Raabe, D. The effect of grain size and grain orientation on deformation twinning in a Fe-22 wt.% Mn-0.6 wt.% C TWIP steel. *Mater. Sci. Eng. A* **2010**, *527*, 3552–3560. [CrossRef]

23. Kearns, W.H. Metals and their weldability. In *AWS Welding Handbook*, 7th ed.; American Welding Society: St. Doral, FL, USA, 1982; Volume 4, pp. 76–146.

24. Santella, M.L.; Babu, S.S.; Riemer, B.W.; Feng, Z. Influence of microstructure on the properties of resistance spot welds. In Proceedings of the 5th International Conference on Trends in Welding Research, Pine Mountain, GA, USA, 1–5 June 1998.

25. Chao, Y.; Miller, K.; Wang, P.C. Impact strength of resistance spot welded joints. In Proceedings of the WS Sheet Metal Welding Conference VIII, Detroit, MI, USA, 13–16 October 1998; pp. 3–12.

26. Kearns, W.H. Welding Processes. In *AWS Welding Handbook*, 7th ed.; American Welding Society: St. Doral, FL, USA, 1978; Volume 2, pp. 1–55.

27. Pouranvari, M.; Asgari, H.R.; Mosavizadch, S.M.; Marashi, P.H.; Goodarzi, M. Effect of weld nugget size on overload failure mode of resistance spot welds. *Sci. Technol. Weld. Join.* **2007**, *12*, 217–225. [CrossRef]

28. Pouranvari, M.; Marashi, S.P. Failure mode transition in AHSS resistance spot welds. Part I. Controlling factors. *Mater. Sci. Eng. A* **2011**, *528*, 8337–8343. [CrossRef]

29. Marashi, P.; Pouranvari, M.; Sanaee, S.M.H.; Abedi, A.; Abootalebi, H.; Goodarzi, M. Relationship between failure behavior and weld fusion zone attributes of austenitic stainless steel resistance spot welds. *Mater. Sci. Technol.* **2008**, *24*, 1506–1512. [CrossRef]

30. Razmpoosh, M.H.; Shamanian, M.; Esmailzadeh, M. The microstructural evolution and mechanical properties of resistance spot welded Fe-31Mn-3Al-3si TWIP steel. *Mater. Des.* **2014**, *67*, 571–576. [CrossRef]

31. Pouranvari, M.; Marashi, S.P. Critical review of automotive steels spot welding: Process, structure and properties. *Sci. Technol. Weld. Join.* **2013**, *18*, 361–403. [CrossRef]

32. Kumar Pal, T.; Bhowmick, K. Resistance spot welding characteristics and high cycle fatigue behavior of DP 780 steel sheets. *J. Mater. Eng. Perform.* **2012**, *21*, 280–285.

33. Vural, M.; Akkuş, A. On the resistance spot weldability of galvanized interstitial free steel sheets with austenitic stainless steel sheets. *J. Mater. Process. Technol.* **2004**, *16*, 53–156. [CrossRef]

34. Sharma, P.; Ghosh, P.K.; Nath, S.K. Fatigue behavior of resistance spot welded Mn-Cr-Mo dual phase steels. *Z. Metallkd.* **1993**, *84*, 513–517.

35. Gupta, P.; Ghosh, P.K.; Nath, K.; Ray, S. Resistance spot weldability of plain carbon and low alloy dual phase steels. *Z. Metallkd.* **1990**, *81*, 502–508.

36. Lin, S.H.; Pan, J.; Wu, S.R.; Tyan, T. *Spot Weld Failure Loads under Combined Mode Loading Conditions*; SAE Technical Paper No. 2001-01-0428; Society of Automotive Engineers: Warrendale, PA, USA, 2001.

37. Kou, S. *Welding Metallurgyü*, 2nd ed.; Wiley: Hoboken, NJ, USA, 2003; pp. 341–352.

38. Higgins, R.A. *Engineering Metallurgy Applied Physical Metallurgy*, 6th ed.; Elsevier: Amsterdam, The Netherlands, 1993; pp. 79–99.

17

Comparison of Hydrostatic Extrusion between Pressure-Load and Displacement-Load Models

Shengqiang Du [1], Xiang Zan [1,2,*], Ping Li [1], Laima Luo [1,2], Xiaoyong Zhu [2] and Yucheng Wu [1,2,*]

[1]   School of Materials Science and Engineering, Hefei University of Technology, Hefei 230009, China;
      15255150652@163.com (S.D.); li_ping@hfut.edu.cn (P.L.); luolaima@126.com (L.L.)
[2]   National–Local Joint Engineering Research Centre of Nonferrous Metals and Processing Technology,
      Hefei 230009, China; zhuxiaoyong@hfut.edu.cn
*   Correspondence: zanx@hfut.edu.cn (X.Z.); ycwu@hfut.edu.cn (Y.W.)

Academic Editors: Myoung-Gyu Lee and Yannis P. Korkolis

Abstract: Two finite element analysis (FEA) models simulating hydrostatic extrusion (HE) are designed, one for the case under pressure load and another for the case under displacement load. Comparison is made of the equivalent stress distribution, stress state ratio distribution and extrusion pressure between the two models, which work at the same extrusion ratio (R) and the same die angle (2α). A uniform Von-Mises equivalent stress gradient distribution and stress state ratio gradient distribution are observed in the pressure-load model. A linear relationship is found between the extrusion pressure (P) and the logarithm of the extrusion ratio ($\ln R$), and a parabolic relationship between P and 2α, in both models. The P-value under pressure load is smaller than that under displacement load, though at the same R and α, and the difference between the two pressures becomes larger as R and α grow.

Keywords: hydrostatic extrusion; FEA; pressure load; die angle; extrusion ratio

---

1. Introduction

Hydrostatic extrusion (HE) is a unique forming method that was presented by Robertson in 1893 [1]. During the process, the material is surrounded by a high-pressure medium, which forms hydrostatic pressure conditions that improve the material's formability; thereby, larger amounts of deformation can be achieved as compared to the conventional extrusion process. The medium also ensures good lubricant conditions, and even generates dynamic lubrication between the die and the billet [2], and hence great surface quality. HE as a special severe plastic deformation (SPD) method has great advantages for large deformation processes and the forming processes of difficult-to-form materials.

By using the HE process, Ozaltin et al. [3] improved the strength of Ti-45Nb by 45% and also attained good plasticity by refining the grain. Also, by HE, Yu et al. [4] realized the deformation of AZ31 at 200–300 °C, at a maximum R of 31.5. Xue et al. [5] improved the properties of Zr-based metallic glass/porous tungsten phase composite; the breaking strength reaching 2112 MPa and the fracture strain reaching 53%. Kaszuwara et al. [6] densified Nd-Fe-B powder to the theoretical maximum density by HE. Kováč et al. [7] prepared MgB_2 wires by internal magnesium diffusion and HE. Skiba et al. [8] deformed GJL250 grey cast iron and GJS500 nodular cast iron by improved HE equipment with back pressure. Hydrostatic extrusion is widely used in the preparation of materials which are hard to deform. Finite element analysis (FEA) has also been used for investigations of the HE processes. Zhang et al. [9] simulated the HE process with tungsten alloy; the displacement load on the upper surface and a rigid boundary on the lateral surface of the billet were used instead of the pressure load of the pressure medium. Li et al. [10] simulated the HE process of W-40 wt. % Cu at

650–800 °C with simplified boundary conditions and calculated the linear relationship between the extrusion pressure and temperature, and proved the simulation results with experiments. Replacing the pressure of a pressure medium with a displacement load, accompanied by near-zero friction between the billet surface and the virtual rigid container, it is easy to model the HE process and improve the convergence rate effectively. However, without hydrostatic pressure, the simulation results reduce the accuracy and differ from real hydrostatic extrusion. In Li's work [10], a large gradient of equivalent stress distribution at the un-deformed region surrounded by the pressure medium was found, which was different from the real HE process where that region was in a hydrostatic state and the equivalent stress should be almost near zero. Thus, using the simplified displacement-load mode may introduce inaccuracy into the simulation results. Manafi and Saeidi [11] simulated 93 tungsten alloy by HE with a pressure boundary condition and found the optimized die angle. Peng et al. [12], by calculating the stress distribution in Nb/Cu composited by HE, investigated its interface bonding status. Manifi et al. [13] improved conventional backward extrusion by employing HE principles to reduce the extrusion load; the maximum load was reduced by 80% compared to the conventional back extrusion process. Kopp and Barton [14] improved the model of HE and analyzed the differences between experimentation and simulation.

The comparisons between the simulation and experiment were discussed in [10–14], but the comparison between the different simulation models has not been discussed in detail yet. In the present study, the pressure-load mode and displacement-load mode are used to simulate the HE process. In addition, the main work of this paper is comparing the deviation of the calculated results of the two modes under the same conditions and finding out the influence of the pressure. The judgments of the comparison are made through the theories of HE.

2. FEM Methods and Materials

The biggest difference between hydrostatic extrusion and conventional extrusion lies in the way of transferring loads from the punch to the billet. In HE process, the billet is tapered to match the die geometry, the gap between the billet and the container is filled with pressure medium, which surrounds the billet and conveys the extrusion force of the moving punch onto it, and the pressure medium is forced by its inherent pressure into the gap between the die and the billet, generating excellent lubrication on the contact surface (Figure 1a). In the present study, castor oil is used as the liquid pressure medium. The billet is tapered to match the die geometry before the extrusion in order to ensure the pressure medium staying in the container. In the real experiment, the gaps between the punch, container and the die were sealed by rubber and pure copper seal rings to ensure the system is under good sealing state. While, in the conventional extrusion, the billet is pressed by the punch directly (Figure 1b) and so deformed.

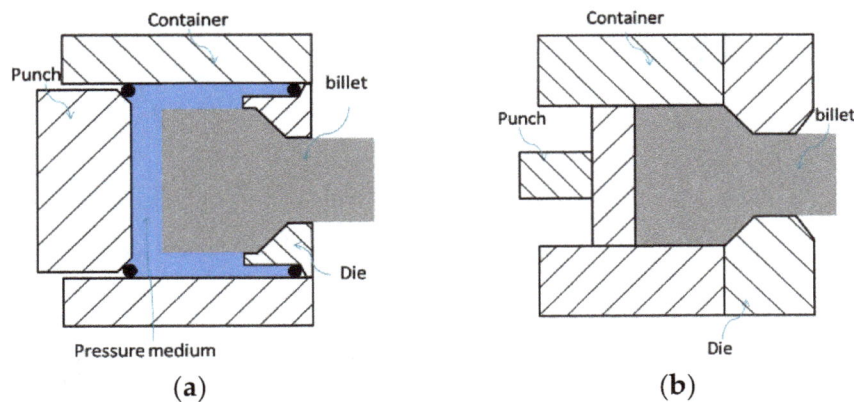

Figure 1. Principle of (**a**) hydrostatic extrusion and (**b**) conventional extrusion.

To simulate the HE process accurately, it has to take the fluid-structure interaction mode, which, however, is too complicated for large-scale calculation. So, the model developed in the present study is a partly simplified one, which improves the calculation efficiency and ensures the calculation precision. The numerical software ANSYS (V15.0, ANSYS Inc., Canonsburg, PA, USA) was used for the simulation. In the model, the pressure medium is replaced by uniform pressure loads, the die is partly replaced by rigid lines, and no friction is set between the billet and the fluid while friction between the billet and the die is in agreement with Coulomb's friction law. The model could be further simplified to 2-D because of the axial symmetry of both billet and die as columns. An eight-node plane element (PLANE 183) is used.

The pressure load mode is the mode replacing the pressure medium as a boundary condition of the billet, modeling only its pressure properties. The pressure is set to increase linearly with the time, replacing the effect of the punch pushing the pressure medium. So, in this model, the punch is not needed because the billet is deformed by the increased pressure. In the pressure load model, the central axis of the billet is the symmetrical axis of both billet and die, and the pressure load only exists over the un-deformed outer surface of the billet. The fillet at upper right of the billet is built to verify the uniform distribution of the pressure load. The coefficient of friction between the billet and the die is set as 0.05 (Figure 2a). A displacement load model which is commonly used is also established, for the sake of comparison (Figure 2b). In the displacement load mode, the displacement load with even speed is directly applied on the upper surface of the billet. The lateral surface of the billet is constrained by displacement constraint to ensure the materials cannot flow along the positive direction of the radius. Thus, the extrusion force can be calculated by the reaction force. Rigid lines are placed outside the un-deformed region instead of the displacement constraint and the coefficient of friction over this region is set as 0 which replaces the zero friction between billet and pressure medium. So, the displacement load mode is essentially a conventional extrusion mode without friction between the billet and the die. The material used to be deformed in both models is AA2024, which is a typical ideal elastoplastic material, whose specific parameters are given in Table 1. The two models are simulated at room temperature (298 K), and the parameters used for the simulations presented in this paper are given in Table 2.

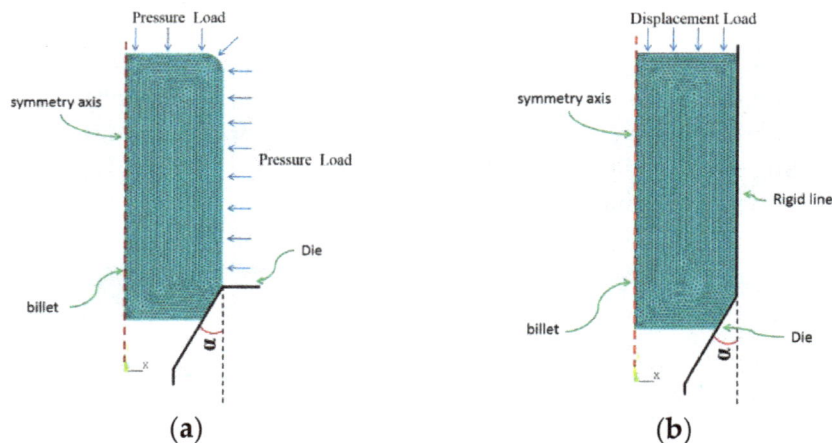

Figure 2. Models of (**a**) pressure load and (**b**) displacement load.

Table 1. Material properties of AA2024.

Material	Density	Poisson Ratio	Elastic Modulus	Yield Stress
AA2024	2.79 g/cm^3	0.3	71.7 GPa	340 MPa

Table 2. Simulation parameters of pressure load mode and displacement load mode.

Load Mode	Extrusion Ratio	Die Angle (2α)	Initial Height	Initial Diameter
Pressure load	2.25, 2.78, 4.00, 6.25	25°, 30°, 35°, 40°, 60°, 90°, 120°	80 mm	30 mm
Displacement load				

3. Results and Discussion

As shown in Figure 3, fluid pressure changed with time as the set pressure was distributed uniformly over the outer surface of the un-deformed region of the billet and decreases gradually and finally disappeared near the entrance of the die. The distributions of the pressure at different stages all proved that the hydrostatic pressure property was perfectly represented. The pressure changed linearly with time and reached a certain amount when the billet was deformed. The distribution of the fluid pressure, again as shown in Figure 3, proved that the pressure load model fits the real HE process well.

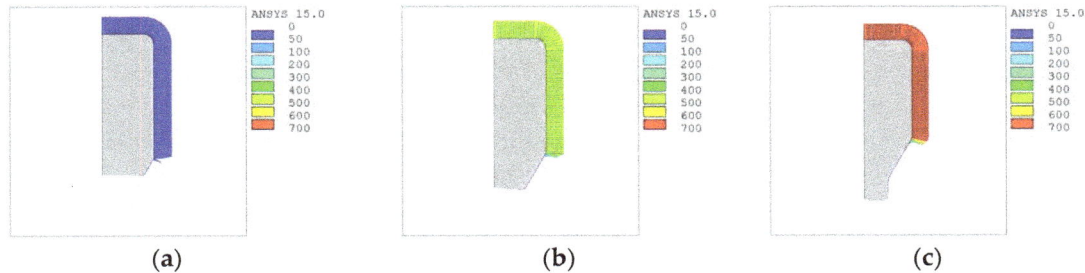

(a)   (b)   (c)

Figure 3. Fluid pressure distribution at (**a**) initial stage; (**b**) pressure-up stage and (**c**) stable stage.

3.1. Comparison of Distribution of Stress and Strain Field

The distribution information, including the equivalent strain, equivalent stress, etc., as calculated, was compared under the same scale bar, between the two models, thus making the difference much more obvious. The comparison was conducted at $R = 4.00$ and $2\alpha = 60°$.

The Von-Mises equivalent strain distributions of the two models (Figure 4) were found to be basically the same, proving that the comparison was conducted as the two billets were experiencing the same degree of deformation.

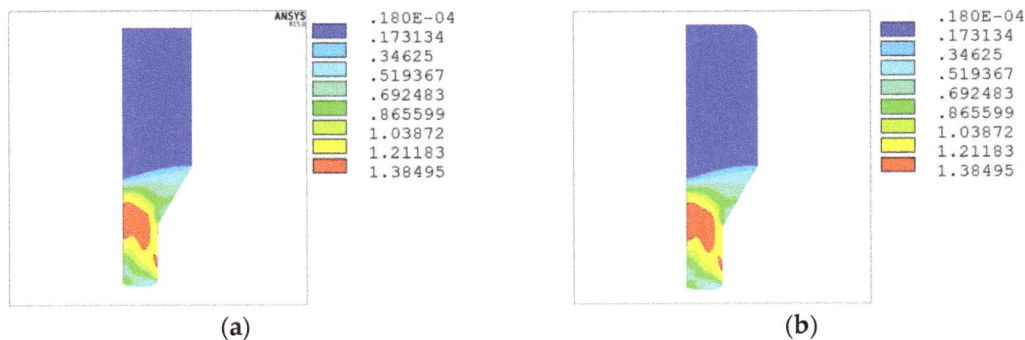

(a)   (b)

Figure 4. Von-Mises equivalent strain distribution of (**a**) displacement-load model and (**b**) pressure-load model.

The axial stress distributions (Figure 5) indicate that, on the surface of the billet under pressure load, the area of the compressed stress–concentrated region in the inlet region was larger and that of the tensile stress–concentrated region was smaller in the outlet region than those in the case of displacement load. These differences can affect the material's formability. So in the pressure-load model, there was less of a tendency to generate cracks on the surface of the billet when the material went through the inlet and outlet regions, and hence there was good surface quality, which is an important characteristic of HE.

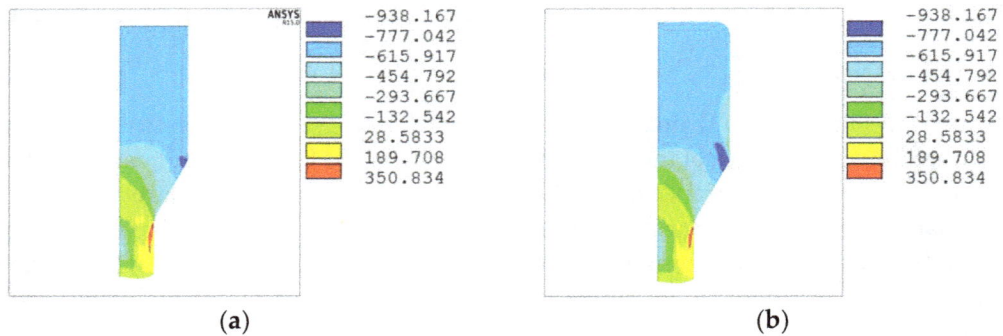

Figure 5. Axial distribution of (**a**) displacement-load model and (**b**) pressure-load model.

The Von-Mises equivalent stress distributions of the two models were obviously different, just as shown in Figure 6. It was found that, under displacement load, the material was pressed by unequal σ_1, σ_2, σ_3, generating an exorbitant Von-Mises stress in the un-deformed region and a tiny Von-Mises stress in a small region only in the core of the billet at the inlet of the die, which is totally different from the situation of the real HE process (Figure 6a). The Von-Mises equivalent stress was extremely small in the un-deformed region under pressure load, because the billet was surrounded by hydrostatic pressure, which made the primary stress (σ_1, σ_2, σ_3) nearly equal. The value of the equivalent stress gradually reached the yield value as the deformation went on, and so there exists a gradient distribution of the equivalent stress in the inlet region, where the deformed and un-deformed regions are clearly demarcated (Figure 6b). In Reference [10], the stress distribution with a large value was found in the un-deformed region, proving the mode in that work was similar to the displacement load. In addition, this equivalent stress difference indicates that the pressure-load model is more suitable for HE analysis.

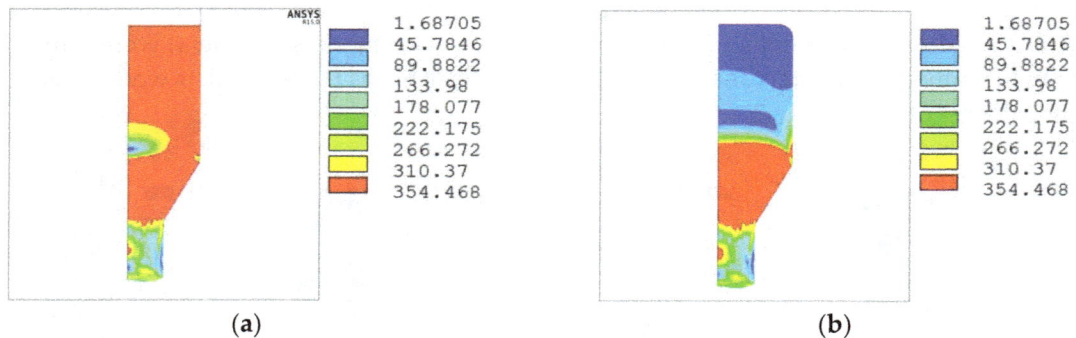

Figure 6. Von-Mises equivalent stress distribution of (**a**) displacement load and (**b**) pressure load.

The equivalent strain and stress distributions of the pressure-load model proved the un-deformed region was under hydrostatic pressure, ensuring no deformation was happening. Some experiments [15,16] were conducted to verify the materials' deformation behavior through HE. The billet was cut through the center along the extrusion axis and a grid was printed on the cut surface. Finally, the two parts were put together and extruded. After extrusion, no deformation was found at the grid in the part surrounded by the pressure medium, the un-deformed region. The experiments fit the simulation results well.

The deformed region can be distinguished by the stress state ratio distribution. The boundary between the deformed and un-deformed regions under displacement load was not stable, lower in the core and higher near the surface (Figure 7a). In the case of pressure load, the boundary was parallel to the top surface, shaped like the Von-Mises equivalent stress distribution, thus proving that the material flowed uniformly under the hydrostatic pressure (Figure 7b). The die limited the material's movement during the deformation, making the material flow more easily in the core but less near the surface. The hydrostatic pressure can effectively improve the material's flow, for the material flowed uniformly even where the die exerts its limitation. However, the displacement load cannot benefit the material's flow, hence the uneven flow and the deformation near the surface lagging behind that in the core.

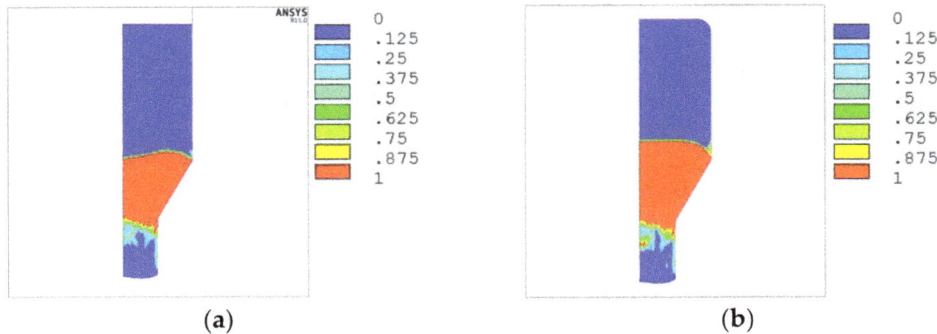

Figure 7. Stress state ratio distribution of (**a**) displacement load (**b**) and pressure load.

As can be seen in the contact pressure distribution, the material's flow near the surface differed between the two models. The contact pressure in the inlet region was higher under displacement load (Figure 8a) because uneven material flow results in more redundant work, so the load to achieve the same deformation is higher, hence the higher contact pressure. A lower contact pressure can be found under pressure load because the material deforms uniformly in this region and so lower redundant work is needed, hence the lower contact pressure (Figure 8b). Hydrostatic pressure can make the material flow uniformly, as was obviously shown in the pressure-load model.

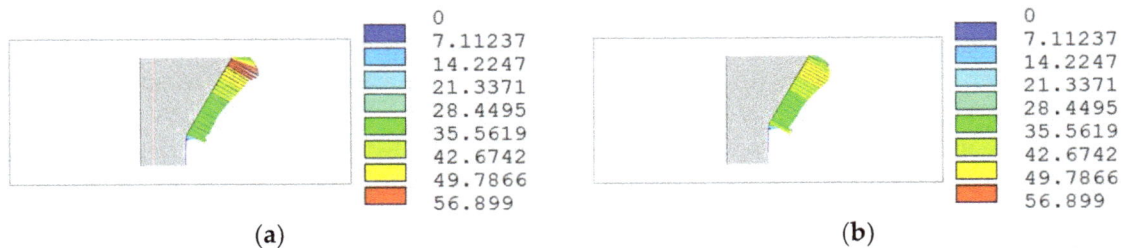

Figure 8. Contact pressure distribution of (**a**) displacement load and (**b**) pressure load.

3.2. Comparison of Extrusion Pressure

By analyzing the difference in HE, a further comparison was made between the two models. R is the deformation ratio, written as $R = D^2/d^2$. In the pressure-load model, the pressure increased linearly with the time; meanwhile, the billet was deformed. The pressure corresponding to the position when the bottom of the billet is pressed out of the die is the extrusion pressure (P), and this is the minimum pressure to complete the extrusion process. In the displacement-load model, the position with same deformation ratio can be found, and the extrusion pressure was calculated by the reaction force and the area of the contact surface. A linear relationship exists between P and $\ln R$ in both models; for instance, at $\alpha = 45°$, $P = 424\ln R + 197$ for displacement load and $P = 347\ln R + 160$ for pressure load, respectively (Figure 9a). The pressure gap between the two load models can be found under the same working conditions. The P-value is higher in the displacement-load model, because the material flows less uniformly, so that a higher P is required to overcome the redundant work. But under a pressure load, the P-value is lower because the hydrostatic pressure load can maintain uniform deformation during the process. The gap becomes larger as the R value increases, because the deformation uniformity decreases as the deformation ratio grows. However, the gap grows only a bit, indicating that the deformation ratio is not the main cause for the redundant work.

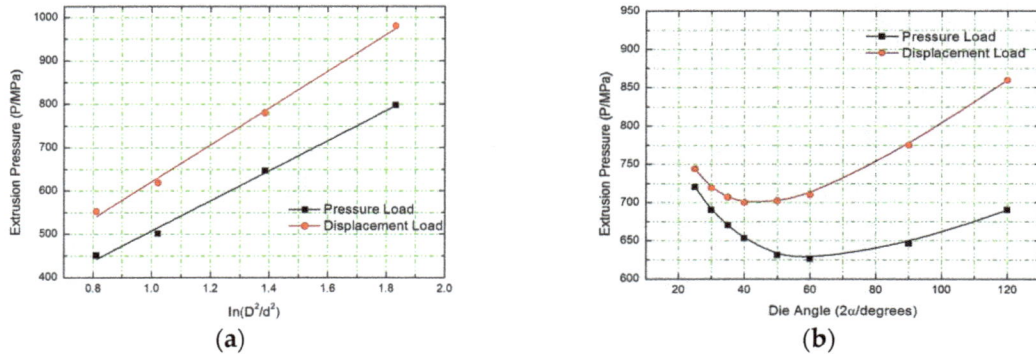

Figure 9. Relationship between (a) P and $\ln R$; (b) P and 2α.

In both models, there exists a parabolic relationship between P and 2α (Figure 9b). Extrusion pressure first declined and then increase when 2α increased from 30° to 120°. The optimized die angle (2α), corresponding to the smallest P-value, was 40° and 60° in the displacement-load and the pressure-load model, respectively. The size of the optimized die angle depends on the redundant work and friction work during the deformation. The redundant work, resistant to non-uniform deformation, increased rapidly when the die angle grew. Meanwhile, the friction work also changed because both the contact pressure and contact area changed. As the die angle grew, the contact pressure increased whereas the contact area decreased, and the friction work first declined and then increased. The redundant work increase was lower than the friction work decrease at first, so the P-value declined, but as the die angle grew, the redundant work increase gradually grew faster than the friction work decrease, leading to the rise of the P-value. The pressure gap between the two models became larger rapidly as 2α grew, indicating that the redundant work increased faster in the displacement-load model. The gap grew rapidly with the die angle, indicating the angle was the main cause for the redundant work.

The relationship between P and 2α can be well explained by the stress state ratio distribution (Figure 10). As the die angle grew from 30° to 120°, the deformation region boundary in the displacement-load model increased more rapidly than in the pressure-load model and the shape of the boundary changed more drastically at the same time. The irregularity of the boundary indicated a non-uniform deformation, so more redundant work was generated, which in turn resulted in greater extrusion pressure to achieve the same deformation. So the extrusion pressure was lower in the pressure load model at the same die angle where the boundary was more stable. In addition, because of the hydrostatic pressure, little redundant work was generated in the pressure-load model, so the extrusion pressure value fluctuated in a small range at different die angles.

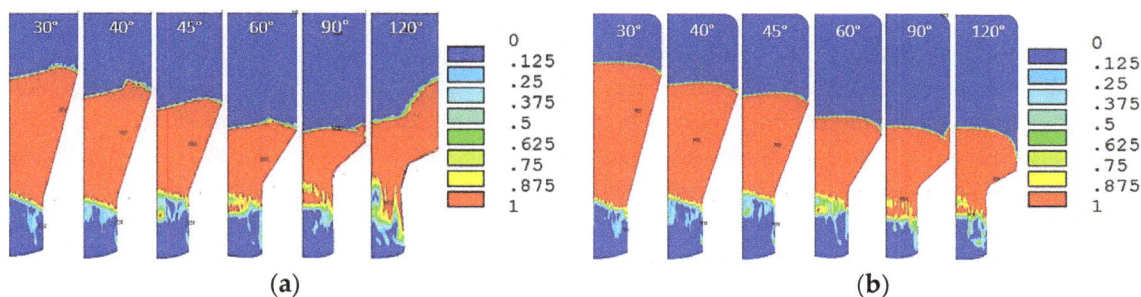

Figure 10. Stress state ratio distribution of (**a**) displacement-load model and (**b**) pressure-load model at $2\alpha = 30°, 40°, 45°, 60°, 90°$ and $120°$.

4. Conclusions

Through numerical simulation, two models, the displacement-load model and pressure-load model, are designed and compared. Both models can simulate the hydrostatic extrusion process to a certain extent, gaining similar results of the strain calculation. Under the pressure load, the value of the Von-Mises equivalent stress in the un-deformed region is very small, which proves that this region is under uniform hydrostatic pressure. The deformation boundary in the stress state ratio distribution is almost horizontal, which proves that the material is pressed by hydrostatic pressure. However, in the displacement-load model, the Von-Mises stress value is large, and the deformation boundary is irregular, proving that, with no hydrostatic pressure, the deformation is non-uniform. The relationship between P and R was found as $P = 347\ln R + 160$ or $P = 424\ln R + 197$ in the pressure-load and displacement-load model. In addition, the optimized die angle was $60°$ or $40°$, respectively. It can be proved that in the displacement-load model, the non-uniform material flow generates more redundant work, resulting in a higher extrusion pressure to achieve the same deformation. With the increase of the die angle, the abnormal growth of the extrusion pressure under displacement load indicates that the redundant work increases rapidly, which will lead to the deviation of the calculation results from the actual HE process. By comparing the data above, it is found that the numerical model with pressure load can simulate the hydrostatic extrusion process more accurately.

Acknowledgments: The authors would like to acknowledge the financial support from the National Magnetic Confinement Fusion Program with Grant No. 2014GB121001, the National Natural Science Foundation of China No. 51675154, and the research support from the Laboratory of Nonferrous Metal Material and Processing Engineering of Anhui Province.

Author Contributions: Shengqiang Du, Xiang Zan and Ping Li designed the simulation models; Shengqiang Du performed the simulation; Xiang Zan, Laima Luo and Xiaoyong Zhu contributed to analyze the simulation results; Yucheng Wu provided support and contributed to the discussions; Shengqiang Du wrote the paper.

Conflicts of Interest: The authors declare no conflict of interest.

References

1. Robertson, J. Method of and Apparatus for Forming Metal Articles. British Patent No. 19 356, 14 October 1894.
2. Wilson, W.R.D.; Walowit, J.A. An isothermal hydrodynamic lubrication theory for hydrostatic extrusion and drawing processes with conical dies. *J. Lubr. Technol.* **1971**, *93*, 69–74. [CrossRef]
3. Ozaltin, K.; Chrominski, W.; Kulczyk, M.; Panigrahi, A.; Horky, J.; Zehetbauer, M.; Lewandowska, M. Enhancement of mechanical properties of biocompatible Ti–45Nb alloy by hydrostatic extrusion. *J. Mater. Sci.* **2014**, *49*, 6930–6936. [CrossRef]
4. Yu, Y.; Zhang, W.C.; Duan, X.R. Study on microstructure and properties of thin tube of AZ31 magnesium alloy by extrusion technology. *Powder Metall. Technol.* **2013**, *31*, 201–206. [CrossRef]
5. Xue, Y.F.; Cai, H.N.; Wang, L.; Wang, F.C.; Zhang, H.F. Strength-improved Zr-based metallic glass/porous tungsten phase composite by hydrostatic extrusion. *Appl. Phys. Lett.* **2007**, *90*, 081901. [CrossRef]

6.   Kaszuwara, W.; Kulczyk, M.; Leonowicz, M.K.; Gizynski, T.; Michalski, B. Densification of Nd-Fe-B powders by hydrostatic extrusion. *IEEE Trans. Magn.* **2014**, *50*, 1–5. [CrossRef]

7.   Kováč, P.; Hušek, I.; Melišek, T.; Kopera, L.; Kováč, J. Critical currents, I_c-anisotropy and stress tolerance of MgB_2 wires made by internal magnesium diffusion. *Sci. Technol.* **2014**, *27*, 88–93. [CrossRef]

8.   Skiba, J.; Pachla, W.; Mazur, A.; Przybysz, S.; Kulczyk, M.; Przybysz, M.; Wróblewska, M. Press for hydrostatic extrusion with back-pressure and the properties of thus extruded materials. *J. Mater. Process. Technol.* **2014**, *214*, 67–74. [CrossRef]

9.   Zhang, Z.H.; Wang, F.C.; Sun, M.Y.; Yang, R.; Li, S.K. Finite element analysis and experimental investigation of the hydrostatic extrusion process of deforming two-layer Cu/Al composite. *J. Beijing Inst. Technol.* **2013**, *22*, 544–549. (In Chinese).

10.   Li, D.R.; Liu, Z.Y.; Yu, Y.; Wang, E.D. Numerical simulation of hot hydrostatic extrusion of W-40 wt. % Cu. *Mater. Sci. Eng. A* **2009**, *499*, 118–122. [CrossRef]

11.   Manafi, B.; Saeidi, M. Deformation behavior of 93 Tungsten alloy under hydrostatic extrusion. *Elixir Mech. Eng.* **2014**, *76*, 28487–28492.

12.   Peng, X.; Sumption, M.D.; Collings, E.W. Finite element modeling of hydrostatic extrusion for mono-core superconductor billets. *IEEE Trans. Appl. Supercond.* **2003**, *13*, 3434–3437. [CrossRef]

13.   Manafi, B.; Shatermashhadi, V.; Abrinia, K.; Faraji, G.; Sanei, M. Development of a novel bulk plastic deformation method: Hydrostatic backward extrusion. *Int. J. Adv. Manuf. Technol.* **2016**, *82*, 1823–1830. [CrossRef]

14.   Kopp, R.; Barton, G. Finite element modeling of hydrostatic extrusion of magnesium. *J. Technol. Plast Technol.* **2003**, *28*, 1–12.

15.   Barton, G. Finite-Elemente Modellierung des Hydrostatischen Strangpressens von Magnesiumlegierungen. Ph.D. Thesis, Rheinisch-Westfaelische Technische Hochschule Aachen, Aachen, Germany, January 2009. Available online: http://publications.rwth-aachen.de/record/51193/files/Barton_Gabriel.pdf (accessed on 1 March 2017).

16.   Kulczyk, M.; Przybysz, S.; Skiba, J.; Pachla, W. Severe plastic deformation induced in Al, Al-Si, Ag and Cu by hydrostatic extrusion. *Arch. Metall. Mater.* **2014**, *59*, 59–64. [CrossRef]

18

Assessment of the Residual Life of Steam Pipeline Material beyond the Computational Working Time

Marek Sroka [1,*], Adam Zieliński [2], Maria Dziuba-Kałuża [2], Marek Kremzer [1], Magdalena Macek [1] and Artur Jasiński [3]

[1] Division of Material Processing Technology, Management and Computer Techniques in Materials Science, Institute of Engineering Materials and Biomaterials, Silesian University of Technology, Konarskiego 18a, 44-100 Gliwice, Poland; marek.kremzer@polsl.pl (M.K.); magdalena.macek@gmail.com (M.M.)

[2] Institute for Ferrous Metallurgy, K. Miarki 12-14, 44-100 Gliwice, Poland; azielinski@imz.pl (A.Z.); mkaluza@imz.pl (M.D.-K.)

[3] Energopomiar, J. Sowińskiego 3, 44-100 Gliwice, Poland; ajasinski@energopomiar.com.pl

* Correspondence: marek.sroka@polsl.pl

Academic Editor: Robert Tuttle

Abstract: This paper presents the evaluation of durability for the material of repair welded joints made from (13HMF) 14MoV6-3 steel after long-term service, and from material in the as-received condition and after long-term service. Microstructure examinations using a scanning electron microscope, hardness measurements and creep tests of the basic material and welded joints of these steels were carried out. These tests enabled the time of further safe service of the examined repair welded joints to be determined in relation to the residual life of the materials. The evaluation of residual life and disposable life, and thus the estimation and determination of the time of safe service, is of great importance for the operation of components beyond the design service life. The obtained test results are part of the materials' characteristics developed by the Institute for Ferrous Metallurgy for steels and welded joints made from these steels to work under creep conditions.

Keywords: creep; degradation; welded joint; Cr-Mo-V steel; residual life

1. Introduction

Pressure components working at an elevated temperature are designed for a definitive working time. This time is based on temporary creep strength used for calculations. It is 100,000 h for old units, while, for those with supercritical working parameters designed and operated at present, it is 200,000 h. Most of the units operated in Poland have significantly exceeded the design service life of 100,000 h, reaching the actual operation time of more than 200,000 h. The extension of the operation time beyond the design one of 100,000 h is made by using the calculation methods based on data concerning the average temporary creep strength for 200,000 h and positive results of comprehensive investigations and diagnostic measurements. Usually, the critical components in the pressure part of boilers and turbines are subject to these investigations and evaluation. Out of these components, those working above the limit temperature, i.e., under creep conditions, are crucial (Figure 1).

The above-mentioned operation of steam boilers for much more than 200,000 h requires a new approach in the materials diagnostics. For safety reasons, a particularly important issue to be solved is creep strength of the welded joints of the steam pipelines working under creep conditions [1–6].

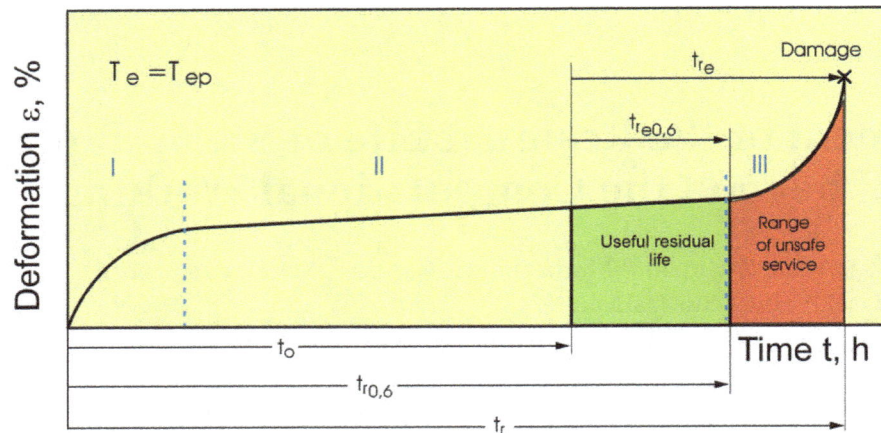

Figure 1. Schematic approach of the definition of residual and disposable durability.

In the evaluation of these components, it is important and necessary to evaluate the condition of their material [7–13]. This evaluation is carried out based on non-destructive or destructive materials tests. In the case of components working for more than 150,000 h, the estimate of residual life by non-destructive testing is not sufficient. It needs to be determined based on destructive tests performed on a sampled representative test specimen [14].

As part of the diagnostics, not only the basic material of the operated component but also the material of welded joints is subject to evaluation [15]. It is necessary to evaluate the condition of the material of welded joints to determine the component's ability to carry the required operating loads during its further service. If there is a need for repair to or replacement of a part or the entire component with a new one, the ability of the basic material under operation to carry out such a repair or replacement must be determined. When the condition of such material after service allows a repair to be made, it is necessary to develop a technology for its performance. The repair welded joint is defined as a new weld made to join a material after service with another material after service, and also to join a new material with a material after service (for replacement of a part of a structural component with a new one). Such repair welded joints are made during the renovation and modernisation works on pressure elements including, but not limited to, steam pipelines.

The subject-matter of the investigations, including the materials and their repair welded joints after long-term service made with materials after service or new materials, is an important issue overseen by the Institute for Ferrous Metallurgy. The selected results of investigations with regard to condition evaluation of the material of repair welded joints are the subject of this study. They mainly concern the elements of primary steam pipelines made from 13HMF (14MoV6-3) steel, which in the majority of Polish power plants exceeded the design service life of 100,000 h. Therefore, an important issue in the evaluation of the safe operation of these devices is to provide a numerical value of the time of their further operation and determine creep strength not for the material pipeline itself, but rather for the welded joints of these materials made during repairs.

2. Material for Investigations

The material for investigations was tested specimens of repair welded joints made from (13HMF) 14MoV6-3 steel after long-term service, and of material in the as-received condition and after long-term service. The summary of the material for investigations, including their steel grades, geometrical dimensions, working parameters, the current time of service and macrophotography of the test specimen is presented in Table 1.

Table 1. Material for investigations.

Repair Welded Joint Made from Pipeline Sections after Long-Term Service, Marked ZS1	
Steel grade: 14MoV6-3 Service time: material after 169,000 h service, marked ZS1	
Dimensions: 273×32 ($D_n \times g_n$)	
Working parameters of sections after service $t_0 = 538\,°C$; $p_0 = 13.0$ Mpa	
Material for investigations: repair circumferential welded joint made under industrial conditions	

Repair Welded Joint Made from Pipeline Sections in the As-Received Condition (before Service) and after Long-Term Service, Marked ZS2	
Steel grade: 14MoV6-3 Service time: material in the as-received condition/material after 169,000 h service, marked ZS2	
Dimensions: 273×32 ($D_n \times g_n$)	
Working parameters of sections after service $t_0 = 538\,°C$; $p_0 = 13.0$ Mpa	
Material for investigations: repair circumferential welded joint made under industrial conditions	

The check analysis of chemical composition of the examined materials of repair welded joints from low-alloy Cr-Mo-V steels after long-term service and a material in the as-received condition and after long-term service was performed in accordance with the following procedures: 3/CHEM,4 "Determination of C, Mn, Si, P, S, Cr, Ni, Cu, Mo, V, Ti, Al, Nb, B and Sn contents in low- and medium-alloy carbon steel by the spark optical emission spectrometry method using natural standards" with the optical emission spectrometer Magellan Q8 by Bruker, Germany. For the chemical composition of the examined steels with regard to the requirements of standard specification [16], see Table 2.

Table 2. Check analysis of chemical composition.

Grade of Material	Content of Elements (%)									
	C	Mn	Si	P	S	Cu	Cr	Ni	Mo	Others
14MoV6-3 according to [16]	0.10 0.18	0.40 0.70	0.15 0.35	max 0.04	max 0.04	max 0.25	0.30 0.60	max 0.30	0.50 0.65	V 0.22–0.35 Al max 0.02
14MoV6-3 169,000 h service Designation ZS1-PM1	0.16	0.58	0.35	0.017	0.018	0.20	0.46	0.23	0.62	V 0.29 Al 0.026
14MoV6-3 169,000 h service Designation ZS1-PM2	0.16	0.58	0.35	0.017	0.020	0.20	0.46	0.23	0.63	V 0.29 Al 0.024
14MoV6-3 in the as-received condition Designation ZS2-PM1	0.17	0.51	0.22	0.008	0.006	0.11	0.53	0.11	0.52	V 0.26 Al 0.013
14MoV6-3 169,000 h service Designation ZS2-PM2	0.16	0.59	0.34	0.018	0.018	0.21	0.48	0.22	0.59	V 0.28 Al 0.023

The analysis results of the check of chemical composition show that the materials of the examined test specimens of repair welded joints meet the requirements of the standard with regard to the chemical composition of the examined steel grade, i.e., 13HMF (14MoV6-3) [16].

3. Research Scope and Methodology

As part of the investigations, the properties of the material of the repair welded joints were evaluated. In the evaluation of the material condition and the level of required utility properties for repair welded joints, the following was subject to investigation:

- The microstructure of repair circumferential welded joints of components in the pressure part of a boiler was examined, with tests were carried out using a scanning electron microscope (SEM, FEI, Hillsboro, OR, USA) Inspect F on nital-etched metallographic microsections;
- Analysis of precipitation processes was carried out using X-ray analysis of isolated carbides, with the use of a Empyrean diffractometer (XRD, Panalytical, Almelo, Netherlands) and selective diffraction of electrons;
- The level of hardness for individual joint components and its nature in the course from the parent material through the heat-affected zone and weld was obtained, taken by Vickers method using a Future—Tech FM—7 machine (Kawasaki, Japan) at the indenter load of 10 kG;
- The material's residual life was determined based on abridged creep tests at a constant test stress corresponding to the operating one $\sigma_b = \sigma_r = $ const and at a constant test temperature T_b for each test. The tests were performed using Instron single-sample machines (Norwood, MA, USA) with an accuracy of temperature during the test of ± 1 °C.

The obtained results of the investigations are part of the study, which is under preparation for verification of the proposed method for evaluating and predicting the time of further safe service of homogeneous circumferential welded joints from low-alloy Cr-Mo-V steels. In the case of its positive result, this test method will be used in materials diagnostics to be performed for the power industry.

4. Results

4.1. Microstructure Investigations: Structure of Steel in the As-Received Condition

The microstructure of 14MoV6-3 steel in the as-received condition is a mixture of bainite and ferrite, sometimes with a slight amount of pearlite. Moreover, very fine MC carbide precipitates that occur inside the ferrite grains are observed within the structure. In the bainite areas, there are small spheroidal cementite precipitates, while in the pearlite colonies, cementite lamellas exist. An example of the characteristic microstructure of 14MoV6-3 steel in the as-received condition is shown in Figure 2.

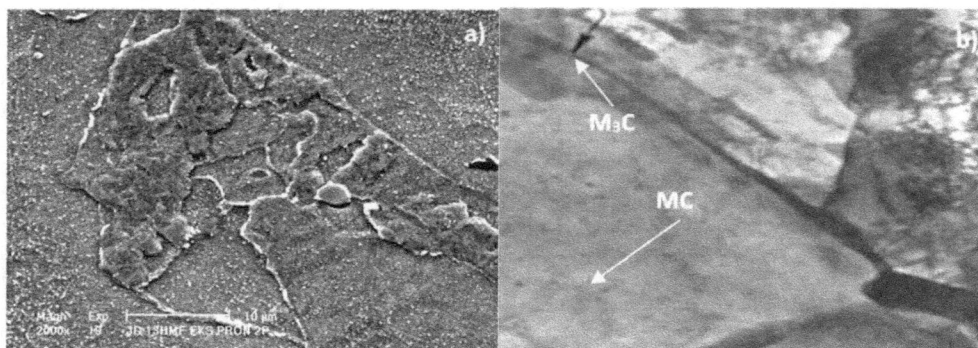

Figure 2. Structure image of 14MoV6-3 ferritic-bainitic steel in the as-received condition (**a**) SEM; (**b**) TEM.

4.2. Evaluation of the Microstructure of Repair Welded Joints

The investigations of microstructure were carried out on metallographic microsections. The microsections were made on the cross-section of test specimens of the examined components in the area of the weld and prepared by mechanical grinding and polishing as well as etching. The observations were performed with magnifications of 500 to 5000×. For the repair welded joint made from materials after long-term service, marked ZS1, the results of the investigations are presented as photographs of the microstructure of the materials of the circumferential welded joint components, in particular: parent material, heat affected zone of the joint and weld, respectively, in Figure 3.

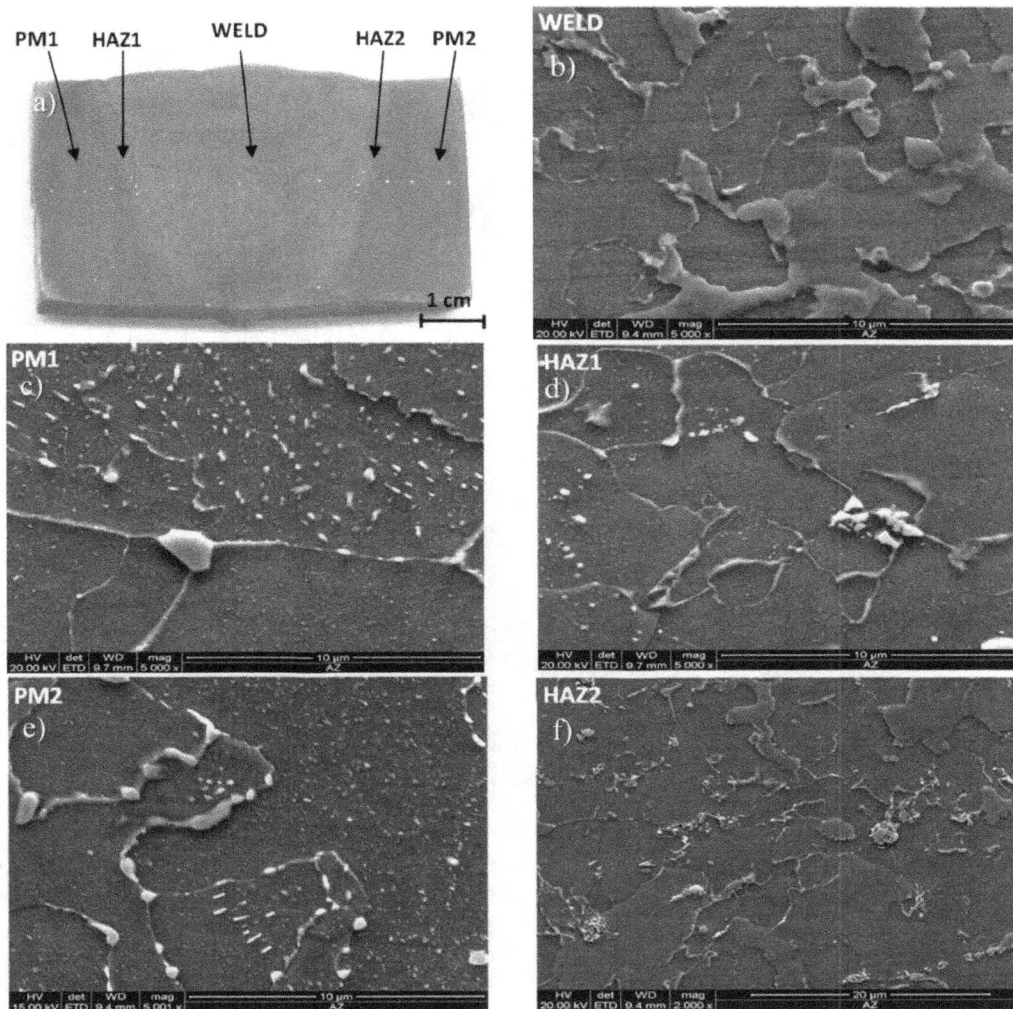

Figure 3. Structure of the material of components of the repair welded joint marked ZS1 made from 14MoV6-3 steel after 169,000 h service; microstructure investigation locations (**a**): parent material marked (**c**) PM1, (**e**) PM2; heat affected zone marked (**d**) HAZ1, (**f**) HAZ2; weld marked (**b**) WELD.

The results of the microstructure investigations for the components of the repair welded joint made from a material in the as-received condition and after long-term service, marked ZS2, in particular: parent material, heat affected zone and weld, are provided in Figure 4.

The classification of the microstructure including the evaluation and exhaustion extent t_e/t_r estimated based on the Institute for Ferrous Metallurgy's own classification [1] is provided in Table 3.

The parent material of the welded joints marked ZS1 (PM1, PM2) and ZS2 (PM2) after service was characterised by a ferritic microstructure with partially coagulated bainite areas. At the ferrite grain boundaries, there are precipitates of different size, mostly fine ones, whereas inside the ferrite grains, mostly very fine precipitates distributed evenly within the structure were observed.

The microstructure of the parent material of the welded joint marked ZS2 (PM1) in the as-received condition was characterised by ferritic-bainitic microstructure, which is typical for this type of steel.

In the microstructure of the materials of the examined repair welded joints, no discontinuities or microcracks, nor initiation of internal damage processes due to creep, were observed.

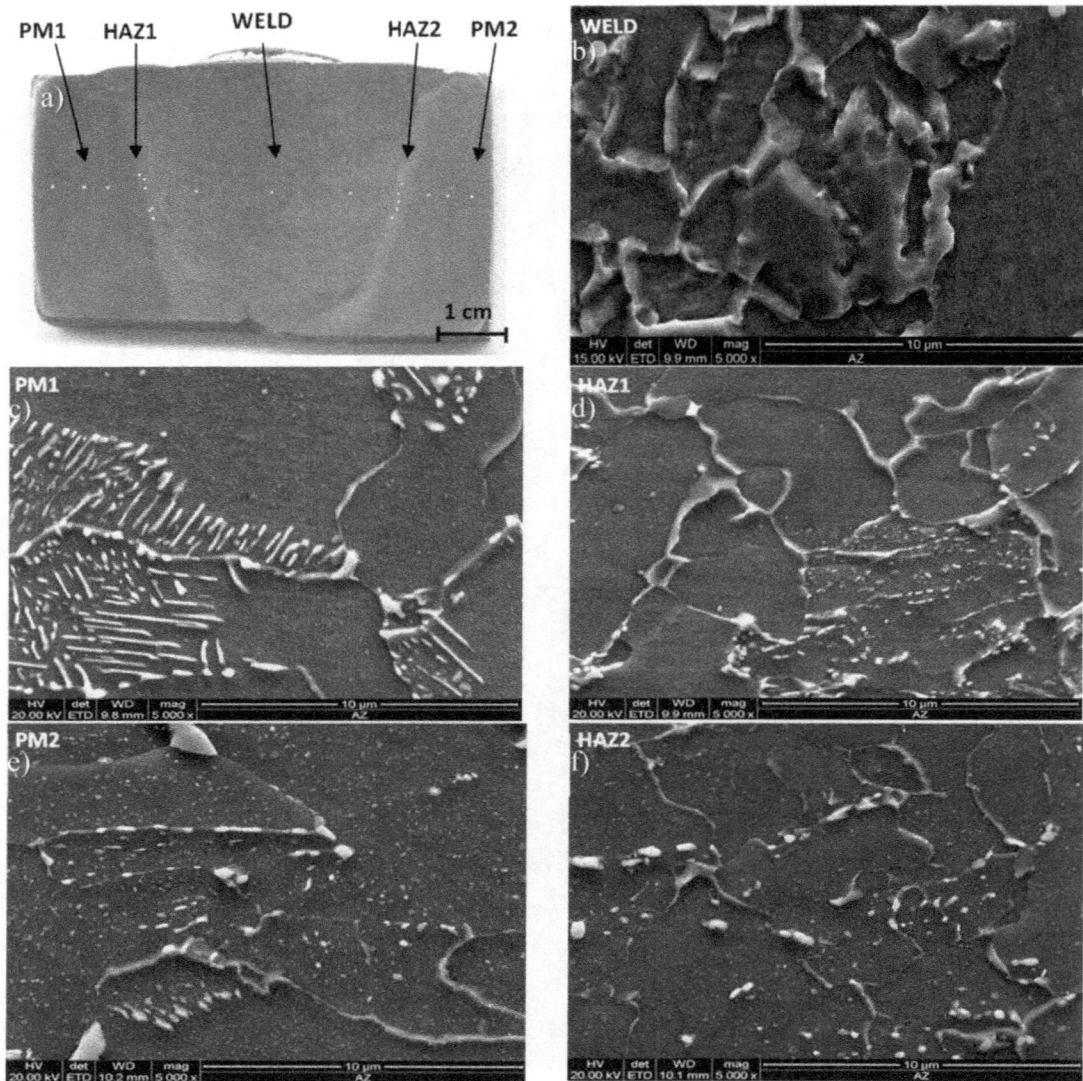

Figure 4. Structure of the material of components of the repair welded joint marked ZS2 made from 14MoV6-3 steel in the as-received condition and after 169,000 h service; investigation performance locations (**a**): parent material marked (**c**) PM1, (**e**) PM2; heat affected zone marked (**d**) HAZ1, (**f**) HAZ2; weld marked (**b**) WELD.

Table 3. Review of the results of microstructure investigations and hardness tests on the material of components of repair welded joints.

Material for Investigations	Figure No.	Description of Microstructure Material Condition—Exhaustion Degree	Hardness HV10
Repair welded joint Designation ZS1 — PM1	Figure 3	Ferritic-bainitic structure. No discontinuities or micro-cracks are observed in the structure. Bainitic areas: class I/II, precipitates: class A. Damaging processes: class 0. CLASS 2, EXHAUSTION DEGREE: approx. 0.3 ÷ 0.4.	173
PM2			169
HAZ1		No discontinuities or micro-cracks are found in the structure.	247
WELD			240
HAZ2			247
Repair welded joint Designation ZS2 — PM1	Figure 4	Ferritic-bainitic structure. No discontinuities or micro-cracks are observed in the structure. Bainitic areas: class 0; precipitates: class 0; Damaging processes: class 0. MATERIAL CONDITION: CLASS 0; EXHAUSTION DEGREE: ~0.	160
PM2		Ferritic-bainitic structure. No discontinuities or micro-cracks are observed in the structure. Bainitic areas: class I/II, precipitates: class A. Damaging processes: class 0. CLASS 2, EXHAUSTION DEGREE: approx. 0.3 ÷ 0.4.	168
HAZ1		No discontinuities or micro-cracks are found in the structure.	247
WELD			242
HAZ2			168

4.3. X-ray Analysis of Phase Composition of Precipitated Carbide Isolates

As a result of dissolving the matrix of the material of the examined test specimens of repair welded joints by the electrolytic method, the existing carbides were isolated. The X-ray phase analysis was carried out on the obtained carbide isolate, and the existing carbides were identified. The obtained results of the investigations of the material of test repair welded joints are summarised in Table 4.

Table 4. Phase composition of carbides in repair welded joints.

Material Condition	Phase Composition of Carbides	Precipitation Sequence
As-received condition 14MoV6-3 steel	M_3C MC	$M_3C + MC$
14MoV6-3 steel 169,000 h service Designation ZS1-PM1	Isovit $Cr_{23}C_6$—main phase Cementite Fe_3C VC	$M_{23}C_{6\ main\_ph.} + M_3C_{av} + MC_{nw}$
14MoV6-3 steel 169,000 h service Designation ZS2-PM2	Isovit $Cr_{23}C_6$—main phase Cementite Fe_3C VC	$M_{23}C_{6\ main\_ph.} + M_3C_{av} + MC_{nw}$

The type and contribution of the revealed precipitates correspond to the exhaustion degree estimated based on the microstructure image of the examined materials of repair welded joints from low-alloy steels after operation beyond the design service time (Table 3).

The sequences of carbides (Table 4) within the examined materials formulated based on the X-ray diffraction analysis of electrolytically isolated carbide deposits confirm the class of microstructure as determined based on the analysis of its observed images.

4.4. Hardness Evaluation of Repair Welded Joints

Hardness measurement was taken by Vickers HV10 method on the transverse metallographic microsection of the examined repair circumferential welded joints made from materials after long-term service, marked ZS1, and material in the as-received condition and after long-term service marked ZS2. The HV10-hardness measurement results against the background of the macro photograph showing a cross-section of the repair welded joints in the examined test specimens are presented in Figure 5.

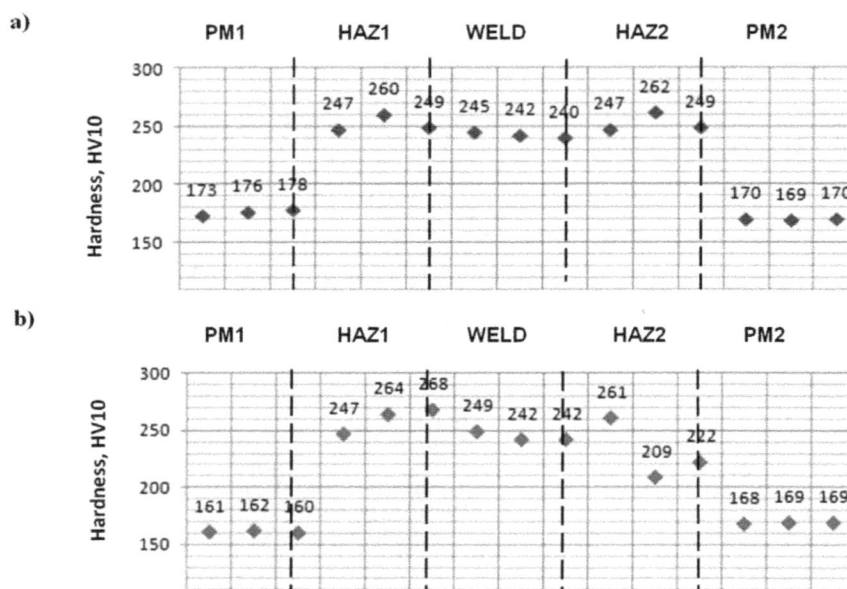

Figure 5. Distribution of the results of HV10 hardness measured on transverse microsections of the repair welded joint made from 14MoV6-3 steel. Test location—transverse microsection: (**a**) ZS1; (**b**) ZS2.

Hardness for all the examined components of repair welded joints is lower than the maximum permitted one, which is 350 HV10 for joints in the as-received condition and ranges from 160 to 179 HV10 for the parent material, from 209 to 268 for the heat-affected zone and from 240 to 249 HV10 for the weld material. This suggests that the examined welded joints were properly heat-treated after welding and will be able to transfer the required considerable loads, including those that occur during water pressure tests, shut-downs and start-ups. Hardness measurements have also shown no sudden changes when passing through the individual zones of the joint. Hardness for the 14MoV6 steel repair circumferential welded joint made from materials after 169,000 h service is, on average, 173 HV10 for the parent material, while in the weld it increases up to 262 HV10. Hardness for the 14MoV6 steel repair circumferential welded joint made from a material in the as-received condition and after 169,000 h service is, on average, 165 HV10 for the parent material, while in the weld it increases up to 268 HV10.

4.5. Abridged Creep Tests

The abridged creep tests were carried out for five test temperature levels ranging between 600 °C and 680 °C at 20 °C intervals with constant test stress σ_b = const corresponding to the operating one, which allows for obtaining test results within several months. This provides a good estimate of residual life t_{re} as it was verified in [17,18].

The method used to reduce the duration of creep tests involves accelerating the creep process by increasing the test temperature T_b well over the temperature level T_e suitable for operation in the samples maintained at a constant test stress corresponding to the operating one $\sigma_b = \sigma_r$ = const. They allow for the plotting of a straight line inclined at the time to rupture the t_r axis. The residual life is determined by extrapolation of the obtained straight line towards a lower temperature corresponding to the operating one T_e.

The results of creep tests for the examined repair welded joints marked ZS1 and ZS2 are summarised in Table 5 and presented in comparative graphs (Figure 6) as $\log t_z = f(T_b)$ at $\sigma_b = $ const, where t_r is the time to rupture in the creep test.

Table 5. Results of abridged creep tests.

Test Specimen Designation	Working Parameters		Test Stress σ_b (Mpa)	Test Temperature, T_b (°C)				
	Pressure P_r (MPa)	Temperature T_r (°C)		600	620	640	660	680
				Time to Rupture, t_r (h)				
Repair welded joint made from materials after 169,000 h service Designation ZS1	-	-	50	(3127)	1197	559	234	120
Repair welded joint made from material in the as-received condition and material after 169,000 h service Designation ZS2	-	-		(3161)	1178	822	179	103
Parent material after 169,000 h service Designation PM	13.0	538		(286)	(1365)	559	429	196
Repair welded joint made from materials after 169,000 h service Designation ZS1	-	-	55	2834	672	373	189	97
Repair welded joint made from material in the as-received condition and material after 169,000 h service Designation ZS2	-	-		(2592)	1297	481	191	84

(-)—tests in progress.

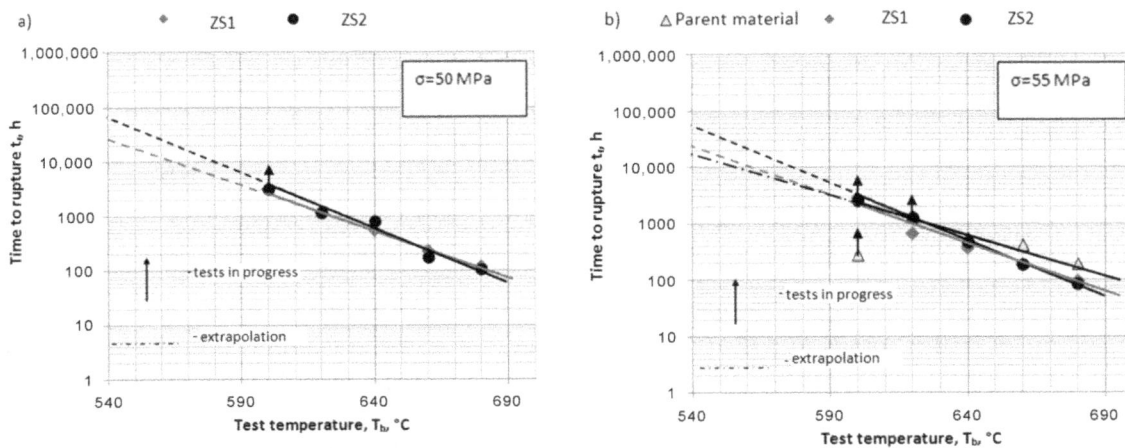

Figure 6. Result of abridged creep tests on the examined ZS1 and ZS2 repair joints and parent material in the form of $\log t_{re} = f(T_b)$ for the adopted stress level of further service (**a**) $b = 50$ MPa; (**b**) $b = 55$ MPa.

The comparison of the results of abridged creep tests in the form of $\log t_r = f(T_b)$ at $\sigma_b \approx \sigma_r = 50$ MPa for the repair welded joint made from 14MoV6-3 steel after 169,000 h service and the repair welded joint made from 14MoV6-3 material in the as-received condition and 14MoV6-3 material after 169,000 h service is presented in Figure 6a.

The comparison of the results of abridged creep tests in the form of $\log t_r = f(T_b)$ at $\sigma_b \approx \sigma_r = 55$ MPa for the parent material of 14MoV6-3 after 169,000 h service and the repair welded joint made from materials of 13HMF (14MoV6-3) after 169,000 h service and the repair welded joint made from 14MoV6-3 steel in the as-received condition and 14MoV6-3 steel after 169,000 h service is presented in Figure 6b.

On the basis of the previously completed creep tests, based on the extrapolation method used, the residual life (interpreted as the time to failure) was determined and the disposable residual life (being the safe time of service, which is about 0.6 of the residual life, Figure 1) was estimated as the safe time of service for the examined parent material, repair welded joint made from materials after long-term service and repair welded joint made from material in the as-received condition and after long-term service. The obtained results of extrapolation based on creep tests are summarised in Table 6 for two values of stress—50 and 55 MPa.

Table 6. Residual life determined and disposable residual life estimated by abridged creep tests of the parent material, repair welded joint made from materials after long-term service and repair welded joint made from material in the as-received condition and after long-term service.

Test Specimen Designation	Adopted Operating Stress σ_r (MPa)	Adopted Further Operation Temperature T_r (°C)	Estimated Life Time (h)	
			Residual t_{re}	Disposable Residual Life t_b (about 0.6 t_{re})
Joint from materials after long-term service Designation ZS1	50	538	25,000	15,000
Joint from material in the as-received condition and material after long-term service Designation ZS2			60,000	36,000
Native material Designation PM1			20,000	12,000
Joint from materials after long-term service Designation ZS1	55		23,000	13,800
Joint from material in the as-received condition and material after long-term service Designation ZS2			58,000	34,800

The residual life determined by extrapolation of creep results obtained in abridged tests, for the temperature of further service and the adopted stress level of further operation of the parent materials and repair welded joints, has allowed the disposable residual life, which is the time of further safe service, to be determined.

The residual life t_{re} determined for the adopted stress level of 50 MPa for the repair welded joint of the materials after service, marked ZS1, is 25,000 h and its estimated disposable life t_b is 15,000 h (Figure 6, Table 4), while the residual life t_{re} determined for the repair welded joint of the material after service and the material in the as-received condition, marked ZS2, is 60,000 h and the estimated disposable life is 36,000 h.

The residual life t_{re} determined for the adopted stress level of 55 MPa for the repair welded joint of the materials after service marked ZS1, is 23,000 h and the estimated disposable life t_b is approx. 14,000 h, while the residual life t_{re} determined for the repair welded joint of the material after service and the material in the as-received condition, marked ZS2, is 58,000 h and the estimated disposable life is approx. 35,000 h. For the parent material after service marked PM1, the residual life is 20,000 h, and the estimated disposable life is t_b 12,000 h.

The time of further safe service of the examined new repair welded joints may be assumed to be 15,000 h for the ZS1 joint and 36,000 h for the ZS2 joint at the further service stress $\sigma_e = 50$ MPa, while for the adopted further service stress $\sigma_e = 55$ MPa the time of further safe service of the examined new repair welded joints may be assumed to be approximately 14,000 h for the ZS1 joint and approximately 35,000 h for the ZS2 joint.

5. Conclusions

1. The set of destructive materials tests presented in this paper allows for the evaluation of material condition and determination of suitability for service of repair. It is of particular importance for the operation of steam pipelines beyond the design service time.

2. The evaluation of the material condition of repair welded joints is made based on a comprehensive summary of the results of investigations on mechanical properties, microstructure and abridged creep tests. These results are in turn a part of the database of the materials' characteristics for steels and their welded joints with materials showing varying degrees of degradation. This database is used in diagnostic tests for pressure parts of boiler elements.

3. The quantitative dimension of suitability for service of the material of repair welded joints is achieved by extrapolating the straight line obtained in abridged creep tests from $\log t_r = f(T_b)$ at $\sigma_b = $ const towards the temperature of assumed operation, which allows residual life t_{re} and disposable residual life t_b to be determined for the working temperature.

4. The knowledge of the share of disposable residual life t_{be} in residual life t_{re} (t_{be}/t_{re}) allows the safe time of service of the examined joints to be determined for the required performance parameters.

5. The examined repair welded joints are suitable for operation for a limited time resulting from the disposable residual life determined for defined temperature and stress parameters of further service.

The completed tests of the material of steam pipeline and welded joints have shown that long-term operation beyond the design service time does not disqualify the material from service. It has been demonstrated that the modernisation and repair works carried out on the steam pipeline materials by making welded joints show lower creep strength than the basic material. The lower strength of repair welded joints in relation to the parent material should be taken into account in design calculations while extending the service time beyond the design service life.

It has also been demonstrated that, in contrast to the microstructural investigations and the basic investigations of mechanical properties, the abridged creep tests allow the real determination of the time of further safe operation of the elements of power equipment working beyond the design service life to be obtained.

The analysis of the research results of abridged creep tests shows that, independently, of the values of the stress, creep resistance of repair welded joint ZS2 is twice as high as welded joints marked ZS1. This difference is probably related to the higher creep resistance of the parent material resulting in a higher creep resistance of joints marked ZS2.

Acknowledgments: The publication was partially financed by the statutory grant from Faculty of Mechanical Engineering, the Silesian University of Technology for the year 2016. The results in this publication were obtained as a part of research co-financed by the National Centre for Research and Development under contract PBS3/B5/42/2015—Project: "Methodology, evaluation and forecast of operation beyond the analytical operation of welded joints in pressure elements of power boilers beyond the design work time".

Author Contributions: M.S., A.Z. and M.D.-K. conceived and designed the experiments; M.D.-K. and A.Z. performed the experiments; M.M., A.J. and M.K. analysed the data; M.S. wrote the paper.

Conflicts of Interest: The authors declare no conflict of interest.

References

1. Dobrzański, J. Materials science interpretation of the life of steels for power plants. *Open Access Libr.* **2011**, *3*, 1–228.

2. Golański, G.; Zieliński, A.; Słania, J.; Jasak, J. Mechanical properties of Vm12 steel after 30,000 hrs of ageing at 600 °C temperature. *Arch. Metall. Mater.* **2014**, *59*, 1351–1354.

3. Zieliński, A.; Golański, G. The influence of repair welded joint on the life of steam pipeline made of Cr-Mo steel serviced beyond the calculated working time. *Arch. Metall. Mater.* **2015**, *60*, 1045–1049. [CrossRef]

4. Cao, J.; Gong, Y.; Yang, Z.G.; Luo, X.M.; Gu, F.M.; Hu, Z.F. Creep fracture behavior of dissimilar weld joints between T92 martensitic and HR3C austenitic steels. *Int. J. Press. Vessels Pip.* **2011**, *88*, 94–98. [CrossRef]

5. Laha, K. Integrity Assessment of Similar and Dissimilar Fusion Welded Joints of Cr-Mo-W Ferritic Steels under Creep Condition. *Procedia Eng.* **2014**, *86*, 195–202. [CrossRef]

6. Sawada, K.; Tabuchi, M.; Hongo, H.; Watanabe, T.; Kimura, K. Z-Phase formation in welded joints of high chromium ferritic steels after long-term creep. *Mater. Charact.* **2008**, *59*, 1161–1167. [CrossRef]

7. Zieliński, A.; Sroka, M.; Miczka, M.; Śliwa, A. Forecasting the particle diameter size distribution in P92 (X10CrWMoVNb9-2) steel after long-term ageing at 600 and 650 °C. *Arch. Metall. Mater.* **2016**, *61*, 753–760.

8. Golański, G.; Zieliński, A.; Zielińska-Lipiec, A. Degradation of microstructure and mechanical properties in martensitic cast steel after ageing. *Materialwiss. Werkst.* **2015**, *46*, 248–255. [CrossRef]

9. Zieliński, A.; Golański, G.; Sroka, M.; Tański, T. Influence of long-term service on microstructure, mechanical properties, and service life of HCM12A steel. *Mater. High Temp.* **2016**, *33*, 24–32. [CrossRef]

10. Zieliński, A.; Golański, G.; Sroka, M.; Dobrzański, J. Estimation of long-term creep strength in austenitic power plant steels. *Mater. Sci. Technol.-Lond.* **2016**, *32*, 780–785. [CrossRef]

11. Dobrzański, J.; Hernas, A.; Moskal, G. Microstructural degradation in power plant steels. In *Power Plant Life Management and Performance Improvement*; Oakey, J.E., Ed.; Woodhead Publishing Limited: Sawston, UK, 2011.

12. Falat, L.; Svoboda, M.; Výrostková, A.; Petryshynets, I.; Sopko, M. Microstructure and creep characteristics of dissimilar T91/TP316H martensitic/austenitic welded joint with Ni-based weld metal. *Mater. Charact.* **2012**, *72*, 15–23. [CrossRef]

13. Kim, M.-Y.; Kwak, S.-C.; Choi, I.-S.; Lee, Y.-K.; Suh, J.-Y.; Fleury, E.; Jung, W.-S.; Son, T.-H. High-temperature tensile and creep deformation of cross-weld specimens of weld joint between T92 martensitic and Super304H austenitic steels. *Mater. Charact.* **2014**, *97*, 161–168. [CrossRef]

14. Sroka, M.; Zieliński, A.; Mikuła, J. The service life of the repair welded joint of Cr-Mo/Cr-Mo-V. *Arch. Metall. Mater.* **2016**, *61*, 969–974. [CrossRef]

15. Zieliński, A.; Golański, G.; Sroka, M. Influence of long-term ageing on the microstructure and mechanical properties of T24 steel. *Mater. Sci. Eng. A* **2017**, *682*, 664–672.

16. *PN-EN 10216-2. Pipes for Pressure Purposes with Specified Elevated Temperature Properties*; Boiler Tubes: Brussels, Belgium, 2014.

17. Zieliński, A.; Dobrzański, J.; Purzyńska, H.; Golański, G. Properties, structure and creep resistance of austenitic steel Super 304H. *Mater. Test.* **2015**, *57*, 859–865. [CrossRef]

18. Zieliński, A.; Sroka, M.; Hernas, A.; Kremzer, M. The effect of long-term impact of elevated temperature on changes in microstructure and mechanical properties of HR3C steel. *Arch. Metall. Mater.* **2016**, *61*, 761–766. [CrossRef]

Permissions

All chapters in this book were first published in METALS, by MDPI; hereby published with permission under the Creative Commons Attribution License or equivalent. Every chapter published in this book has been scrutinized by our experts. Their significance has been extensively debated. The topics covered herein carry significant findings which will fuel the growth of the discipline. They may even be implemented as practical applications or may be referred to as a beginning point for another development.

The contributors of this book come from diverse backgrounds, making this book a truly international effort. This book will bring forth new frontiers with its revolutionizing research information and detailed analysis of the nascent developments around the world.

We would like to thank all the contributing authors for lending their expertise to make the book truly unique. They have played a crucial role in the development of this book. Without their invaluable contributions this book wouldn't have been possible. They have made vital efforts to compile up to date information on the varied aspects of this subject to make this book a valuable addition to the collection of many professionals and students.

This book was conceptualized with the vision of imparting up-to-date information and advanced data in this field. To ensure the same, a matchless editorial board was set up. Every individual on the board went through rigorous rounds of assessment to prove their worth. After which they invested a large part of their time researching and compiling the most relevant data for our readers.

The editorial board has been involved in producing this book since its inception. They have spent rigorous hours researching and exploring the diverse topics which have resulted in the successful publishing of this book. They have passed on their knowledge of decades through this book. To expedite this challenging task, the publisher supported the team at every step. A small team of assistant editors was also appointed to further simplify the editing procedure and attain best results for the readers.

Apart from the editorial board, the designing team has also invested a significant amount of their time in understanding the subject and creating the most relevant covers. They scrutinized every image to scout for the most suitable representation of the subject and create an appropriate cover for the book.

The publishing team has been an ardent support to the editorial, designing and production team. Their endless efforts to recruit the best for this project, has resulted in the accomplishment of this book. They are a veteran in the field of academics and their pool of knowledge is as vast as their experience in printing. Their expertise and guidance has proved useful at every step. Their uncompromising quality standards have made this book an exceptional effort. Their encouragement from time to time has been an inspiration for everyone.

The publisher and the editorial board hope that this book will prove to be a valuable piece of knowledge for researchers, students, practitioners and scholars across the globe.

List of Contributors

Sefika Kasman
Izmir Vocational School, Dokuz Eylul University, Izmir 35160, Turkey

Fatih Kahraman
Department of Mechanical Engineering, Faculty of Engineering, Dokuz Eylul University, Izmir 35140, Turkey

Anıl Emiralioğlu
Department of Mechanical Engineering, Faculty of Engineering, Dokuz Eylul University, Izmir 35140, Turkey
Graduate School of Natural and Applied Sciences, Dokuz Eylul University, Izmir 35140, Turkey

Haydar Kahraman
Graduate School of Natural and Applied Sciences, Dokuz Eylul University, Izmir 35140, Turkey
Department of MetallurgicalandMaterialsEngineering, Faculty of Engineering, Dokuz Eylul University, Izmir 35140, Turkey

Sebastián F. Medina , Inigo Ruiz-Bustinza and José Robla
National Centre for Metallurgical Research (CENIM-CSIC), Av. Gregorio del Amo 8, 28040 Madrid, Spain

Jessica Calvo
Technical University of Catalonia (ETSEIB — UPC), Av. Diagonal 647, 08028 Barcelona, Spain

Andrei Shishkin
Rudolfs Cimdins Riga Biomaterials Innovations and Development Centre of Riga Technical University (RTU), Institute of General Chemical Engineering, Faculty of Materials Science and Applied Chemistry, Riga Technical University, Pulka 3, LV-1007 Riga, Latvia

Maria Drozdova and Irina Hussainova
Department of Mechanical and Industrial Engineering, Tallinn University of Technology, 19086 Tallinn

Viktor Kozlov
Sidrabe Inc., LV-1073 Riga, Latvia

Dirk Lehmhus
MAPEX Center for Materials and Processes, University of Bremen, 28359 Bremen, Germany

El-Sayed M. Sherif
Deanship of Scientific Research, Advanced Manufacturing Institute (AMI), King Saud University, P.O. Box 800, Al-Riyadh 11421, Saudi Arabia
Electrochemistry and Corrosion Laboratory, Department of Physical Chemistry, National Research Centre, El-Behoth St. 33, Dokki, 12622 Cairo, Egypt

Hany S. Abdo
Deanship of Scientific Research, Advanced Manufacturing Institute (AMI), King Saud University, P.O. Box 800, Al-Riyadh 11421, Saudi Arabia
Mechanical Design and Materials Department, Faculty of Energy Engineering, Aswan University, Aswan 81521, Egypt

Ehab A. El Danaf
Mechanical Engineering Department, College of Engineering, King Saud University, P.O. Box 800, Al Riyadh 11421, Saudi Arabia

Hasan Al-Khazraji
Former Ph.D. Student at Institute of Materials Science and Engineering, Clausthal University of Technology, Clausthal-Zellerfeld 38678, Germany

Shujun Qiu, Xingyu Ma, Errui Wang, Yongjin Zou, Cuili Xiang, Fen Xu and Lixian Sun
Guangxi Key Laboratory of Information Materials, Guangxi Collaborative Innovation Center of Structure and Property for New Energy and Materials, School of Materials Science and Engineering, Guilin University of Electronic Technology, Guilin 541004, China

Hailiang Chu
Guangxi Key Laboratory of Information Materials, Guangxi Collaborative Innovation Center of Structure and Property for New Energy and Materials, School of Materials Science and Engineering, Guilin University of Electronic Technology, Guilin 541004, China
Key Laboratory of Advanced Energy Materials Chemistry (Ministry of Education), Nankai University, Tianjin 300071, China

Mingxing Zhou, Guang Xu , Haijiang Hu, Qing Yuan and Junyu Tian
The State Key Laboratory of Refractories and Metallurgy, Key Laboratory for Ferrous Metallurgy and Resources Utilization of Ministry of Education ,Wuhan University of Science and Technology, Wuhan 430081, China

Qing Fang, Hongwei Ni , Bao Wang, Hua Zhang and Fei Ye
The State Key Laboratory of Refractories and Metallurgy, Wuhan University of Science and Technology, Wuhan 430081, China

Jhih You Chen and Ming Tsung Sun
Department of Mechanical Engineering, Chang Gung University, Taoyuan 33302, Taiwan

Ching An Huang
Department of Mechanical Engineering, Chang Gung University, Taoyuan 33302, Taiwan
Department of Mechanical Engineering, Ming Chi University of Technology, New Taipei 24301, Taiwan
Bone and Joint Research Center, Chang Gung Memorial Hospital, Taoyuan 33305, Taiwan

Ramona Sola, Roberto Giovanardi and Paolo Veronesi
Department of Engineering "E. Ferrari", University of Modena and Reggio Emilia, Modena 41125, Italy

Giovanni Parigi
Stav, Barberino del Mugello, Florence 50031, Italy

Yu Liu and Xianwei Reng
Research Institute of Light Alloy, Central South University, Changsha 410083, China
Nouferrous Metal Oriented Advanced Structural Materials and Manufacturing Cooperative Innovation Center, Central South University, Changsha 410083, China

Yuanchun Huang and Zhengbing Xiao
Research Institute of Light Alloy, Central South University, Changsha 410083, China
Nouferrous Metal Oriented Advanced Structural Materials and Manufacturing Cooperative Innovation Center, Central South University, Changsha 410083, China

College of Mechanical and Electrical Engineering, Central South University, Changsha 410083, China

Aytekin Hitit and Hakan Şahin
Department of Materials Science and Engineering, Afyon Kocatepe University, Afyonkarahisar 03200, Turkey

Ruixuan Li, Haiying Hao, Yangyong Zhao and Yong Zhang
State Key Laboratory for Advanced Metals and Materials, University of Science and Technology Beijing, No. 30, Xueyuan Road, Beijing 100083, China

Xue Bai
School of Materials Science and Engineering, Beihang University, Beijing 100191, China

Department of Materials Science and Engineering, The University of Tennessee, Knoxville, TN 37996, USA
Beijing Key Laboratory of Advanced Nuclear Materials and Physics, Beihang University, Beijing 100191, China

Sujun Wu
School of Materials Science and Engineering, Beihang University, Beijing 100191, China
Beijing Key Laboratory of Advanced Nuclear Materials and Physics, Beihang University, Beijing 100191, China

Peter K. Liaw
Department of Materials Science and Engineering, The University of Tennessee, Knoxville, TN 37996, USA

Lin Shao and Jonathan Gigax
Department of Nuclear Engineering, Texas A&M University, College Station, TX 77843, USA

Chun-Liang Chen and Chen-Han Lin
Department of Materials Science and Engineering, National Dong Hwa University, Hualien 97401, Taiwan

Qingli Qi ,Xiaoqian Bao and Xuexu Gao
State Key Laboratory for Advanced Metals and Materials, University of Science and Technology Beijing, Beijing 100083, China

Jiheng Li
State Key Laboratory for Advanced Metals and Materials, University of Science and Technology Beijing, Beijing 100083, China

Department of Chemistry-Ångström Laboratory, Uppsala University, Uppsala 75121, Sweden

Chao Yuan
Grirem Advanced Materials Co., Ltd., Beijing 100088, China

Havva Kazdal Zeytin
TUBITAK MAM, Gebze, Kocaeli 41470, Turkey

Hayriye Ertek Emre and Ramazan Kaçar
Department of Manufacturing Engineering, Karabuk University, Karabuk 78050, Turkey

Shengqiang Du and Ping Li
School of Materials Science and Engineering, Hefei University of Technology, Hefei 230009, China

Xiang Zan Laima Luo and Yucheng Wu
School of Materials Science and Engineering, Hefei University of Technology, Hefei 230009, China
National–Local Joint Engineering Research Centre of Nonferrous Metals and Processing Technology, Hefei 230009, China

Xiaoyong Zhu
National–Local Joint Engineering Research Centre of Nonferrous Metals and Processing Technology, Hefei 230009, China

Marek Sroka, Marek Kremzer and Magdalena Macek
Division of Material Processing Technology, Management and Computer Techniques in Materials Science, Institute of Engineering Materials and Biomaterials, Silesian University of Technology, Konarskiego 18a, 44-100 Gliwice, Poland

Adam Zieliński and Maria Dziuba-Kałuża
Institute for Ferrous Metallurgy, K. Miarki 12-14, 44-100 Gliwice, Poland

Artur Jasiński
Energopomiar, J. Sowińskiego 3, 44-100 Gliwice, Poland

Index

www.ingramcontent.com/pod-product-compliance
Lightning Source LLC
Chambersburg PA
CBHW050452200326
41458CB00014B/5157

* 9 7 8 1 6 8 2 8 5 5 7 5 1 *